WITHDRAWN

EPITHELIAL TRANSPORT AND BARRIER FUNCTION

PATHOMECHANISMS IN GASTROINTESTINAL DISORDERS

DATE DUE

ANNALS OF THE NEW YORK ACADEMY OF SCIENCES
Volume 915

EPITHELIAL TRANSPORT AND BARRIER FUNCTION

PATHOMECHANISMS IN GASTROINTESTINAL DISORDERS

*Edited by Jörg-Dieter Schulzke, Michael Fromm,
Ernst-Otto Riecken, and Henry J. Binder*

The New York Academy of Sciences
New York, New York
2000

Tennessee Tech Library
Cookeville, TN

Copyright © 2000 by the New York Academy of Sciences. All rights reserved. Under the provisions of the United States Copyright Act of 1976, individual readers of the Annals *are permitted to make fair use of the material in them for teaching or research. Permission is granted to quote from the* Annals *provided that the customary acknowledgment is made of the source. Material in the* Annals *may be republished only by permission of the Academy. Address inquiries to the Permissions Department (editorial@nyas.org) at the New York Academy of Sciences.*

Copying fees: *For each copy of an article made beyond the free copying permitted under Section 107 or 108 of the 1976 Copyright Act, a fee should be paid through the Copyright Clearance Center, Inc., 222 Rosewood Drive, Danvers, MA 01923 (www.copyright.com).*

⊚ *The paper used in this publication meets the minimum requirements of the American National Standard for Information Sciences—Permanence of Paper for Printed Library Materials, ANSI Z39.48-1984.*

Library of Congress Cataloging-in-Publication Data

Epithelial transport and barrier function : pathomechanisms in gastrointestinal disorders / edited by Jörg-Dieter Schulzke ... [et al.].
    p. cm. — (Annals of the New York Academy of Sciences, ISSN 0077-8923 ;v.915)
    Includes bibliographical references and index.
    ISBN 1-57331-259-2 (cloth : alk. paper) — ISB N1-57331-260-6 (pbk. : alk. paper)
    1. Gastrointestinal system—Pathophysiology—Congresses. 2. Gastrointestinal mucosa—Congresses. 3. Epithelium–Congresses. 4. Tight junctions (Cell biology)—Congresses. 5. Biological transport—Congresses. 6. Pathology, Molecular—Congresses. I. Schulzke, Jörg-Dieter. II. Series.

Q11.N5 vol. 915
[RC817]
500 s—dc21
[616.3'307]
                                        00-051126

GYAT/B-M
Printed in the United States of America
ISBN 1-57331-259-2 (cloth)
ISBN 1-57331-260-6 (paper)
ISSN 0077-8923

ANNALS OF THE NEW YORK ACADEMY OF SCIENCES

Volume 915

December 2000

EPITHELIAL TRANSPORT AND BARRIER FUNCTION

PATHOMECHANISMS IN GASTROINTESTINAL DISORDERS

Editors

JÖRG-DIETER SCHULZKE, MICHAEL FROMM, ERNST-OTTO RIECKEN, AND HENRY J. BINDER

This volume is the result of a conference entitled **Epithelial Transport and Barrier Function: Pathomechanisms in Gastrointestinal Disorders,** held on March 26–27, 1999 in Berlin, Germany.

CONTENTS

Financial assistance was received from:

- ASTRA GMBH
- BAXTER IMMUNO DEUTSCHLAND GMBH
- BAYER VITAL GMBH & CO. KG
- BECKMAN COULTER GMBH
- BERLIN-CHEMIE AG
- BIOCHROM KG
- BOEHRINGER INGELHEIM PHARMA KG
- BRISTOL-MYERS SQUIBB GMBH
- BRAUN MELSUNGEN
- BYK GULDEN
- CLONTECH LABORATORIES GMBH
- DR. RENTSCHLER ARZNEIMITTEL GMBH & CO.
- EPPENDORF-NETHELER-HINZ GMBH
- ESSEX PHARMA GMBH
- DR. FALK PHARMA GMBH
- FERRING ARZNEIMITTEL GMBH
- FRESENIUS AG
- GRÜNENTHAL GMBH
- HOECHST MARION ROUSSEL DEUTSCHLAND GMBH
- HOFFMANN-LA ROCHE AG
- INTERSAN, INSTITUT FÜR PHARMAZEUTISCHE UND
  KLINISCHE FORSCHUNG GMBH
- IN VITRO SYSTEMS AND SERVICES
- DR. WERNER JANSSEN NACHF. CHEM.-PHARM. PRODUKTE GMBH
- LEDERLE ARZNEIMITTEL GMBH & CO
- MEDAC GESELLSCHAFT FÜR KLINISCHE SPEZIALPRÄPARATE GMBH
- MERCK KGAA
- MERCKLE GMBH
- MERCK SHARP & DOHME LIMITED
- MILLIPORE GMBH
- NOVARTIS PHARMA GMBH
- OLYMPUS
- PHARMACIA & UPJOHN GMBH
- PROMEGA GMBH
- QIAGEN
- RAYTEST ISOTOPENMEßGERÄTE GMBH
- ROCHE DIAGNOSTICS GMBH
- ROCHE NICHOLAS DEUTSCHLAND GMBH
- SARTORIUS AG
- SMITH KLINE BEECHAM PHARMA GMBH
- THIEMANN ARZNEIMITTEL GMBH
- WALLAC-ADL GMBH

The New York Academy of Sciences believes it has a responsibility to provide an open forum for discussion of scientific questions. The positions taken by the participants in the reported conferences are their own and not necessarily those of the Academy. The Academy has no intent to influence legislation by providing such forums.

Preface

Physiological and pathophysiological knowledge forms the basis of rational decisions in daily clinical work. This volume, which is the result of a conference entitled Epithelial Transport and Barrier Function: Pathomechanisms in Gastrointestinal Disorders, held on March 26–27, 1999 in Berlin, is an attempt to learn from basic research about epithelial transport and barrier function in the small and large intestine for diagnostic and therapeutic approaches in the clinic. Thus, this conference brought together clinicians and basic scientists for a fruitful discussion on pathophysiological mechanisms.

The intestinal epithelium is the site of vectorial transport of solutes and water and a functional barrier between the intestinal lumen and circulation. The net effect of transport is influenced by the tightness (or leakiness) of the barrier and *vice versa*. Both transport and barrier functions are physiologically regulated, and both can be dramatically altered in disease conditions, leading to symptoms of malabsorption, diarrhea, and inflammation. It is the purpose of this volume to describe the interrelationship between intestinal transport and barrier function in health and disease.

In Part I of this book, the molecular structure of active ion and nutrient transporters as well as their intracellular signal transduction mechanisms and extracellular regulatory mechanisms—including the enteric nervous and the immune system—are described. This discussion includes the molecular structure and the cellular expression of epithelial transporters as well as the route of ions, substrates, and water transport through the epithelium.

In Part II, the focus is on function of the tight junction domain of the epithelium in determining epithelial barrier properties; such phenomena as apoptosis, the contribution of M cells to epithelial permeability, and epithelial remodeling are also incorporated. Much attention has been given to the role of neurotransmitters and paracrine and endocrine factors, including cytokines that activate or inactivate these agents. This part also contains original work on the molecular structure of occludins and claudins. Pathological factors that can interfere with the epithelial barrier function are also discussed.

In Part III, clinical conditions—for example, intestinal inflammation—are elucidated by animal models of inflammation or by studies using epithelial cell culture models. This section also includes direct intestinal studies of patients with inflammatory bowel disease, HIV enteropathy, and infectious types of diarrhea, including cholera and *Clostridium difficile*–induced diarrhea.

Part IV covers diarrheal mechanisms that arise in gastrointestinal diseases. Two forms are focused upon here: malabsorptive diarrhea, due to a reduced absorptive area or due to defective transporters; and leak flux diarrhea, due to an impaired epithelial barrier and a passive loss of ions and water into the intestinal lumen. Leak flux diarrhea has only just gained attention, after it was found that several enterotoxins are able to impair the intestinal barrier.

Taken together, the present volume is the result of a very productive conference in which participants exchanged the latest results from basic research and recent insights from clinically oriented work.

— JÖRG-DIETER SCHULZKE
— MICHAEL FROMM

EPITHELIAL TRANSPORT AND BARRIER FUNCTION

PATHOMECHANISMS IN GASTROINTESTINAL DISORDERS

Expression and Function of $Na^+HCO_3^-$ Cotransporters in the Gastrointestinal Tract

URSULA SEIDLER,[a] HEIDI ROSSMANN, PETRA JACOB, OLIVER BACHMANN, STEPHANIE CHRISTIANI, GEORG LAMPRECHT, AND MICHAEL GREGOR

Medizinische Klinik der Universität Tübingen, 72076 Tübingen, Germany

ABSTRACT: The stomach, duodenum, colon, and pancreas secrete HCO_3^- ions into the lumen. Although the importance of HCO_3^- secretion for the maintenance of mucosal integrity, a normal digestion, and the reabsorption of Cl^- has been well established, the molecular nature of the apical and basolateral HCO_3^- transporting proteins has remained largely unknown. Functional studies have suggested that a $Na^+HCO_3^-$ cotransport system, similar but not identical to the well-characterized $Na^+HCO_3^-$ cotransporter in the basolateral membrane of the kidney proximal tubule, is present in duodenal and colonic enterocytes, pancreatic ducts cells, and gastric cells and involved in HCO_3^- uptake from the interstitium. This report describes our work towards understanding the molecular nature, cellular origin, and functional relevance of the $Na^+HCO_3^-$ cotransporter(s) in the stomach and intestine and reviews work by others on the function and localization of $Na^+HCO_3^-$ cotransport processes in the gastrointestinal tract.

INTRODUCTION

Virtually all cellular mechanisms are affected by changes in the pH of their surroundings. One fundamental reason for this is that the amino acids that make up every protein contain weak acid or weak base moieties, and a pH change could have substantial effects on the conformational state of the protein. In many instances the pH dependency of biological processes has been demonstrated experimentally. Because pH is so important for such a wide variety of biological processes, cells have evolved multiple mechanisms to transport protons and base equivalents across their membranes, both those of intracellular organelles and the plasma membrane.

All these mechanisms are crucially dependent on the pH of the extracellular fluid, and the major pH regulator for the long-term regulation of extracellular fluid pH is the kidney. Therefore, this organ was of particular interest for the elucidation of pH_i- regulatory mechanisms. Using pH- and ion-sensitive microelectrodes, Boron and Boulpaep first described an electrogenic Na^+- and HCO_3^--dependent cotransport system in the basolateral membrane of the salamander proximal tubule.[1] This $Na^+HCO_3^-$ cotransporter was shown to be the major base transport system in the basolateral membrane of proximal tubule cells in all species studied, and, together with

[a]Address for correspondence: Dr. U. Seidler, Abteilung Innere Medizin I, Universitätsklinikum Schnarrenberg, Eberhard-Karls Universität Tübingen, Otfried-Müller Str. 10, 72076 Tübingen, Germany. Voice: +49-7071-2983187; fax: +49-7071-295221.
ursula.seidler@uni-tuebingen.de

1

a Na^+/H^+ exchanger in the luminal membrane, was responsible for HCO_3^- reabsorption from the tubule lumen into the blood.[2,3] This renal electrogenic $Na^+HCO_3^-$ cotransport system has been extensively characterized in renal cortex basolateral membrane vesicles,[4] isolated and perfused proximal tubules,[3,5–7] renal cell lines,[8–10] and immortalized proximal tubules cells[11,12] by a variety of techniques long before the successful cloning of the salamander and human renal $Na^+HCO_3^-$ cotransporter by Romero et al.[13] and Burnham et al.[14] in 1997. The coupling ratio of this transporter was found to be most likely a 3:1 coupling of HCO_3^- to Na^+, facilitating HCO_3^- efflux under resting conditions.[11,15] Transport was found to be inhibited by stilbenes and regulated by a variety of hormones and transmitters[16–19] and its activity to be modified by intracellular pH, similar to what is described for the Na/H exchangers.[20] Recent developments on the molecular characterization of the electrogenic $Na^+HCO_3^-$ cotransporter are reviewed in Reference 21.

Similar Na^+- and HCO_3^--dependent, stilbene-sensitive base transporters have been identified in a variety of organs and cell types, including the gastrointestinal tract. In some instances, such transport was Cl^- dependent, thus making it likely that it acts as a base importer under physiological ion concentrations gradients.[22–25] However, even Cl^--independent Na^+- and HCO_3^--dependent base transporters were found to mediate base influx, suggesting that $Na^+HCO_3^-$ cotransporters may exist with different coupling ratios when compared to the renal one. The biological function of these $Na^+HCO_3^-$ cotransporters was thought to lie in the regulation of the pH_i on a cellular level but also to serve as mechanisms to transport HCO_3^- and—to a lesser degree—Na^+ across epithelia.

FUNCTIONAL AND MOLECULAR EVIDENCE FOR NA+HCO3− COTRANSPORT SYSTEMS IN THE STOMACH AND THEIR POTENTIAL BIOLOGICAL ROLES

Early studies with amphibian gastric tissues in Ussing chamber setups suggested that a Na^+- and HCO_3^--dependent base importer may provide a part of the HCO_3^- destined for secretion into the lumen.[26] Also, a large body of experimental evidence has accumulated that demonstrated the paramount importance of HCO_3^- availability in the blood or serosal fluid for the maintenance of gastric epithelial integrity (reviewed in Refs. 27–29). The molecular mechanisms behind this phenomenon remained unknown, but these findings led to an interest in, and an investigation of, HCO_3^--dependent ion transport mechanisms immediately after techniques became available to study such transport systems.

Using a conventional microelectrode technique, Curci et al. described the existence of an electrogenic $Na^+HCO_3^-$ cotransporter in the basolateral membrane of frog oxyntic cells, which they thought was importing HCO_3^- into the cell.[30] The same group developed a technique to micropuncture those cells with pH-sensitive microelectrodes and provided evidence to suggest that frog oxyntic cells secrete bicarbonate in the non–acid secreting state and use a $Na^+HCO_3^-$ cotransporter for base uptake.[31,32] Yanaka et al. found that in bullfrog oxyntopeptic mucosa, pH_i recovery after intracellular acidification by a luminal acid load was stilbene sensitive and HCO_3^- dependent, but Caroppo et al. failed to detect evidence for a $Na^+HCO_3^-$ cotransporter in the basolateral membrane of the oxyntopeptic cells of this species.[33,34] Kiviluoto *et al.* found that in *Necturus* antrum, the ability of the cells to maintain a neutral pH_i in the presence of a luminal acid load was sensitive to stilbenes in the antrum, whereas it was amiloride sensitive in the fundus.[35] Results regarding the existence of a $Na^+HCO_3^-$ cotransporter in mammalian parietal cells have also yielded controversial results: Although Townley *et al.* found evidence for Na^+- and HCO_3^--dependent, stilbene-sensitive pH_i recovery in rabbit parietal cells, Thomas *et al.* did not.[36,37] We found no difference in parietal cell proton extrusion rates in the absence and presence of HCO_3^- after intracellular acidification to a pH_i range between 6.4 and 7.3,[38] and only a relatively small enhancement of ^{22}Na uptake into acid-loaded parietal-cell basolateral membrane vesicles by HCO_3^-.[39] However, in intact frog oxyntic mucosa, we observed Na^+- and HCO_3^--dependent base influx into the epithelial cells that was not fully inhibited by inhibitors of NaH exchange, suggesting that a $Na^+HCO_3^-$ cotransporter may be involved in base uptake (Stumpf *et al.*, unpublished observations). Inasmuch as it was known that a number of pitfalls existed that could falsely identify a $Na^+HCO_3^-$ cotransporter when none existed, the question about the existence of a $Na^+HCO_3^-$ cotransporter in the stomach appeared still open despite substantial experimental effort.

When the salamander renal $Na^+HCO_3^-$ cotransporter (NBC) was cloned, we were able to amplify a rabbit NBC cDNA fragment, and we used this probe to study the expression of NBC in the different cell types of rabbit gastric mucosa. Compared to NBC expression in the pancreas (not shown), kidney, or colon, NBC expression is low in the stomach (FIG. 1). However, in gastric surface cells, which show the highest expression of NBC among the different gastric cell types, NBC expression was as high or higher than that of other proton extrusion mechanisms expressed in this cell type, such as the isoforms NHE1 and NHE2 of the Na^+/H^+ exchanger gene fam-

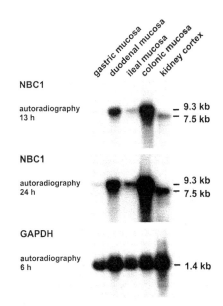

FIGURE 1. High-stringency Northern blot analysis of ~10 μg once-purified poly (A⁺) RNA from rabbit gastric mucosa, duodenal mucosa, ileal mucosa, colonic mucosa, and kidney cortex probed with ^{32}P-labeled homologous NBC1 cDNA fragment (435 bp; rabbit NBC1 nucleotide 661–1095) and exposed to X-ray film. After 13 h autoradiography bands are detected in all lanes, which intensify after 24 h autoradiography: a 9.3-kb transcript was found in gastric and intestinal mucosa, and a 7.5-kb transcript was found in kidney cortex. Rehybridization with a GAPDH probe confirms the loading of intact RNA in all lanes.

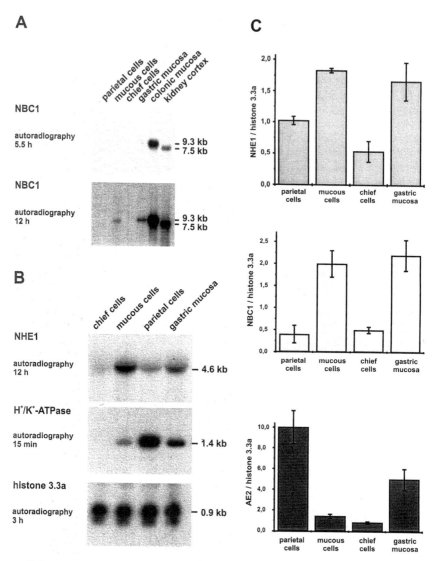

FIGURE 2. Northern analysis and semiquantitative PCR to study expression levels of NHE1, NBC1, and AE2 in the different gastric cell types (modified from Ref. 51, with permission). (**A**) High-stringency Northern blot analysis of ~15 μg once-purified poly (A+) RNA from the tissues as indicated, probed with ^{32}P-labeled homologous NBC1 cDNA. After 12-h autoradiography similar transcripts as in colonic mucosa appear in gastric mucosa and in the mucous cell fraction. (**B**) High-stringency Northern blot analysis of 30 μg total RNA from rabbit gastric epithelial cell types and gastric mucosa probed with ^{32}P-labeled homologous NHE1 (5′ untranslated region and nucleotides 1-1524 of the coding sequence, kindly provided by C.M. Tse and M. Donowitz), H+/K+-ATPase β-subunit (nucleotide 6-676[73]) and histone 3.3a (nucleotide 40-562[74]) cDNA fragments. The blot demonstrates that histone 3.3a is evenly distributed in the different gastric cell types, making it a suitable internal standard for semiquantitative PCR for evaluation of relative expression levels. (**C**)

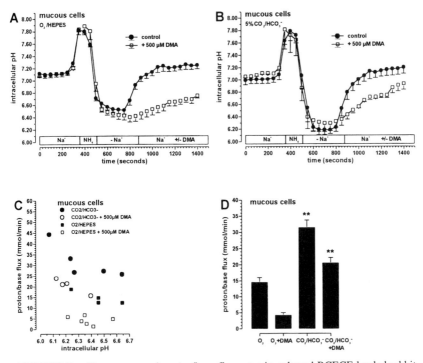

FIGURE 3. pH$_i$ recovery and proton/base flux rates in cultured BCECF-loaded rabbit mucous cells in the absence and presence of CO_2/HCO_3^- and dimethyl-amiloride (DMA) modified from Ref. 51, with permission). (**A, B**) pH$_i$ recovery after acidification by NH₄Cl prepulse in Hepes-buffered, O₂-gassed (**A**) and 5% CO_2/HCO_3^- (**B**), in the absence and presence of 500 μM DMA, which inhibits all NHE isoforms expressed in rabbit gastric cells. C) Initial proton/base flux rates in the first minute of acidification. For it is obvious that both Na⁺/H⁺ exchange rates (*closed squares*) and Na⁺HCO₃⁻ cotransport rates (*open circles*) are strongly dependent on pH$_i$. (**D**) Summary of proton/base flux rates. It is evident that a HCO₃⁻-dependent base importer is the major pH$_i$ recovery mechanism in mucous cells. For details of the technique see Ref. 75.

ily (FIG. 2). Correspondingly, a Na⁺HCO₃⁻ cotransport system was the major base importer in cultured rabbit surface cells (FIG. 3, A–D). Because the ability of the gastric mucosa to maintain its integrity in the face of a strong luminal acid load depends primarily on an intact surface layer (reviewed in Ref. 28), these findings explain, in

---

Semiquantitative RT-PCR analysis of NHE1 (**top**), NBC1, and AE2 in the different cell types of rabbit stomach. The integrated optical density of the ethidium bromide–stained bands was calculated and the ratio gene of interest/histone 3.3a was determined during the exponential phase of both reactions. The amplification curves of NHE1, NBC1, AE2, and histone 3.3a from rabbit gastric mucosa (0.8 μg total RNA) are not shown, but were parallel. For the bar graph the ratio of gene of interest/histone 3.3a (representing the relative expression level of AE2, see Ref. 51) was plotted for the stomach and the different epithelial cell types ($n = 3$). To ascertain the identity of the PCR amplimers, NBC1, AE2, and histone 3.3a were cloned and sequenced, and NHE1 was analyzed by restriction digest.

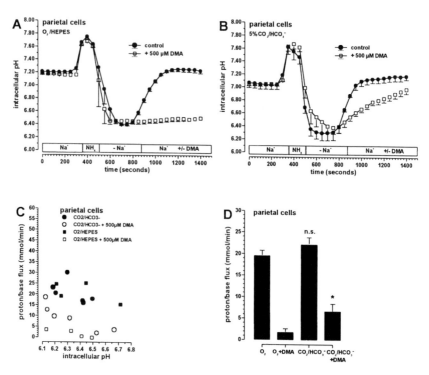

FIGURE 4. Identical experiments to those in FIG. 3 but performed in cultured rabbit parietal cells. Proton/base flux rates were not significantly different in the absence and presence of CO_2/HCO_3^-, but although DMA completely inhibited pH_i recovery in the absence of CO_2/HCO_3^-, it did not do so in its presence. This indicates that in CO_2/HCO_3^-, an alternative base importer is operative.

part, why the presence of HCO_3^- is of paramount importance for the stomach's ability to maintain a near-neutral pH during even very high luminal acid concentrations.

In cultured parietal cells acidified to various pH_i values, base import rates were identical in the absence and presence of CO_2/HCO_3^-, as observed years ago in isolated parietal cells.[38] However, after inhibition of all Na^+/H^+ exchanger isoforms expressed in this cell type, pH_i recovery was completely inhibited in the absence, but not the presence, of CO_2/HCO_3^- (FIG. 4, A–D), findings compatible with the action of a Na^+- and HCO_3^--dependent base importer active under these conditions. Interestingly, in the absence of Cl^-, the base import rates were clearly higher in the presence of CO_2/HCO_3^- compared to its absence even in parietal cells (FIG. 5). We speculate that the concomitant action of the Cl^-/HCO_3^- exchanger, which is very highly expressed in parietal cells (FIG. 2C), is also dependent on the presence of HCO_3^- but acts as an acid loader and may counteract the base-importing action of the $Na^+HCO_3^-$ cotransporter during recovery from an acid load in parietal cells. Curci et al. and Debellis et al. observed $Na^+HCO_3^-$ cotransport into frog oxyntic cells only when acid secretion was completely inhibited,[31,40] suggesting that indeed the

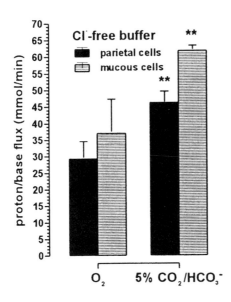

FIGURE 5. pH_i recovery from acidification by an $(NH_4)_2SO_4$ prepulse in Cl-depleted parietal (*full bars*) and mucous (*hatched bars*) cells in the absence and presence of $CO_2/$ HCO_3^-. Initial pH_i after acidification was 5.8–6.2, $n = 4–7$. It is evident that the presence of CO_2/HCO_3^- significantly enhanced the proton/base flux rates both in parietal and mucous cells (in contrast to the results in FIG. 4). Possibly, the inhibition of the Cl^-/HCO_3^- exchanger, which acidifies parietal cells, unmasks the alkalinizing effect of the $Na^+HCO_3^-$ cotransporter in parietal cells.

Cl^-/HCO_3^- exchanger must be quiescent to observe $Na^+HCO_3^-$ cotransport in parietal cells.

In summary, a $Na^+HCO_3^-$ cotransporter protein is both highly expressed and highly active in rabbit gastric surface cells. It is the major mediator of base uptake into acidified surface cells under physiological CO_2/HCO_3^- concentrations and therefore of immense importance in the pH_i maintenance of these cells during an acid load, to which these cells are continuously exposed in normal gastric physiology. Moreover, this $Na^+HCO_3^-$ cotransport system could also be involved in other biological functions, such as epithelial restitution, a term used to describe the phenomenon that superficial wounds in the gastrointestinal tracts are rapidly closed by migration of cells from the isthmus of the gastric glands.[41–43] This phenomenon is exquisitely pH sensitive and needs serosal CO_2/HCO_3^- to take place under physiologic acid concentrations in the lumen.[44] In parietal cells, expression levels of NBC is markedly lower than in surface cells, and an involvement in base uptake is seen only after inhibition of either the Cl^-/HCO_3^- or the Na^+/H^+ exchanger.[45] A biological role for pH_i regulation is envisioned during nighttime acid inhibition, when the gastric lumen can still contain very high luminal proton concentrations but when parietal cells are in the resting state.

FUNCTIONAL AND MOLECULAR EVIDENCE FOR THE NA⁺HCO₃⁻ COTRANSPORTER IN THE DUODENUM, PANCREAS, AND COLON

Duodenum

Duodenocytes are far more sensitive to damage by acid than gastric epithelial cells.[46] In the proximal duodenum of several species, including humans, particularly high rates of HCO_3^- secretion have been measured.[47] Duodenal HCO_3^- secretion has been shown to be stimulated by luminal acid[48,49] and to represent a major protective factor of the epithelium against gastric acid.[50] Duodenal ulcer patients have been shown to have a reduced HCO_3^- secretory response to luminal acid,[51] which is related to their infection with *helicobacter pylori*.[52]

Similar to HCO_3^- secretion in the stomach, duodenal HCO_3^- secretion *in vitro* has been found to be dependent on the availability of Na^+ and HCO_3^- and to be sensitive to the presence of stilbenes in the serosal compartment, suggesting that a part of the HCO_3^- ions destined for apical secretion may be imported through the basolateral enterocyte membrane via a $Na^+HCO_3^-$ cotransporter. Functional evidence for the possible presence of such a $Na^+HCO_3^-$ cotransport system in isolated rat duodenocytes was first provided by Isenberg *et al.*[53] We have isolated rabbit duodenal apical and basolateral membranes and clearly demonstrated the presence of HCO_3^--dependent $^{22}Na^+$ uptake into duodenal basolateral membrane vesicles (FIG. 6). This

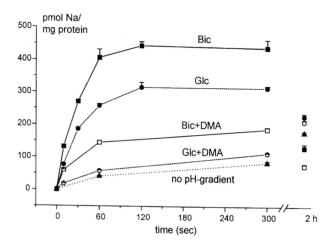

FIGURE 6. pH-gradient ($pH_i = 5.5$, $pH_o = 7.5$) driven $^{22}Na^+$ uptake into rabbit duodenal basolateral membrane vesicles (BLMvs) in the absence (glc = Na^+ gluconate) and presence of CO_2/HCO_3^- (bic = $Na^+HCO_3^-$), with or without the Na^+/H^+ exchange inhibitor dimethyl-amiloride (DMA). Duodenal BLMvs were isolated using a combination of differential and density gradient centrifugation. It is evident that the presence of HCO_3^- enhances $^{22}Na^+$ uptake into duodenal BLMvs and that DMA fully inhibits pH-driven $^{22}Na^+$ uptake in the absence, but not in the presence, of HCO_3^-. This demonstrates the presence of $Na^+HCO_3^-$ cotransport, which is Cl^- independent, inasmuch as no Cl^- was present in any of the buffers. For details of the technique, see Ref. 39.

^{22}Na$^+$ uptake was stilbene sensitive, not inhibited by even high concentrations of amiloride or analogues, which inhibit all Na$^+$/H$^+$ exchanger isoforms expressed in the intestine, and independent of Cl$^-$, thus representing Na$^+$HCO$_3^-$ cotransport. We have also studied the expression of NBC1 mRNA in the rabbit duodenum and found a high expression of this mRNA in the duodenal mucosa, both when analyzed by Northern blotting (FIG. 1) and when analyzed by semiquantitative PCR with the 18sRNA as an internal standard. As seen in FIGURE 1, the size of the NBC mRNA throughout the intestinal tract differs from that in the kidney. We have then cloned and sequenced the full-length rabbit duodenal NBC1 cDNA (GenBank accession no. AF 149418) and cloned and sequenced cDNA fragments from the N-terminal ends of the NBC1 from rabbit stomach, duodenum, pancreas, colon, and kidney. The N-terminal sequence of the NBC1 cloned from the duodenum was identical to that in other parts of the GI tract and the pancreas, demonstrating that a gastrointestinal NBC1 subtype exists with a different N-terminal sequence to that of the kidney, and that this subtype is expressed throughout the GI tract. In situ hybridization of NBC1 mRNA has demonstrated a predominant expression in villus cells in the rat (Urs Berger, personal communication). These are also the cells with the highest expression of carboanhydrase II, the enzyme involved in providing the HCO$_3^-$ ions that are generated intracellularly by CO$_2$ hydration (Herbert Helander, personal communication). We have functional evidence that rat duodenal villus cells, when stimulated to secrete anions, have a higher percentage of HCO$_3^-$ to Cl$^-$ ions in the secretion than crypt cells.[54] Therefore, in the duodenum, as well as in the stomach, the cells known to secrete HCO$_3^-$ are also the site of the highest expression of the NBC1.

Ileum and Colon

MacLeod et al. have described Na$^+$HCO$_3^-$ cotransport in guinea pig crypt cells in 1996, looking at ^{22}Na$^+$ fluxes from cells and Na$^+$- and HCO$_3^-$-related pH$_i$ and volume changes in crypt cells.[55] Compared to colon and duodenum, NBC1 expression in the ileum is quite low in the rabbit, and so is HCO$_3^-$ secretion by rat, mouse, and rabbit ileum when compared to other segments of the GI tract (Jacob et al., manuscript in preparation).

The colon is an organ known to secrete HCO$_3^-$, and the major biological significance of this is thought to be the reabsorption of Cl$^-$ in exchange for HCO$_3^-$. In 1991, Rajendran et al. have described a stilbene-sensitive Na$^+$-anion cotransporter in rat distal colon,[56] and in 1993 Teleky et al. described a Na$^+$HCO$_3^-$ cotransport in isolated human crypt cells, studying pH$_i$ recovery of BCECF-loaded crypt cells in the absence and presence of HCO$_3^-$ and stilbenes.[57] We find both basal and stimulated HCO$_3^-$ secretion in the rat distal colon that is, in part, sensitive to inhibition by basolateral stilbenes (Seidler et al., unpublished observations). Consistent with these functional data, we find a high expression of NBC1 in the colon of several species (see FIG. 1).

Pancreas

The pancreas is the organ with the highest bicarbonate secretory rates. Various transport systems have been implicated to mediate the basolateral HCO$_3^-$ uptake into the pancreatic duct cells.[58,59] Although Novak and Greger ruled out the existence of a $Na^+HCO_3^-$ contransporter in the basolateral membrane of rat pancreatic ducts on the basis of electrophysiological studies, Zhao *et al.* found evidence for its presence in the same cell type using fluorescence microscopy.[60,61] Moreover, it became clear that in some species, CO_2 hydration and basolateral extrusion of the generated protons via a Na^+/H^+ exchanger may not be the main mechanism of HCO_3^- ion generation destined for secretion into the ducts.[62] Using fluorescence microscopy and pH- and Na^+-sensitive dyes for the measurement of pH_i and $[Na^+]_i$ and the intraluminal pH in microdissected guinea pig pancreatic ducts, Ishiguro *et al.* demonstrated a HCO_3^--dependent, stilbene-sensitve accumulation of intracellular HCO_3^-, as well as a marked inhibition of intraluminal HCO_3^- secretion after basolateral stilbene application. Thus it was demonstrated that in this species, the majority of the secreted HCO_3^- is taken up via a $Na^+HCO_3^-$ cotransporter.[63,64] When the renal $Na^+HCO_3^-$ cotransporter was cloned, it became evident that a strong hybridization signal was present when pancreatic mRNA was probed with the renal NBC cDNA. Abuladze *et al.* first demonstrated that the NBC expressed in the the human pancreas had a unique N-terminal end not expressed in the kidney.[65] As mentioned above, we find the same situation in the rabbit. The intestinal (including the pancreatic) NBC1 subtype and the renal NBC1 subtype share 914 C-terminal amino acids, and the intestinal NBC1 has an additional 84. The kidney NBC1 40 N-terminal amino acids have very little sequence homology. The functional significance of this finding remains to be determined, but Abuladze *et al.* have pointed out that in the human pancreatic NBC1 the N terminus contains protein kinase phosphorylation sites that are missing in the kidney NBC1, suggesting a possibility of differential regulation by second messengers.[65] Boron and collaborators are currently investigating whether the different transport direction in kidney (basolateral HCO_3^- efflux) versus pancreatic and intestinal $Na^+HCO_3^-$ cotransport (basolateral HCO_3^- influx) may be explained by their different N termini (Boron, personal communication).

NBC1 antibodies developed by Bernhard Schmitt in the laboratory of Walter Boron have delineated a basolateral localization of NBC1 in both kidney, stomach, and intestine.[66,67] Antibodies made against the pancreatic NBC1 demonstrated basolateral staining in pancreatic acini and ducts and also some apical staining in the ducts.[68]

FUNCTIONAL AND MOLECULAR EVIDENCE FOR THE EXISTENCE OF ADDITIONAL NA$^+$ HCO$_3^-$ COTRANSPORT SYSTEMS IN THE GASTROINTESTINAL TRACT

Recently, a gene was cloned from human retina with a 53% homology to the human NBC sequence and called NBC2.[69] Because no expression studies were performed, the functional proof that this gene codes for an $Na^+HCO_3^-$ cotransporter is still lacking; however, the high homology to NBC (then called NBC1) does indeed suggest that it represents a gene encoding a HCO_3^--transporting protein. In the retina, apical as well as basolateral $Na^+HCO_3^-$ cotransport processes have been functionally described.[70] Because the NBC2 message was found in low levels in other organs, we studied the expression of NBC2 throughout the gastrointestinal tract by semiquantitative PCR. Overall NBC2 expression was markedly lower than NBC1 expression and was found only in certain cell types, such as gastric chief cells,[45] or certain intestinal segments, such as the ileum.[71] NBC2 expression levels were similar or slightly higher than NBC1 expression levels. These cell types or organs are not among those known to secrete high rates of HCO$_3^-$ secretion, and we are currently further investigating the potential role of NBC2 in the GI tract.

It is likely that additional Na$^+$- and HCO$_3^-$-dependent ion transport systems, such as a Cl$^-$-dependent one, which are as yet unidentified on a molecular level, are present in the gastrointestinal tract.[72] The successful cloning of NBC cDNAs from various tissues and the fact that the NBC shares a homology of approximately 40% with the AE anion exchanger family make it very likely that the molecular nature of other ion transporters with a high HCO$_3^-$ affinity will be established in the near future. This will greatly facilitate our understanding of the structure–function relationships and the cellular regulation of this group of anion transport proteins.

ACKNOWLEDGMENTS

The authors thank Walter Boron, Rosella Caroppo, Bernhard Schmitt, Lucantonio Debellis, Silvana Curci, Eberhard Frömter, Gunnar Flemström, Eero Kivilaakso, Tuula Kiviluoto, Gary Shull, and Urs Berger for helpful discussions.

REFERENCES

1. BORON, W.F. & E.L. BOULPAEP. 1983. Intracellular pH regulation in the renal proximal tubule of the salamander. Basolateral HCO$_3^-$ transport. J. Gen. Physiol. **81**: 53–94.
2. ALPERN, R.J. 1990. Cell mechanisms of proximal tubule acidification. Physiol. Rev. **70**: 79–114.
3. MOE, O.W., P.A. PREISIG & R.J. ALPERN. 1990. Cellular model of proximal tubule NaCl and NaHCO$_3$ absorption. Kidney Int. **38**: 605–611.
4. ARONSON, P.S., M. SOLEIMANI & S.M. GRASSL. 1991. Properties of the renal Na(+)-HCO$_3^-$ cotransporter. Semin. Nephrol. **11**: 28–36.
5. KRAPF, R., R.J. ALPERN, F.C. RECTOR JR. & C.A. BERRY. 1987. Basolateral membrane Na/base cotransport is dependent on CO$_2$/HCO$_3$ in the proximal convoluted tubule. J. Gen. Physiol. **90**: 833–853.
6. PREISIG, P.A. & R.J. ALPERN. 1989. Basolateral membrane H-OH-HCO$_3$ transport in the proximal tubule. Am. J. Physiol. **256**: F751–F765.
7. SEKI, G., S. COPPOLA, K. YOSHITOMI *et al.* 1996. On the mechanism of bicarbonate exit from renal proximal tubular cells. Kidney Int. **49**: 1671–1677.
8. JENTSCH, T.J., H. MATTHES, S.K. KELLER & M. WIEDERHOLT. 1986. Electrical properties of sodium bicarbonate symport in kidney epithelial cells (BSC-1). Am. J. Physiol. **251**: F954–F968.
9. JENTSCH, T.J., B.S. SCHILL, P. SCHWARTZ *et al.* 1985. Kidney epithelial cells of monkey origin (BSC-1) express a sodium bicarbonate cotransport. Characterization by ^{22}Na$^+$ flux measurements. J. Biol. Chem. **260**: 15554–15560.
10. JENTSCH, T.J., P. SCHWARTZ, B.S. SCHILL *et al.* 1986. Kinetic properties of the sodium bicarbonate (carbonate) symport in monkey kidney epithelial cells (BSC-1). Interactions between Na$^+$, HCO$_3^-$, and pH. J. Biol. Chem. **261**: 10673–10679.
11. GROSS, E. & U. HOPFER. 1996. Activity and stoichiometry of Na$^+$:HCO$_3^-$ cotransport in immortalized renal proximal tubule cells. J. Membr. Biol. **152**: 245–252.
12. GROSS, E. & U. HOPFER. 1998. Voltage and cosubstrate dependence of the Na-HCO3 cotransporter kinetics in renal proximal tubule cells. Biophys. J. **75**: 810–824.
13. ROMERO, M.F., M.A. HEDIGER, E.L. BOULPAEP & W.F. BORON. 1997. Expression cloning and characterization of a renal electrogenic Na$^+$/HCO$_3^-$ cotransporter. Nature **387**: 409–413.

14. BURNHAM, C.E., H. AMLAL, Z. WANG et al. 1997. Cloning and functional expression of a human kidney Na⁺:HCO₃⁻ cotransporter. J. Biol. Chem. **272:** 19111–19114.
15. SOLEIMANI, M., S.M. GRASSI & P.S. ARONSON. 1987. Stoichiometry of Na⁺-HCO-3 cotransport in basolateral membrane vesicles isolated from rabbit renal cortex. J. Clin. Invest. **79:** 1276–1280.
16. RUIZ, O.S. & J.A. ARRUDA. 1992. Regulation of the renal Na-HCO3 cotransporter by cAMP and Ca-dependent protein kinases. Am. J. Physiol. **262:** F560–F565.
17. RUIZ, O.S., Y.Y. QIU, L.J. WANG & J.A. ARRUDA. 1995. Regulation of the renal Na-HCO3 cotransporter: IV. Mechanisms of the stimulatory effect of angiotensin II. J. Am. Soc. Nephrol. **6:** 1202–1208.
18. RUIZ, O.S., Y.Y. QIU, L.R. CARDOSO & J.A. ARRUDA. 1997. Regulation of the renal Na-HCO3 cotransporter: VII. Mechanism of the cholinergic stimulation. Kidney Int. **51:** 1069–1077.
19. RUIZ, O.S., Y.Y. QIU, L.R. CARDOSO & J.A. ARRUDA. 1998. Regulation of the renal Na-HCO3 cotransporter: IX. Modulation by insulin, epidermal growth factor and carbachol. Regul. Pept. **77:** 155–161.
20. SOLEIMANI, M., Y.J. HATTABAUGH & G.L. BIZAL. 1992. pH sensitivity of the Na⁻:HCO₃⁻ cotransporter in basolateral membrane vesicles isolated from rabbit kidney cortex. J. Biol. Chem. **267:** 18349–18355.
21. ROMERO, M.F. & W.F. BORON. 1999. Electrogenic Na⁺/HCO₃⁻ cotransporters: cloning and physiology. Annu. Rev. Physiol. **61:** 699–723.
22. ALPERN, R.J. & M. CHAMBERS. 1987. Basolateral membrane Cl/HCO₃ exchange in the rat proximal convoluted tubule. Na-dependent and -independent modes. J. Gen. Physiol. **89:** 581–598.
23. OLSNES, S., J. LUDT, T.I. TONNESSEN & K. SANDVIG. 1987. Bicarbonate/chloride antiport in Vero cells: II. Mechanisms for bicarbonate-dependent regulation of intracellular pH. J. Cell Physiol. **132:** 192–202.
24. OLSNES, S., T.I. TONNESSEN & K. SANDVIG. 1986. pH-regulated anion antiport in nucleated mammalian cells. J. Cell Biol. **102:** 967–971.
25. TONNESSEN, T.I., J. LUDT, K. SANDVIG & S. OLSNES. 1987. Bicarbonate/chloride antiport in Vero cells: I. Evidence for both sodium-linked and sodium-independent exchange. J. Cell Physiol. **132:** 183–191.
26. TAKEUCHI, K., A. MERHAV & W. SILEN. 1982. Mechanism of luminal alkalinization by bullfrog fundic mucosa. Am. J. Physiol. **243:** G377–G388.
27. KIVILAAKSO, E. & W. SILEN. 1979. Pathogenesis of experimental gastric-mucosal injury. N. Engl. J. Med. **301:** 364–369.
28. SILEN, W. 1987. Gastric mucosal defense and repair. In Physiology of the Gastrointestinal Tract. L.R. Johnson, Ed.: 1055–1069. Raven Press. New York.
29. ALLEN, A., G. FLEMSTROM, A. GARNER & E. KIVILAAKSO. 1993. Gastroduodenal mucosal protection. Physiol. Rev. **73:** 823–857.
30. CURCI, S., L. DEBELLIS & E. FROMTER. 1987. Evidence for rheogenic sodium bicarbonate cotransport in the basolateral membrane of oxyntic cells of frog gastric fundus. Pfluegers Arch. **408:** 497–504.
31. CURCI, S., L. DEBELLIS, R. CAROPPO & E. FROMTER. 1994. Model of bicarbonate secretion by resting frog stomach fundus mucosa. I. Transepithelial measurements. Pfluegers Arch. **428:** 648–654.
32. DEBELLIS, L., R. CAROPPO, E. FROMTER & S. CURCI. 1998. Alkaline secretion by frog gastric glands measured with pH microelectrodes in the gland lumen. J. Physiol. (Lond.) **513:** 235–241.
33. YANAKA, A., K.J. CARTER, P.J. GODDARD & W. SILEN. 1991. Effect of luminal acid on intracellular pH in oxynticopeptic cells in intact frog gastric mucosa. Gastroenterology **100:** 606–618.
34. CAROPPO, R., S. COPPOLA & E. FROMTER. 1994. Electrophysiological investigation of microdissected gastric glands of bullfrog. I. Basolateral membrane properties in the resting state. Pfluegers Arch. **429:** 193–202. (Abstract).
35. KIVILUOTO, T., H. MUSTONEN, J. SALO & E. KIVILAAKSO. 1995. Regulation of intracellular pH in isolated Necturus gastric mucosa during short-term exposure to luminal acid. Gastroenterology **108:** 999–1004.

36. TOWNSLEY, M.C. & T.E. MACHEN. 1989. Na-HCO3 cotransport in rabbit parietal cells. Am. J. Physiol. **257:** G350–356.
37. THOMAS, H.A. & T.E. MACHEN. 1991. Regulation of Cl/HCO$_3$ exchange in gastric parietal cells. Cell Regul. **2:** 727–737.
38. SEIDLER, U., M. HUBNER, S. ROITHMAIER & M. CLASSEN. 1994. pH$_i$ and HCO$_3$-dependence of proton extrusion and Cl(-)-base exchange rates in isolated rabbit parietal cells. Am. J. Physiol. **266:** G759–G766.
39. LAMPRECHT, G., U. SEIDLER & M. CLASSEN. 1993. Intracellular pH-regulating ion transport mechanisms in parietal cell basolateral membrane vesicles. Am. J. Physiol. **265:** G903–910.
40. DEBELLIS, L., C. IACOVELLI, E. FROMTER & S. CURCI. 1994. Model of bicarbonate secretion by resting frog stomach fundus mucosa. II. Role of the oxyntopeptic cells. Pfluegers Arch. **428:** 655–663.
41. SVANES, K., S. ITO, K. TAKEUCHI & W. SILEN. 1982. Restitution of the surface epithelium of the *in vitro* frog gastric mucosa after damage with hyperosmolar sodium chloride. Morphologic and physiologic characteristics. Gastroenterology **82:** 1409–1426.
42. ITO, S., E.R. LACY, M.J. RUTTEN *et al.* 1984. Rapid repair of injured gastric mucosa. Scand. J. Gastroenterol. Suppl. **101:** 87–95.
43. PAIMELA, H., P.J. GODDARD & W. SILEN. 1995. Present views on restitution of gastrointestinal epithelium. Dig. Dis. Sci. **40:** 2495–2496.
44. SVANES, K., K. TAKEUCHI, S. ITO & W. SILEN. 1983. Effect of luminal pH and nutrient bicarbonate concentration on restitution after gastric surface cell injury. Surgery **94:** 494–500.
45. ROSSMANN, H., O. BACHMANN, W.-K. SIEGEL *et al.* 1999. Na$^+$HCO$_3^-$ cotransport and expression of NBC1 and NBC2 in cultured rabbit parietal and surface cells. Gastroenterology **116:** 1389–1398.
46. PAIMELA, H., T. KIVILUOTO, H. MUSTONEN *et al.* 1990. Tolerance of rat duodenum to luminal acid. Dig. Dis. Sci. **35:** 1244–1248.
47. AINSWORTH, M.A., M.A. KOSS, D.L. HOGAN & J.I. ISENBERG. 1995. Higher proximal duodenal mucosal bicarbonate secretion is independent of Brunner's glands in rats and rabbits. Gastroenterology **109:** 1160–1166.
48. HEYLINGS, J.R., B.C. HURST & A. GARNER. 1984. Effect of luminal acid on gastric and duodenal bicarbonate transport. Scand. J. Gastroenterol. Suppl. **92:** 59–62.
49. KNUTSON, L. & G. FLEMSTROM. 1989. Duodenal mucosal bicarbonate secretion in man. Stimulation by acid and inhibition by the alpha 2-adrenoceptor agonist clonidine. Gut **30:** 1708–1715.
50. AINSWORTH, M.A., C. FENGER, P. SVENDSEN & O.B. SCHAFFALITZKY DE MUCKADELL. 1993. Effect of stimulation of mucosal HCO3$^-$ secretion on acid-induced injury to porcine duodenal mucosa. Scand. J. Gastroenterol. **28:** 1091–1097.
51. ISENBERG, J.I., J.A. SELLING, D.L. HOGAN & M.A. KOSS. 1987. Impaired proximal duodenal mucosal bicarbonate secretion in patients with duodenal ulcer. N. Engl. J. Med. **316:** 374–379.
52. HOGAN, D.L., R.C. RAPIER, A. DREILINGER *et al.* 1996. Duodenal bicarbonate secretion: eradication of *Helicobacter pylori* and duodenal structure and function in humans [see comments]. Gastroenterology **110:** 705–716.
53. ISENBERG, J.I., M. LJUNGSTROM, B. SAFSTEN & G. FLEMSTROM. 1993. Proximal duodenal enterocyte transport: evidence for Na(+)-H+ and Cl(-)-HCO3$^-$ exchange and NaHCO3 cotransport. Am. J. Physiol. **265:** G677–685.
54. GUBA, M., M. KUHN, W.G. FORSSMANN *et al.* 1996. Guanylin strongly stimulates rat duodenal HCO3$^-$ secretion: proposed mechanism and comparison with other secretagogues [see comments]. Gastroenterology **111:** 1558–1568.
55. MACLEOD, R.J., F. REDICAN, P. LEMBESSIS *et al.* 1996. Sodium-bicarbonate cotransport in guinea pig ileal crypt cells. Am. J. Physiol. **270:** C786–C793.
56. RAJENDRAN, V.M., M. OESTERLIN & H.J. BINDER. 1991. Sodium uptake across basolateral membrane of rat distal colon. Evidence for Na-H exchange and Na-anion cotransport. J. Clin. Invest. **88:** 1379–1385.
57. TELEKY, B., G. HAMILTON, E. COSENTINI *et al.* 1994. Intracellular pH regulation of human colonic crypt cells. Pfluegers Arch. **426:** 267–275.

58. CASE, R.M. & B.E. ARGENT. 1990. Pancreatic secretion: *in vivo*, perfused gland, and isolated duct studies. Methods Enzymol. **192:** 256–271.
59. VILLANGER, O., T. VEEL & M.G. RAEDER. 1995. Secretin causes $H^+/HCO_3 HCO_3^-$ secretion from pig pancreatic ductules by vacuolar-type H(-)-adenosine triphosphatase. Gastroenterology **108:** 850–859.
60. NOVAK, I. & R. GREGER. 1988. Electrophysiological study of transport systems in isolated perfused pancreatic ducts: properties of the basolateral membrane. Pfluegers Arch. **411:** 58–68.
61. ZHAO, H., R.A. STAR & S. MUALLEM. 1994. Membrane localization of H^+ and HCO_3^- transporters in the rat pancreatic duct. J. Gen. Physiol. **104:** 57–85.
62. GROTMOL, T., T. BUANES, O. BRORS & M.G. RAEDER. 1986. Lack of effect of amiloride, furosemide, bumetanide and triamterene on pancreatic NaHCO3 secretion in pigs. Acta Physiol. Scand. **126:** 593–600.
63. ISHIGURO, H., M.C. STEWARD, A.R. LINDSAY & R.M. CASE. 1996. Accumulation of intracellular HCO_3^- by Na(+)-HCO_3^- cotransport in interlobular ducts from guinea-pig pancreas. J. Physiol. (Lond.) **495:** 169–178.
64. ISHIGURO, H., M.C. STEWARD, R.W. WILSON & R.M. CASE. 1996. Bicarbonate secretion in interlobular ducts from guinea-pig pancreas. J. Physiol. (Lond.) **495:** 179–191.
65. ABULADZE, N., I. LEE, D. NEWMAN *et al.* 1998. Molecular cloning, chromosomal localization, tissue distribution, and functional expression of the human pancreatic sodium bicarbonate cotransporter. J. Biol. Chem. **273:** 17689–17695.
66. SCHMITT, B.M., D. BIEMESDERFER, M.F. ROMERO *et al.* 1999. Immunolocalization of the electrogenic Na$^+$-HCO_3^- cotransporter in mammalian and amphibian kidney. Am. J. Physiol. Renal Physiol. **45:** F27–F38.
67. SCHMITT, B.M., D. BIEMESDERFER, S. BRETON *et al.* 1998. Immunolocalization of the electrogenic Na$^+/HCO_3^-$ cotransporter (NBC) in mammalian epithelia. Pflügers Arch. **435:** R59. (Abstract).
68. ROUSSA, E., B.M. SCHMITT, M.F. ROMERO & F. THEVENOD. 1999. Molecular cloning, biochemical characterization and immunolocalization of the rat pancreatic Na$^+$/HCO_3^- cotransporter. Pfluegers Arch. **437:** R106. (Abstract).
69. ISHIBASHI, K., S. SASAKI & F. MARUMO. 1998. Molecular cloning of a new sodium bicarbonate cotransporter cDNA from human retina. Biochem. Biophys. Res. Commun. **246:** 535–538.
70. KENYON, E., A. MAMINISHKIS, D.P. JOSEPH & S.S. MILLER. 1997. Apical and basolateral membrane mechanisms that regulate pH_i in bovine retinal pigment epithelium. Am. J. Physiol. **273:** C456–C472.
71. JACOB, P., S. CHRISTIANI, H. ROSSMANN *et al.* 2000. Role of Na$^+$HCO$_3^-$ cotransporter NBC1, H$^+$/K$^+$ exchanger NHE1 and carbonic anhydrase in rabbit duodenal bicarbonate secretion. Gastroenterology **119**(2): 406–419.
72. BLEICH, M., R. WARTH, I. THIELE & R. GREGER. 1998. pH-regulatory mechanisms in *in vitro* perfused rectal gland tubules of *Squalus acanthias*. Pfluegers Arch. **436:** 248–254.
73. REUBEN, M.A., L.S. LASATER & G. SACHS. 1990. Characterization of the beta subunit of the gastric H$^+$/K$^+$ transporting ATPase. Proc. Natl. Acad. Sci. USA **87:** 6767–6771.
74. CHALMERS, M. & M.J. WELSH. 1999. Extreme sequence conservation characterises the rabbit H3.3A histone gene. Nucleic Acid Res. **18:** 3075–3075.
75. BACHMANN, O., T. SONNENTAG, W.K. SIEGEL *et al.* 1998. Different acid secretagogues activate different H$^+$/K$^+$ exchanger isoforms in rabbit parietal cells. Am. J. Physiol. Gastrointestinal Liver Physiol. **38:** G1085–G1093.

Characterization and Molecular Localization of Anion Transporters in Colonic Epithelial Cells

VAZHAIKKURICHI M. RAJENDRAN[a] AND HENRY J. BINDER

Department of Internal Medicine, Yale University School of Medicine, New Haven, Connecticut 06520, USA

ABSTRACT: This study describes the identification and characterization of anion transporters in apical membrane (APM) and basolateral membrane (BLM) of rat distal colon. $Cl\text{-}HCO_3$, Cl-OH, Cl-butyrate, and butyrate-HCO_3 exchanges and Na-HCO_3 cotransporter are present in rat distal epithelial cells. $Cl\text{-}HCO_3$ exchange (1) is present only in APM from surface, but not from crypt cells; (2) is also present in BLM; and (3) of surface cell is encoded by anion exchange (AE)–1 isoform, whereas BLM $Cl\text{-}HCO_3$ is encoded by AE2 isoform. Cl-OH exchange is present only in APM, but not in BLM from surface and crypt cells, and is responsible for regulation of cell functions (i.e., cell pH and cell volume regulation). Butyrate-HCO_3 exchange (1) is also present in apical membrane vesicles (AMV) from surface, but not from crypt cells; (2) is present in BLM; and (3) is responsible for SCFA-dependent HCO_3 secretion. By contrast, Cl-butyrate exchange: (1) is present in APM from both surface and crypt cells; (2) is not present in BLM; and (3) recycles butyrate by absorbing Cl. Na-HCO_3 cotransport: (1) is present only in BLM; (2) is expressed predominantly in midcrypt regions; and (3) may be linked to HCO_3 secretion. A mechanism for HCO_3 movement across the crypt apical membrane has not as yet been identified.

BACKGROUND

Multiple ion transport processes have been identified in colonocytes that are involved both in fluid absorption and secretion and in the regulation of one or more intracellular functions.[1] Many of these investigations have focused on the role of Na-dependent processes, but the identification of the function and regulation of anion transport mechanisms is not as comprehensive. During the past decade we have explored several different aspects of anion transport function primarily in the rat distal colon. The present communication summarizes some of these functions and the initial molecular identification of apical membrane anion exchanges and basolateral membrane Na-HCO_3 cotransport (NBC).

[a]Address for correspondence: V. M. Rajendran, Ph.D., Department of Internal Medicine, Yale University School of Medicine, 333 Cedar St., New Haven, CT 06520. Voice: 203-785-4131; fax: 203-737-1755.
vazhaikkurichi.rajendran@yale.edu

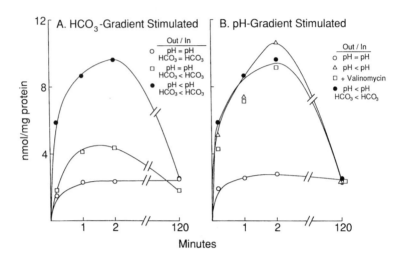

FIGURE 1. Cl-anion exchange in apical membrane: **(A)** Outward HCO_3 gradient in the absence of pH gradient stimulates Cl uptake *(open squares)* compared to in the absence of HCO_3 gradient *(open circles).* Additional imposition of OH-gradient (pH_o/pH_i: 5.5/7.5) enhances the HCO_3 gradient–driven Cl uptake *(closed circles).* **(B)** Outward OH gradient (pH_o/pH_i: 5.5/7.5) stimulates Cl uptake *(open squares)* compared to that in the absence of pH gradient *(open circles).* Additional imposition of HCO_3 gradient did not stimulate the OH gradient–driven Cl uptake *(closed circles).* Voltage clamping with K and valinomycin did not inhibit HCO_3 and OH gradient–driven Cl uptake *(open triangles).* (Reproduced with permission from Rajendran & Binder.[7])

APICAL MEMBRANE ANION TRANSPORTERS

Cl-HCO$_3$ and Cl-OH Exchanges

In vitro studies have established that electroneutral Cl absorption is both Na dependent and HCO_3 dependent and have proposed that Na-H exchange and Cl-HCO_3 exchange coupled via intracellular pH is the mechanism for electroneutral Na-Cl absorption in intestinal epithelial cells including rat distal colon.[1-3] Although several studies characterized Na-H as the mechanism for electroneutral Na absorption,[4-6] limited studies have characterized the mechanism of electroneutral Cl absorption. Thus, to establish the mechanism of Cl uptake across the apical membrane, Cl uptake studies were performed with apical membrane vesicles (AMV) isolated from rat distal colon.[7] Cl uptake was driven by both artificially imposed outward HCO_3 gradient ($HCO_3o < HCO_{3i}$) and OH gradient (pHo/pHi: 5.5/7.5) (FIG. 1). HCO_3 gradient–driven Cl uptake (Cl-HCO_3 exchange) was further stimulated by an additional imposition of an outward OH gradient (FIG. 1A). By contrast, OH gradient–driven Cl uptake (Cl-OH exchange) was not altered by the additional presence of outward HCO_3 gradient (FIG. 1B). These results suggest that OH ion stimulates Cl uptake via both HCO_3– and OH ion–dependent pathways, whereas HCO_3 stimulates Cl uptake only via a HCO_3-dependent pathway.[7]

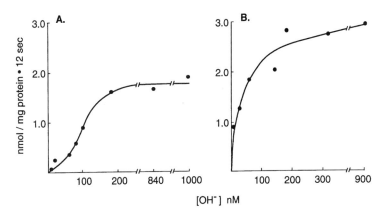

FIGURE 2. Effect of HCO_3 on Cl-OH exchange kinetics: **(A)** Increasing intravesicular [OH] in the nominal absence of HCO_3 stimulates and saturates Cl uptake with a *sigmoidal curve* indicating the presence of two binding sites for Cl. **(B)** Increasing intravesicular [OH] in the presence of 25 mM HCO_3 saturates the Cl uptake, with a *hyperbolic curve*, indicating that Cl uptake in the presence of HCO_3 occurs only via Cl-OH exchange. (Reproduced with permission from Rajendran & Binder.[7])

This hypothesis was confirmed by studies of the effect of [OH] on Cl uptake (FIG. 2). In these kinetic studies increasing intravesicular [OH] in the absence of HCO_3 stimulated Cl uptake with a sigmoidal curve (FIG. 2A), indicating the presence of more than one Cl binding site in the extravesicular membrane. By contrast, in the presence of HCO_3, increasing intravesicular [OH] stimulated Cl uptake with a hyperbolic curve (FIG. 2B), indicating the stimulation of Cl uptake via a single transport system. Additional kinetic analyses indicate (1) K_m for Cl of Cl-HCO_3 and Cl-OH exchanges were approximately 11.9 ± 1.5 and 22.6 ± 2.5 mM, respectively; and (2) K_i for DIDS for Cl-HCO_3 and Cl-OH exchanges were 7.8 ± 0.6 and 106.6 ± 6.5 μM, respectively (TABLE 1). These results establish that (1) Cl-HCO_3 and Cl-OH exchanges of colonic AMV are distinct and separate anion exchange processes; and (2) Cl-HCO_3 exchange has an affinity for OH ion in the nominal absence of HCO_3 ion.

SCFA-HCO₃ Exchange and Cl-SCFA Exchange

Butyrate, a short-chain fatty acid (SCFA), stimulates electroneutral Na-Cl absorption in the absence of HCO_3.[8–10] SCFAs synthesized by colonic bacteria are the predominant anions in the large intestine.[8] Major SCFAs that stimulate electroneutral Na-Cl absorption include butyrate > propionate > acetate.[8,9] As a result, the initial model of SCFA-stimulated Na-Cl absorption proposed (1) nonionic SCFA (SCFAH), formed in the luminal acidic microclimate and maintained by apical Na-H exchange, diffuses into surface epithelial cells across the apical membrane; (2) SCFAH then acidifies these cells by releasing SCFA⁻ and H⁺ with activation of apical Na-H exchange; and (3) SCFA⁻ partially recycles across the apical membrane via Cl-SCFA exchange.[9] To test this proposed model, SCFA uptake studies were per-

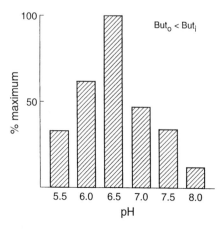

FIGURE 3. Effect of pH on Cl-butyrate exchange. Outward butyrate gradient–driven Cl uptake is maximal at pH 6.5. (Reproduced with permission from Rajendran & Binder.[12])

formed in AMV from rat distal colon.[11] Initial studies were designed to establish the presence of nonionic diffusion of SCFA. In these studies the effect of various pHs with (pHo < pHi) and without (pHo = pHi) gradients on SCFA uptake was examined. Absence of significant stimulation of SCFA uptake at lower pH levels indicated that neither nonionic diffusion nor SCFA-OH exchange in colonic AMV was a major mechanism for apical uptake of SCFA.[11]

SCFA absorption and SCFA-stimulated electroneutral Na-Cl absorption are associated with HCO_3 secretion and SCFA-Cl exchange.[9,10] Thus, the effect of various anions on SCFA uptake was also examined.[11] SCFA uptake was stimulated by an outward HCO_3 gradient; an outward SCFA gradient also stimulated SCFA uptake. By contrast, outward Cl gradients did not stimulate SCFA uptake. Increasing extravesicular [SCFA] saturated HCO_3 gradient–driven SCFA uptake with a K_m for butyrate of 26.9 ± 1.6 mM. SCFA-HCO_3 exchange was inhibited by several SCFAs. These results established that SCFA-HCO_3 exchange is a carrier-mediated process, and apical SCFA uptake is not a result of a diffusion mechanism.

The cellular model of SCFA-stimulated electroneutral Na-Cl absorption includes an SCFA-Cl exchange mechanism.[9] However, an outward Cl gradient did not stimulate SCFA uptake.[11] Because the proposed model includes intracellular SCFA recycling via Cl-SCFA exchange,[9] the effect of an outward SCFA gradient on Cl uptake was examined in AMV from distal colon.[12] An outward SCFA gradient under voltage clamp conditions significantly stimulated DIDS-sensitive Cl uptake. It is possible that SCFAH diffusion could establish an outward alkaline pH gradient (i.e., outward OH gradient) and stimulate Cl uptake via Cl-OH exchange. To establish whether SCFA stimulated Cl uptake via Cl-OH exchange, SCFA gradient–driven Cl uptake was measured at different pHs in the presence of FCCP, a proton ionophore (FIG. 3). SCFA gradient–driven Cl uptake was greatest at pH 6.5. Lower rates of Cl uptake at pHs both lower and higher than pH 6.5 suggest that SCFA-stimulated Cl uptake is not a result of Cl-OH exchange. SCFA gradient–driven Cl uptake is (1) saturated by increasing extravesicular [Cl] with a K_m for Cl of 26.8 ± 3.4 mM; (2) saturated by increasing intravesicular [SCFA] with a K_m for butyrate of 5.9 ± 1.4 mM; and (3) inhibited by DIDS with half-maximal inhibitory constants (IC_{50}) of approx-

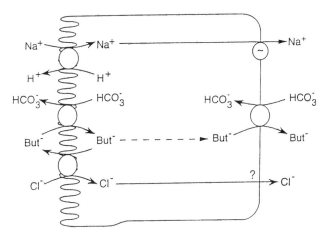

FIGURE 4. Cellular model of butyrate-stimulated Na-Cl absorption. (Reproduced with permission from Rajendran & Binder.[12])

imately 6.8 ± 0.4 μM. These observations establish that SCFA gradient–driven Cl uptake occurs via a carrier-mediated Cl-SCFA exchange. Electroneutral ion exchangers in general are bidirectional and transport ions in both directions. By contrast, although outward SCFA gradient stimulated Cl uptake, an outward Cl-gradient did not stimulate SCFA uptake. Thus, the unidirectional Cl-SCFA exchange is a novel anion exchanger that may be explained by asymmetrical SCFA binding to its transport protein.[12] As a result of these observations, the cellular model of SCFA-stimulated electroneutral Na-Cl absorption was revised: (1) initial SCFA uptake is via an apical SCFA-HCO_3 exchange that acidifies the cells and activates Na-H exchange; and (2) SCFA recycles via an apical Cl-SCFA exchange (FIG. 4).

Spatial Distribution of Apical Membrane Anion Exchanges in Surface and Crypt Cells

Inasmuch as absorptive processes are generally believed present in surface, but not in crypt, cells,[13] the spatial distribution of Cl-HCO_3, Cl-OH, Cl-SCFA, and SCFA-HCO_3 exchanges in surface and crypt cells was examined. FIGURE 5 presents the results of Cl-HCO_3 and Cl-OH exchanges in AMV prepared from both surface and crypt cells. Compared to the DIDS-sensitive Cl-HCO_3 exchange in surface cells, Cl-HCO_3 exchange present in AMV from crypt cells was only 2.4%. By contrast, Cl-OH exchange in crypt AMV was 50% of that in surface AMV (FIG. 5, right panel). SCFA-HCO_3 and SCFA-SCFA exchanges are present only in AMV from surface cells but not in AMV from crypt cells (FIG. 6). By contrast, Cl-SCFA exchange is present in AMV from both surface and crypt cells (FIG. 6). Thus, the distributions of these four apical membrane anion exchanges in surface and crypt cells differ: HCO_3-dependent anion exchanges (Cl-HCO_3 and SCFA-HCO_3 exchanges) are almost exclusively present only in surface cells, whereas HCO_3-independent anion exchanges (Cl-OH and Cl-SCFA exchanges) are present both in surface and in crypt cells.

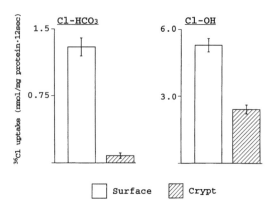

FIGURE 5. Cl-anion exchanges in surface and crypt cells. Cl-HCO$_3$ exchange predominates in apical membranes from surface, but not in crypt, cells (*left panel*). Cl-OH exchange is present in apical membranes of both surface and crypt cells (*right panel*). (Reproduced with permission from Rajendran & Binder.[19])

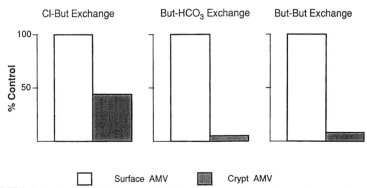

FIGURE 6. Spatial distribution of SCFA exchanges in surface and crypt cells. Cl-butyrate exchange is present in apical membranes of both surface and crypt cells (*left panel*), whereas butyrate-HCO$_3$ and butyrate-butyrate exchanges are present in apical membranes of surface, but not crypt, cells (*middle and right panels*).

The spatial distribution of Cl-HCO$_3$ and SCFA-HCO$_3$ exchanges suggests that their functions are linked to transport events exclusively present in surface epithelial cells. Thus, we have speculated that Cl-HCO$_3$ exchange mediates electroneutral HCO$_3$-dependent Cl absorption, whereas Cl-OH exchange is responsible for one or more cell functions—for example, cell volume and pHi regulation. In addition to the spatial distribution of SCFA-HCO$_3$ exchange to surface and crypt cells, recent studies indicate that certain SCFA-dependent transport events occur in surface but not in crypt cells.[14] Thus, exposure of the crypt lumen to butyrate (SCFA) fails to induce acidification, whereas superfusion of surface cells to butyrate is associated with intracellular acidification.[14] This latter observation is consistent with butyrate-induced HCO_3 extrusion across the apical membrane in surface but not in crypt cells. Because SCFA-HCO_3 exchange in crypt cells is 5% of that in surface cells (FIG. 6), we suspect that SCFA-HCO_3 exchange is responsible for luminal SCFA-induced acidification in surface cells. Similarly, exposure of the basolateral membranes of colonic crypts to butyrate is associated with acidification, an observation consistent with the presence of a butyrate-HCO_3 exchange in basolateral membranes.[15]

Regulation of Apical Membrane Anion Transporters by Aldosterone

Studies with intact tissues established that aldosterone inhibits electroneutral Na-Cl absorption and induces electrogenic Na absorption.[16] Subsequent studies demonstrated that aldosterone inhibits electroneutral Na-H exchange and induces electrogenic Na uptake in apical membranes of rat distal colon.[17] Aldosterone inhibition of Na-H exchange occurs at a pretranslational level, as aldosterone reduces apical Na-H exchange (NHE2 and NHE3) isoform–specific mRNA and protein abundance in distal colon.[18] Cl uptake studies indicate that Cl-HCO_3 exchange is almost completely inhibited in AMV from distal colon of aldosterone-treated animals (FIG. 7A). By contrast, Cl-OH exchange is not inhibited by aldosterone (FIG. 7B). These results confirm the conclusion that Cl-HCO_3 exchange is responsible for Cl absorption, whereas Cl-OH exchange may be involved in one or more cell functions (e.g., pHi, cell volume). These results also confirm that aldosterone inhibits Na-Cl absorption by the downregulation of both Na-H exchange[18] and Cl-HCO_3 exchange.[19]

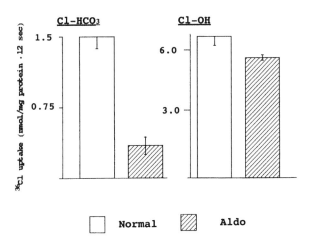

FIGURE 7 Effect of aldosterone on Cl-anion exchanges in apical membrane. Aldosterone substantially inhibits Cl-HCO_3 exchange, but does not inhibit Cl-OH exchange in apical membranes. (Reproduced with permission from Rajendran & Binder.[19])

BASOLATERAL ANION TRANSPORTERS

Cholera toxin and other bacterial enterotoxins induce plasma-like fluid secretion with high HCO_3 concentration.[20] The secreted HCO_3 is derived both from HCO_3 synthesized intracellularly by carbonic anhydrase from the metabolic end product of CO_2 and water, and from extracellular/blood HCO_3. Although several studies have examined the role of carbonic anhydrase in HCO_3 secretion, there is evidence that HCO_3 secretion is in part dependent on serosal HCO_3 and Na.[20] This observation indicates that extracellular HCO_3 is critical for HCO_3 secretion and suggests that HCO_3 uptake across the basolateral membrane is mediated by a Na-dependent process.

Cl-HCO₃ Exchange

The initial experiment of anion transport in basolateral membrane was designed to determine whether a mechanism for anion gradient stimulation of Cl uptake was present in BLMV isolated from rat distal colon.[21] In these studies the effect of outward OH gradient (pH_o/pH_i: 5.5/7.5) and outward HCO_3 gradient on Cl uptake was examined. Similar to the observations in AMV, both an outward OH gradient and an outward HCO_3 gradient stimulated transient Cl accumulation in BLMV.[21] Simultaneous imposition of HCO_3 and OH gradients resulted in a rate of Cl uptake that was greater than that observed in the presence of a HCO_3 gradient or a OH gradient alone. To determine whether Cl uptake occurred as a result of a tightly coupled electroneutral anion exchange and/or an electrodiffusionally coupled potential-dependent (i.e., electrogenic) process, OH and HCO_3 gradient–stimulated Cl uptake was performed under voltage clamped conditions. Voltage clamping (produced by valinomycin and K) almost completely inhibited OH gradient–stimulated Cl uptake. In contrast to OH gradient–driven Cl uptake, voltage clamping only partially inhibited (40%) the initial rate of HCO_3 gradient–driven Cl uptake. This result suggests that an HCO_3 gradient stimulated Cl uptake via both electroneutral (Cl-HCO_3 exchange) and electrodiffusional coupled pathways. Cl-HCO_3 exchange was almost completely inhibited by 1 mM DIDS but was not inhibited by NPPB (10 μM), a Cl channel blocker. By contrast, under non-voltage clamp conditions 10 μM NPPB only partially inhibited HCO_3 gradient–driven Cl uptake. It should be noted that 1 mM DIDS is both an inhibitor of electroneutral anion exchanges and a Cl channel blocker.[22] The observations that voltage clamping inhibited the potential-dependent component of HCO_3 gradient–driven Cl uptake and that NPPB did not affect the HCO_3 gradient–driven Cl uptake under voltage clamp conditions indicate that electroneutral Cl-HCO_3 exchange is present in these vesicles. These results further suggest that an HCO_3 gradient stimulated Cl uptake both via an electroneutral anion exchange and via an NPPB-sensitive, potential-dependent electrogenic process. The relationship of these BLM HCO_3 transport processes and HCO_3 secretion remain to be established.

SCFA-HCO₃ Exchange

Transepithelial SCFA transport is present in ruminal epithelium.[8,15] Although studies in intact tissue suggested nonionic diffusion as the mechanism for SCFA absorption, studies with AMV demonstrated SCFA-HCO_3 exchange as a mechanism for SCFA uptake.[11] Therefore, the effect of different pHs with (pH_o/pH_i: 5.5/7.5; 6.0/8.0) and without (pH_o/pH_i: 5.5/5.5; 6.5/6.5; 7.5/7.5) pH gradients and outward HCO_3 gradients on butyrate (SCFA) uptake was examined.[23] Similarly to AMV, SCFA uptake was not affected by pH, indicating an absence of nonionic diffusion of SCFA in these BLMV. By contrast, SCFA uptake was stimulated by an outward HCO_3 gradient. This HCO_3 gradient–driven SCFA uptake was (1) saturated as a function of increasing extravesicular [butyrate] with a K_m for butyrate of 6.9 ± 0.4 mM; (2) saturated by increasing intravesicular [HCO_3] with a K_m for HCO_3 of 27.4 ± 8.1 mM; (3) inhibited competitively by propionate; and (4) inhibited by DIDS. These results indicate that HCO_3 gradient–driven SCFA uptake is a carrier-mediated anion exchange process and that this BLMV SCFA-HCO_3 exchange is distinct from that in AMV. Thus, the increase in luminal HCO_3 concentration by SCFA may occur as a result of transepithelial HCO_3 transport mediated by the coordinated regulation of both apical and basolateral SCFA-HCO_3 exchanges.

Na-HCO₃ Cotransport

Both *in vivo* and *in vitro* studies have identified HCO_3 secretion in mammalian colon.[24–26] These studies revealed that basolateral Na-dependent, Cl-independent transcellular HCO_3 transport is associated with HCO_3 secretion. However, the mechanism of HCO_3 uptake across basolateral membranes of colonic epithelia is not known. Recently, we demonstrated novel electrogenic HCO_3 gradient–driven Na uptake (Na-HCO_3 cotransport) in BLMV isolated from rat distal colon.[27] Basolateral Na-HCO_3 cotransport is also responsible for HCO_3 secretion in renal proximal tubules.[28] The properties of colonic Na-HCO_3 cotransport, however, differ from those of renal Na-HCO_3 cotransport. The colonic Na-HCO_3 cotransport is sensitive to both DIDS and amiloride (FIG. 8), whereas renal Na-HCO_3 cotransport is sensitive only to DIDS.[28] A recent study with intact tissues described basolateral Na-dependent HCO_3 secretion that is only DIDS sensitive in rat distal colon.[29] The explanation for these different results of Na-dependent HCO_3 between vesicle and intact tissue stud-

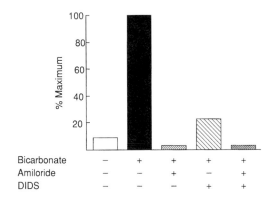

FIGURE 8. Na-HCO_3 cotransport in basolateral membrane vesicles (BLMV). Imposition of inward HCO_3 stimulated Na uptake in basolateral membrane vesicles. Both amiloride and DIDS inhibit HCO_3 gradient–driven Na uptake. (Reproduced with permission from Rajendran *et al.*[27])

ies is not known. Increasing intravesicular [H] activated colonic Na-HCO$_3$ cotransport by increasing its maximal uptake rate (i.e., V_{max} for Na).[30] This observation suggests that Na-HCO$_3$ cotransport is an important mechanism for HCO$_3$ uptake across the basolateral membrane and is involved in HCO$_3$ secretion.

During secretagogue-induced HCO$_3$ secretion, HCO$_3$ movement across apical membrane occurring via Cl-HCO$_3$ exchange will acidify pH$_i$, which in turn should stimulate HCO$_3$ uptake across the basolateral membrane by activating Na-HCO$_3$ cotransport. Further studies are in progress to test these possibilities and whether secretagogues directly activate Na-HCO$_3$ cotransport in colonic epithelial cells.

MOLECULAR IDENTITIES OF COLONIC ANION TRANSPORTERS

Anion Exchange Isoforms

Three different anion exchange (AE) cDNAs (AE1, AE2, and AE3) have been cloned from one or more nonintestinal rat tissues.[31,32] In vitro expression of these AE cDNAs revealed the Cl-HCO$_3$ exchange function.[33] Although at least four physiologically distinct Cl-anion exchanges (Cl-HCO$_3$, Cl-OH, Cl-SCFA, and SCFA-HCO$_3$ exchanges) have been identified in the AMV of rat distal colon, their molecular identities are not known.[7,11,12] As discussed earlier, Cl-HCO$_3$ exchange is present only in surface cells (FIG. 5A), whereas Cl-OH exchange activity is present in AMV from both surface and crypt cells of distal colon (FIG. 5B). Therefore, initial studies were performed to examine whether any of the AE isoform–specific mRNA are expressed in rat distal colon.[34] AE isoform–specific cDNA fragments were amplified using the reverse-transcriptase chain reaction (RT-PCR) technique. cDNA synthesized from mRNA of rat distal colonocytes was used as a template. AE1, AE2, and AE3 isoform–specific primers were designed based on the published sequences. cDNA fragments of the expected size, amplified by RT-PCR, were purified, subcloned, and sequenced. The sequence of AE1 and AE2 isoform–specific cDNA fragments had 100% homology to their respective isoforms. By contrast, an AE3 isoform–specific cDNA fragment was not amplified by RT-PCR, suggesting the absence of the AE3 isoform in rat colonocytes. The RT-PCR–amplified colonic AE1 and AE2 cDNA fragments were then used as probes to analyze mRNA abundance by Northern blot analyses. AE1 and AE2 isoform–specific cDNA probes hybridized with 4.6- and 4.2-kb–sized mRNA of normal rat distal colon, respectively. Because the AE3 isoform–specific cDNA fragment was not amplified by RT-PCR techniques, a full-length AE3 cDNA of heart (provided by Dr. Gary E. Shull, University of Cincinnati, Cincinnati, OH) was used as a probe to screen the Northern blot. This AE3 probe also did not hybridize with mRNA isolated from colonocytes of distal colon (V.M. Rajendran & H.J. Binder, unpublished observations). These observations provide the initial demonstration of the expression of at least two AE gene products (AE1 and AE2) in the rat distal colon.

In Situ Hybridization Studies

Absorptive processes are generally believed to operate in surface, but not in crypt, cells.[13] Cl-HCO$_3$ exchange takes place only in AMV from surface cells, whereas Cl-OH exchange takes place in AMV from both surface and crypt cells (FIGS. 5A and 5B). Thus, we proposed that $Cl-HCO_3$ and Cl-OH exchanges are responsible for Cl absorption and for the regulation of one or more cell functions (e.g., pH_i and cell volume regulation), respectively. Recent studies have also demonstrated that Na-H exchange isoform–3 (NHE3), which is responsible for Na absorption, is expressed exclusively in surface epithelial cells; whereas NHE1 isoform, responsible for housekeeping, is localized in basolateral membranes of both surface and crypt cells.[6,18] Thus, experiments were designed to examine whether, similar to the distribution of $Cl-HCO_3$ and Cl-OH exchange, AE isoforms also exhibit cell-specific expression in the distal colon. The possibility of spatial distribution of AE1 and AE2 isoform–specific mRNA transcripts in surface and crypt cells was assessed by *in situ* hybridization studies.[34] AE1 antisense probe hybridization signal was present within surface cells of the distal colon, but not within crypt cells. In contrast to the distribution of AE1 transcripts, AE2 hybridization signal was detected in both surface and crypt cells of the distal colon. No hybridization signal was observed either with AE1 or with AE2 sense probes in the distal colon. This observation, together with the spatial distribution of Cl-anion exchanges, provides strong evidence suggesting that AE1 mRNA encodes $Cl-HCO_3$ exchange.

Regulation of AE1 and AE2 Isoform–Specific mRNA Abundance by Aldosterone

Aldosterone inhibits electroneutral Na-Cl absorption,[16] Na-H exchange,[17] and $Cl-HCO_3$ exchange (FIG. 7)[18] and reduces apical membrane Na-H exchange isoform (NHE2 and NHE3)–specific mRNA and protein abundance.[18] It is not known whether, similarly to Na-H exchange messages, aldosterone also reduces AE isoform–specific mRNA abundances. Northern blot analyses indicate that aldosterone substantially reduced AE1, but not AE2, isoform–specific mRNA abundance in the rat distal colon. This observation supports the conclusion that AE1 encodes $Cl-HCO_3$ exchange in the rat distal colon.

Peptide Localization Studies

Three distinct AE2 isoforms (AE2a, AE2b, and AE2c) with N terminal variations that are transcribed by different promoters have recently been identified in noncolonic tissue.[35] Although AE2 mRNA expression was localized in both surface and crypt cells by a "common" AE2 cDNA probe (i.e,. common to AE2a, AE2b, and AE2c), it is possible that AE2 isoforms may be expressed in different cell types and/or membranes. Thus, immunofluorescent studies were performed using a monoclonal antibody raised against a 13–amino acid polypeptide that encodes the C terminal end of the AE2 isoform and is common to all three AE2 isoforms.[36] Western blot analyses indicated that AE2 antibody recognized a 170-kDa protein in both apical and basolateral membranes of the rat distal colon. The expression of AE2-specific protein was at least fourfold higher in basolateral than in apical membranes. AE2 expression in apical membrane may represent either contamination from the basolateral membrane or a different AE2 isoform. Cell- and membrane-specific localization established by immunofluorescent studies indicates that AE2 isoform–specific protein was expressed only in basolateral, but not in apical, membranes of surface cells. By contrast, AE2 expression was localized in both apical and basolateral membranes of crypt cells. These observations suggest that different AE2 isoforms are likely expressed in different membranes and cell types of the rat distal colon. Since the antibody used in this study is common to all three AE2 isoforms, further studies with AE2 isoform–specific antibodies will be required to establish the specific expression of AE2 isoforms.

Membrane-specific expression of AE1 isoform has not been identified. However, surface cell–specific mRNA expression is consistent with the possibility that AE1 isoform may be responsible for Cl absorption across apical membranes of the rat distal colon. By contrast, AE2, which is expressed in basolateral membranes of both surface and crypt cells, and apical membranes of crypt cells may be involved in more than one function: (1) AE2, in concert with AE1, may regulate transcellular Cl transport in surface cells; (2) similarly, apical- and basolateral-specific AE2 isoforms may be involved in Cl transport in crypt cells; and (3) basolateral AE2 may be important in the regulation of one or more cell functions (e.g., pH_i, cell volume). This latter possibility is supported by the recent *in vitro* expression studies of AE2 in oocytes in which AE2 was regulated both by pH_i[37] and by volume.[38]

Na-HCO₃ Cotransporter

A cDNA that encodes $Na\text{-}HCO_3$ cotransport has recently been cloned from rat kidney (rNBC).[39] *In vitro* expression of rNBC exhibited DIDS-sensitive electrogenic $Na\text{-}HCO_3$ cotransport activity.[39] In initial studies we used rNBC isoform–specific cDNA probe and polyclonal antibodies to examine the mRNA (Northern blot analyses) and protein (immunofluorescent studies) expression in the rat distal colon, respectively.[40] The rNBC-specific polyclonal antibody used was raised against a fusion protein that encodes a portion of rNBC.[40] rNBC polyclonal antibody identified a 160-kDa protein in basolateral membranes of both kidney and distal colon (unpublished observations). This result indicates the expression of rNBC or an rNBC-like protein in the distal colon. Immunocytochemical analysis localized an rNBC-like protein that is expressed predominantly in basolateral membranes of the midcrypt region in the rat distal colon.

Northern blot analyses were performed to establish whether rNBC cDNA hybridizes with single or multiple mRNA transcripts in the rat distal colon. rNBC cDNA hybridizes with a 7.5-kb transcript in colonic mRNA; by contrast, it hybridizes with a 7.0-kb mRNA in the kidney.[40] This observation indicates the expression of a different NBC isoform in the rat distal colon. However, the colonic isoform may have significant nucleotide and amino acid homology with rNBC. It is possible, similarly to AE2 isoforms, that NBC isoforms of rat colon and kidney may differ primarily at their N terminal end. To confirm this hypothesis, cloning and characterization of the full-length cDNA of colonic NBC will be required.

CONCLUSIONS

Cl- and SCFA-dependent HCO_3 secretion occurs as a result of $Cl\text{-}HCO_3$ and SCFA-HCO_3 exchanges, respectively. Although $Cl\text{-}HCO_3$ and SCFA-HCO_3 exchangers secrete HCO_3, their major function is the absorption of Cl and SCFA, respectively, across apical membranes. Several observations provide the basis for the speculation: (1) The general consensus is that absorptive processes are located primarily in surface cells, whereas secretory processes are in crypt cells. FIGURES 5A and 6A demonstrate that Cl-HCO_3 and SCFA-HCO_3 exchanges are present primarily in AMV prepared from surface cells, not in AMV from crypt cells. By contrast, Cl-OH and Cl-SCFA exchanges are present in both surface and crypt cells (FIGS. 5B and 6B). (2) Aldosterone inhibits NaCl absorption and Na-H exchange; FIGURE 7 presents evidence that aldosterone inhibits Cl-HCO_3 exchange but does not substantially alter Cl-OH exchange. Thus, these observations indicate that Cl-HCO_3 exchange and SCFA-HCO_3 exchange are present in surface cells and are most likely associated with transepithelial electrolyte movement of Cl and SCFA. Conversely, Cl-OH exchange is present in both surface and crypt cells; we propose that Cl-OH exchange is associated not with transepithelial transport, but rather with the regulation of one or more intracellular functions (i.e., pH_i, cell volume).

Specific expression of AE1 mRNA and the presence of Cl-HCO_3 exchange only in surface cells suggests that AE1 isoform encodes Cl-HCO_3 exchange in colonic epithelial cells. In addition, the effect of aldosterone on colon-specific AE isoforms parallels aldosterone alteration of Cl-anion exchange. Thus, AE1 isoform mRNA abundance is substantially reduced in the distal colon from aldosterone animals compared to normal animals. By contrast, AE2 isoform mRNA abundance is comparable in both normal and aldosterone-treated animals and that AE2-specific protein is not expressed in apical membranes of surface cells. This pattern of AE1 isoform mRNA abundance can be correlated with the effect of aldosterone on Cl-HCO_3 exchange. Aldosterone inhibits Cl-HCO_3 exchange, but not Cl-OH exchange. As a result, we speculate that AE1 isoform encodes Cl-HCO_3 exchange, which is responsible for transepithelial Cl absorption. By contrast, AE2 isoform is closely associated with an as-yet-unidentified basolateral Cl-anion exchange function. We also speculate that Cl-HCO_3 exchange is responsible for Cl-dependent HCO_3 secretion in surface cells, as Cl-HCO_3 exchanges are present in both apical and basolateral membranes.

Pharmacologically distinct Na-HCO_3 cotransport that has high homology with renal Na-HCO_3 cotransport is expressed in the rat distal colon. This colonic Na-HCO_3 cotransport is localized predominantly in basolateral membranes of midcrypt cells. As a result, we propose that Na-HCO_3 cotransport is responsible for Na-dependent HCO_3 uptake across basolateral membranes and transepithelial HCO_3 secretion. If confirmed, this would establish a mechanism for HCO_3 uptake across the basolateral membrane. By contrast, a mechanism for HCO_3 movement across apical membranes of colonic crypt cells has not as yet been established.

ACKNOWLEDGMENTS

We thank Ann Thompson for excellent secretarial assistance. This study was supported by USPHS Research Grant DK 14669-27 from the National Institute of Diabetes and Digestive and Kidney Diseases.

REFERENCES

1. BINDER, H.J. & G.I. SANDLE. 1994. Electrolyte transport in the mammalian colon. *In* Physiology of the Gastrointestinal Tract, 3rd edit., Vol. 1. L.R. Johnson, Ed.: 2133–2172. Raven Press. New York.
2. FRIZZELL, R.A., M. FIELD & S.G. SHULTZ. 1979. Sodium-coupled chloride transport by epithelial tissues. Am. J. Physiol. **236:** F1–F8.

3. BINDER, H.J., E S. FOSTER, M E. BUDINGER & J.P. HAYSLETT. 1987. Mechanism of elec-troneutral sodium-chloride absorption in distal colon of the rat. Gastroenterology **93:** 449–455.
4. RAJENDRAN, V.M. & H.J. BINDER. 1990. Characterization of Na-H exchange in apical membrane vesicles of rat colon. J. Biol. Chem. **265:** 8408–8414.
5. KNICKELBEIN, R.G., P.S. ANDERSON, W. ATHERTON & J.W. DOBBINS. 1983. Sodium and chloride transport across rabbit ileal brush border. I. Evidence for Na-H exchange. Am J. Physiol. **245:** G504–G510.
6. YUN, C.H., C.M. TSE, S.K. NATH *et al.* 1995. Mammalian Na^+/H^+ exchanger gene fam-ily: structure and function studies. Am. J. Physiol. **269:** G1–11.
7. RAJENDRAN, V.M. & H.J. BINDER. 1993. $Cl-HCO_3$ and Cl-OH exchanges mediate Cl uptake in apical membrane vesicles of rat distal colon. Am. J. Physiol. **264:** G874–G879.
8. BERGMAN, E.N. 1990. Energy contribution of volatile fatty acids from the gastrointesti-nal tract in various species. Physiol. Rev. **70:** 567–590.
9. BINDER, H.J. & P. MEHTA. 1989. Short-chain fatty acids stimulate active sodium and chloride absorption *in vitro* in the rat distal colon. Gastroenterology **96:** 989–996.
10. ARGENZIO, R.A. & S.C. WHIPP. 1979. Inter-relationship of sodium chloride, bicarbon-ate and acetate transport by the colon of the pig. J. Physiol. (Lond.) **295:** 356–381.
11. MASCOLO, N., V.M. RAJENDRAN & H.J. BINDER. 1991. Mechanism of short-chain fatty acid absorption in apical membrane vesicles of rat distal colon. Gastroenterology **101:** 331–338.
12. RAJENDRAN, V.M. & H.J. BINDER. 1994. Apical membrane Cl-butyrate exchange: mechanism of short-chain fatty acid stimulation of active chloride absorption in rat distal colon. J. Membr. Biol. **141:** 51–58.
13. WELSH, M.J., P.L. SMITH, M. FROMM & R.A. FRIZZELL. 1982. Crypts are the site of intestinal fluid and electrolyte secretion. Science **218:** 1219–1221.
14. SINGH, S.K., H.J. BINDER, A.C. CIMINI & W.F. BORON. 1996. Unusually low apical per-meability to butyric acid in perfused colonic crypts. Gastroenterology **110:** A361 (Abstr.).
15. STEVENS, C.E. 1998. Comparative Physiology of the Vertebrate Digestive System. Cambridge University Press. New York.
16. HALEVY, J., M.E. BUDINGER, J.P. HAYSLETT & H.J. BINDER. 1986. Role of aldosterone in the regulation of sodium and chloride transport in the distal colon of sodium-depleted rats. Gastroenterology **91:** 1227–1233.
17. RAJENDRAN, V.M., M. KASHGARIAN & H.J. BINDER. 1989. Aldosterone induction of electrogenic sodium transport in the apical membrane vesicles of rat distal colon. J. Biol. Chem. **264:** 18638–18644.
18. IKUMA, M., H.J. BINDER, M. KASHGARIAN & V.M. RAJENDRAN. 1999. Differential regu-lation of NHE-isoforms by sodium depletion in proximal and distal segments of rat colon. Am. J. Physiol. **276:** G539–G549.
19. RAJENDRAN, V.M. & H.J. BINDER. 1999. Distribution and regulation of apical Cl/anion exchanges in surface and crypt cells of rat distal colon. Am. J. Physiol. **276:** G132–G137.
20. POWELL, D.W., L.I. SOLBERG, G.R. PLOTKIN *et al.* 1971. Experimental diarrhea 3. Bicarbonate transport in rat salmonella enterocolitis. Gastroenterology **60:** 1076–1086.
21. IKUMA, M., J. GEIBEL, H.J. BINDER & V.M. RAJENDRAN. 1998. $Cl-HCO_3$ and Cl-OH exchanges of basolateral membrane (BLM) regulate intracellular pH (pHi) in crypt cells of rat distal colon. Gastroenterology **114:** A380 (Abstr.).
22. LINSDELL, P. & J.W. HANRAHAN. 1996. Disulfonic stilbene block of cystic fibrosis transmembrane conductance regulator Cl channels expressed in a mammalian cell line and its regulation by a critical pore residue. J. Physiol. **496:** 687–693.
23. REYNOLDS, D.A., V.M. RAJENDRAN & H.J. BINDER. 1993. Bicarbonate-stimulated ^{14}C-butyrate uptake in basolateral membrane vesicles of rat distal colon. Gastroenterol-ogy **105:** 725–732.
24. SHEERIN, H.E. & M. FIELD. 1975. Ileal HCO_3 secretion: relationship to Na and Cl transport and effect of theophylline. Am. J. Physiol. **228:** 1056–1074.

25. SULLIVAN, S.K. & P.L. SMITH. 1986. Bicarbonate secretion by rabbit proximal colon. Am. J. Physiol. **251:** G436–G445.
26. SMITH, P.L., M.A. CASCAIRO & S.K. SULLIVAN. 1985. Sodium dependence of luminal alkalization by rabbit ileal mucosa. Am. J. Physiol. **249:** G358–G368.
27. RAJENDRAN, V.M., M. OESTERLIN & H.J. BINDER. 1991. Sodium uptake across basolateral membrane of rat distal colon. Evidence for Na-H exchange and Na-anion cotransport. J. Clin. Invest. **88:** 1379–1385.
28. BORON, W.F. & E.L. BOULPAEP. 1989. The electrogenic Na/HCO3 cotransporter. Kidney Int. **36:** 392–402.
29. FELDMAN, G.M. 1994. HCO_3 secretion by rat distal colon: effects of inhibitors and extracellular Na. Gastroenterology **107:** 329–338.
30. RAJENDRAN, V.M. & H.J. BINDER. 1994. Differential modulation of Na-HCO_3 cotransport and Na-H exchange by pH in basolateral membrane vesicles of rat distal colon. J. Biol. Chem. **269:** 156–160.
31. KUDRYCKI, K.E. & G.E. SHULL. 1990. Primary structure of the rat kidney band 3 anion exchange protein deduced from a cDNA. J. Biol. Chem. **264:** 8185–8192.
32. KUDRYCKI, K.E., P.R. NEWMAN & G.E. SHULL. 1990. cDNA cloning and tissue distribution of mRNAs for two proteins that are related to the band 3 Cl-HCO_3 exchanger. J. Biol. Chem. **265:** 462–471.
33. ALPER, S. L. 1994. The band 3-related anion exchanger gene family. Cell Physiol. Biochem. **4:** 265–281.
34. RAJENDRAN, V.M., J. BLACK, T.A. ARDITO et al. 2000. Regulation of DRA and AE1 in rat colon by dietary Na depletion. Am. J. Physiol. **279:** G931–G942.
35. WANG, Z., P.J. SCHULTHEIS & G.E. SHULL. 1996. Three N-terminal variants of the AE2 Cl-HCO_3 exchanger are encoded by mRNA transcribed from alternative promoters. J. Biol. Chem. **271:** 7835–7843.
36. RAJENDRAN, V.M., P. SANGAN, A.S. MANN et al. 1998. Tissue specific expression of basolateral anion exchange (AE)–2 isoform in rat distal colon. Gastroenterology **114:** A408 (Abstr.).
37. HUMPHREYS, B.D., L. JIANG, M.N. CHERNOVA & S.L. ALPER. 1994. Functional characterization and regulation by pH of murine AE2 anion exchanger expressed in Xenopus oocytes. Am. J. Physiol. **267:** C1295–C1307.
38. HUMPHREYS, B.D., L. JIANG, M.N. CHERNOVA & S.L. ALPER. 1995. Hypertonic activation of AE2 anion exchanger in Xenopus oocytes via NHE-mediated intracellular alkalinization. Am. J. Physiol. **268:** C201–C209.
39. ROMERO, M.F., P. FONG, U.V. BERGER, M.A. HEDIGER & W.F. BORON. 1998. Cloning and functional expression of rNBC, an electrogenic Na-HCO_3 cotransporter from rat kidney. Am. J. Physiol. **274:** F425–F432.
40. SCHMITT, B.M., D. BIEMESDERFER, S. BRETON et al. 1998. Immunolocalization of the electrogenic Na/HCO3 cotransporter (NBC) in mammalian epithelia. Pflugers Arch. **435** (Suppl.): O5–3 (Abstr.).

Short-Term Regulation of NHE3 by EGF and Protein Kinase C but Not Protein Kinase A Involves Vesicle Trafficking in Epithelial Cells and Fibroblasts

M. DONOWITZ,[a] A. JANECKI, S. AKHTER, M. E. CAVET, F. SANCHEZ,
G. LAMPRECHT, M. ZIZAK, W. L. KWON, S. KHURANA, C. H. C. YUN,
AND C. M. TSE

*Departments of Physiology and Medicine, Johns Hopkins University School of Medicine,
Baltimore, Maryland 21205-2195, USA*

ABSTRACT: NHE3 is an intestinal epithelial isoform Na^+/H^+ exchanger that is present in the brush border of small intestinal, colonic, and gallbladder Na^+-absorbing epithelial cells. NHE3 is acutely up- and downregulated in response to some G protein–linked receptors, tyrosine kinase receptors, and protein kinases when studied in intact ileum, when stably expressed in PS120 fibroblasts, and in the few studies reported in the human colon cancer cell line Caco-2. In most cases this is due to changes in V_{max} of NHE3, although in response to cAMP and squalamine there are also changes in the $K'(H^+)_i$ of the exchanger. The mechanism of the V_{max} regulation as shown by cell surface biotinylation and confocal microscopy in Caco-2 cells and biotinylation in PS120 cells involves changes in the amount of NHE3 on the plasma membrane. In addition, in some cases there are also changes in turnover number of the exchanger. In some cases, the change in amount of NHE3 in the plasma membrane is associated with a change in the amount of plasma membrane. A combination of biochemical studies and transport/inhibitor studies in intact ileum and Caco-2 cells demonstrated that the increase in brush border Na^+/H^+ exchange caused by acute exposure to EGF was mediated by PI 3–kinase. PI 3–kinase was also involved in FGF stimulation of NHE3 expressed in fibroblasts. Thus, NHE3 is another example of a transport protein that is acutely regulated in part by changing the amount of the transporter on the plasma membrane by a process that appears to involve vesicle trafficking and also to involve changes in turnover number.

PHYSIOLOGY OF INTESTINAL NHE3

Na^+/H^+ exchanger isoform 3 (NHE3) is an Na^+/H^+ exchanger isoform that is present in the apical membrane of small intestinal, colonic, and gallbladder Na^+-absorbing cells.[1,2] In the small intestine and colon of human, dog, rabbit, rat, mouse, and chicken, NHE3 is present in the brush border (BB) of the villus (small intestinal)

[a]Address for correspondence: Mark Donowitz, M.D., GI Division, Ross 925, Johns Hopkins University School of Medicine, 720 Ruland Avenue, Baltimore, MD 21205-2195. Voice: 410-955-9675; fax: 410-955-9677.
mdonowit@welch.jhu.edu

and surface (colon and gallbladder) epithelial cells and also in the BB of epithelial cells of the upper crypt. These are the epithelial cells that are involved in Na^+ absorption.

The Na^+ absorptive process, which involves NHE3, is called neutral NaCl absorption. In neutral NaCl absorption, one molecule of Na^+ and one molecule of Cl^- are absorbed together. In the GI tract, this process is not believed to be related to the NaCl cotransporter gene family and rather is thought to be made up of an apical membrane Na^+/H^+ exchanger linked to an apical membrane $Cl^-/HCO3^-$ exchanger.[3] The identity of the Na^+/H^+ exchanger involved in NaCl absorption appears to vary between NHE3 and another epithelial isoform, NHE2, based on the species, segment of intestine, and neurohormonal secretory status under which the study is conducted. NHE3 and NHE2 both appear able to take part in neutral NaCl absorption. In dog ileum, all basal BB Na^+/H^+ exchange is due to NHE3.[4,5] In rat ileum the great majority is NHE3, whereas in rabbit ileum and chicken small intestine both NHE2 and NHE3 make up approximately 50% of basal BB Na^+/H^+ exchange.[6,7] In chicken colon, the great majority of BB Na^+/H^+ exchange is NHE2,[6] and in rabbit descending colon there is no NHE3, but NHE2 is present in the surface cell BB.[7] In rat colon, in contrast, at least 75% of BB Na^+/H^+ exchange is NHE3.[8] How the contributions of NHE2 and NHE3 to BB Na^+/H^+ exchange change with regulation of NaCl absorption remains to be characterized in more detail.

The molecular identify of the intestinal BB $Cl^-/HCO3^-$ exchanger linked to Na^+/H^+ exchange has not been definitely determined but is likely to be the gene product of the downregulated in adenoma gene (DRA). In addition, the method of linking of Na^+/H^+ and $Cl^-/HCO3^-$ exchangers is unknown. The only suggestion is that it is small changes in intracellular pH induced close to the BB by changes in Na^+/H^+ or $Cl^-/HCO3^-$ exchange that provide the linkage by changes in pH_i of these two exchangers.[3]

The neutral NaCl absorptive process is variable in rate based on the status of digestion. Under basal (fasting) conditions neutral NaCl absorption occurs as does BB Na^+/H^+ exchange. It has been assumed, but not shown, that in the immediate postprandial state, there is transient inhibition of neutral NaCl absorption accompanying secretion of water in the small intestine. In the later postprandial state, there is a marked increase in Na^+ absorption in the small intestine (what occurs in the colon has not been defined). The cause of the increased Na^+ absorption in the small intestine appears to differ in the jejunum and ileum. In the jejunum the increase in Na^+ absorption appears to be due to stimulation by end products of digestion of Na^+-linked transporters, including SGLT1 and Na^+-L-amino acid transporters. This increased absorption occurs in the proximal small intestine, where most of the digestion-generated substrates are absorbed in the upper third of the small intestine. There is also stimulation of BB Na^+/H^+ exchange in the ileum, but this appears to occur mostly independently of digestion-generated luminal substrates.[5] At least in the dog, this is due to neurohumoral-induced stimulation of NHE3. In fact, by current understanding, NHE3 is the major component of the intestinal Na^+-absorptive processes, which are regulated over short time periods by changes in neurohumoral regulators that are released as part of digestion.

As reviewed,[9,10] multiple neurohumoral mediators have been shown to rapidly (minutes) regulate small intestinal and colonic neutral NaCl absorption. This regulation consists of agents that stimulate and those that inhibit NaCl absorption. Many of these agonists similarly regulate BB Na^+/H^+ exchange. It has not been established whether in intact intestine both NHE2 and NHE3, or only NHE3, are regulated by these agonists. Studies of effects of neurohumoral regulation on SGLT1, on the intestinal Na^+-linked L-amino acid transporters (many of which have not been molecularly identified), and on the BB $Cl^-/HCO3^-$ exchanger(s) DRA have not been completed. Thus the extent of their neurohumoral factor regulation is not known. For SGLT1 exposed in oocytes, there is evidence of protein kinase regulation.[11,12] Also unknown is the relative effects of pathophysiologic changes in BB Na^+/H^+ exchange on NHE2 and NHE3. For instance, it is not known what cholera toxin exposure to ileum does to BB Na^+/H^+ exchange, nor is the effect of cholera toxin known on ileal BB NHE2 and NHE3.

MECHANISTIC STUDIES OF RAPID PROTEIN KINASE/GROWTH FACTOR REGULATION OF NHE3

Mechanistic studies of how neurohumoral mediators, protein kinases, and growth factors regulate NHE3 have shown that most regulation of NHE3 occurs by changes in the V_{max}.[13,14] Studies with sheets of ileal mucosa studied via the Ussing chamber/voltage clamp/unidirectional fluxes of $^{22}Na^+$ and $^{22}Cl^-$ were carried out under conditions in which changes in V_{max} were demonstrated. These showed that cAMP, cGMP, and elevating intracellular Ca^{2+} acting via protein kinase C inhibited NaCl absorption and BB Na^+/H^+ exchange, whereas EGF and α_2 adrenergic agonists stimulated NaCl absorption and BB Na^+/H^+ exchange. Our studies in Caco-2 cells have examined endogenous NHE3 17–21 days postconfluency as well as stably transfected NHE3 and NHE3 epitope tagged on the C-terminus with the vesicular stomatitis virus g protein studied 7–10 days postconfluency when a combination of transfected and endogenous NHE3 are present.[15] In these Caco-2 cells, there is no NHE2, based on Western and Northern analysis.[16] Results are similar with only endogenous NHE3 and with stably transfected combinations of endogenous/exogenous NHE3. The up/downregulation shown to occur in the ileum, when tested, also occurs in Caco-2 cells. For instance, the phorbol ester phorbol myristate acetate (PMA) inhibits NHE3 by decreasing initial rates of Na^+/H^+ exchange with an effect on the V_{max} but with no discernible change in $K'(H^+)_i$.[15,16] (FIGS. 1 and 2). Carbachol (data not shown) as well as PMA inhibits NHE3, whereas EGF stimulates NHE3 (FIG. 1). We have not yet carried out detailed studies of the effect of cAMP on NHE3 in Caco-2 cells. However, it has been recently reported that cAMP inhibits NHE3 in Caco-2/NHE3 cells and also in OK cells (opossum kidney cell line, which is a model for the proximal tubule).[17–19]

The most detailed kinetic studies of NHE3 regulation have been carried out in fibroblasts, called PS120 and AP-1, which are null in all Na^+/H^+ exchangers. For the most part, protein kinase and growth factor regulation of NHE3 in fibroblasts have mimicked the regulation in Caco-2 cells and ileum, although the intestinal epithelial studies have been less extensively studied from the point of view of the details of kinetics analysis.[13,14,20,21] For instance, epidermal growth factor (EGF) and fibroblast growth factor (FGF) stimulate NHE3 expressed in PS120 cells, whereas phorbol ester inhibits it.[13] Also, hyperosmolarity inhibits NHE3 in both PS120 and AP-1 fibroblasts and on the apical surface of Caco-2 cells.[20–23] A major difference between

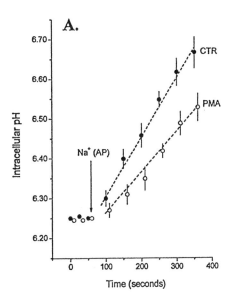

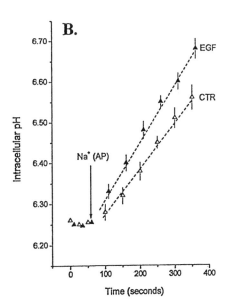

FIGURE 1. Inhibition by PMA (**A**) and stimulation by EGF (**B**) of the activity of endogenous NHE3 in Caco-2 cells 17 days postconfluency. Caco-2 monolayers were preincubated with PMA (1 μM) or EGF (200 ng/mL) for 20 min, and the Na$^+$-dependent rate of intracellular alkalinization was evaluated using a fluorometric method with BCECF. Data shown are means ± SD of initial rates observed in 9 monolayers in three separate experiments for each treatment. *Dotted lines* represent the least square linear fit curves for control conditions (CTR), and for monolayers exposed to PMA (PMA) or monolayers exposed to EGF (EGF). *Arrow* (Na$^+$(AP)) indicates the onset of exposure of the apical surface of the monolayers to 130 mM Na$^+$. HOE694 was present in all solutions to inhibit NHE1. (Reproduced with permission from Janecki *et al.*[15,16])

regulation of NHE3 in ileum and PS120 fibroblasts was identified concerning inhibition by cAMP. cAMP failed to inhibit NHE3 when expressed in PS120 fibroblasts. This was due to the lack in PS120 fibroblasts of two members of the regulatory factor gene family.[24] These two proteins are present in ileum, Caco-2 cells, and another fibroblast cell line, AP-1 cells (only NHERF studied), but are lacking in PS120 cells.

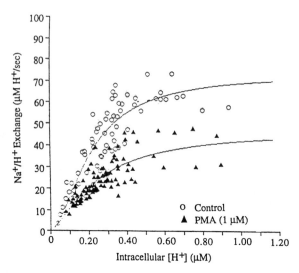

FIGURE 2. Effect of PMA (1 mM) on BB NHE3 expressed in Caco-2 cells. Caco-2/ NHE3 cells studied 17–21 days postconfluency were exposed to PMA for 10 min and then Na^+/H^+ exchange was determined over a range of pH_i, using BCECF to monitor intracellular pH and considering intracellular buffering capacity. PMA treatment inhibited Na^+/H^+ exchange. The effect was an inhibition of NHE3 V_{max} from 85 µM/s to 50 µM/second.

These proteins are E3KARP (NHE3 kinase A regulatory protein) and NHERF (Na$^+$/ H$^+$ exchanger regulatory factor).[24] These are closely related, PDZ domain–containing proteins that act as scaffolds, binding to NHE3 and the BB cytoskeleton protein ezrin.[19,24,25] They appear to act to allow cAMP-dependent protein kinase II to be correctly localized near NHE3 and are necessary to allow cAMP to inhibit NHE3.[20,25]

The kinetics of regulation of NHE3 in PS120 cells indicates that most regulation is through changes in the V_{max} with no change detected in the $K'(H^+)_i$.[13] There are two exceptions. When E3KARP or NHERF were stably transfected into PS120/ NHE3 cells, cAMP inhibition of NHE3 was reconstituted (FIG. 3A and B). Inhibition was due predominantly to a change in $K'(H^+)_i$, with a probable smaller inhibition of V_{max} (FIG. 3B).[19] Other studies have indicated that cAMP regulates NHE3 in OK cells by a V_{max} effect.[18] In addition, studies of the endogenous steroid from the shark liver, squalamine, showed that it inhibited NHE3 by a process that took one hour to become maximum and involved effects both to decrease the V_{max} and to increase the $K'(H^+)_i$.[26] Thus although changes in V_{max} are the predominant way NHE3 is regulated, changes in affinity for intracellular H$^+$ ions play a role in some circumstances.

Even in the kinetic studies of Na$^+$/H$^+$ exchange, performed using NHE3 expressed in fibroblasts, there are limitations to the studies. $^{22}Na^+$ uptake studies of the amiloride-sensitive component have been limited by imperfect estimates of intracellular pH to normalize the total driving force for Na$^+$ uptake. For the measurements made with fluorometry and an intracellular pH–sensitive dye, such as BCECF, the

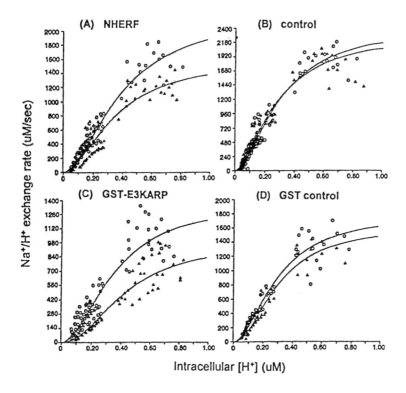

FIGURE 3A. cAMP-induced inhibition of NHE3 in PS120 cells. Stably transfected PS120 fibroblasts acidified with NH_4Cl were either recovered in Na^+ medium (O) or treated with 0.5 mM 8-Br-cAMP for 5–10 min before the recovery in Na^+ medium (▲). Na^+/H^+ efflux rates were calculated at various pH_is, and lines were fitted to the data using an allosteric model. Treatment of **(A)** PS120/NHE3/NHERF with 0.5 mM 8-Br-cAMP (▲) inhibited the Na^+/H^+ exchange activity with a decrease in V_{max} by ~30% with no effect on $K'(H_i^+)$ or n_{app}. **C:** Similarly, PS120/NHE3V/GST-E3KARP was inhibited by 8-Br-cAMP by a decrease in V_{max} by ~28%. By contrast, PS120/NHE3 **(B)** and PS120/NHE3/GST **(D)** were not affected by 8-Br-cAMP. Shown here are data from four or more experiments for each condition. (Reproduced with permission from Yun *et al.*[24]).

high Na^+-driven alkalinization is so rapid that initial rates are difficult to quantitate, leading to underestimates of the V_{max}. Because of these technical limitations, it is possible that the contribution to NHE3 regulation of changes in sensitivity of the intracellular H^+ modifier site has been underestimated. Recently Kinsella *et al.* have described an acid-induced (at level of pH_i 6.0) increase in sensitivity of the H^+ modifier site of BB Na^+/H^+ exchange from rabbit cortex.[27] The changes start over minutes and are long lived, also lasting minutes. Inasmuch as changes in intracellular H^+ occur as part of regulation of NHE3, it is possible that changes in the NHE3 H^+ modifier site contribute more than has been realized to date.

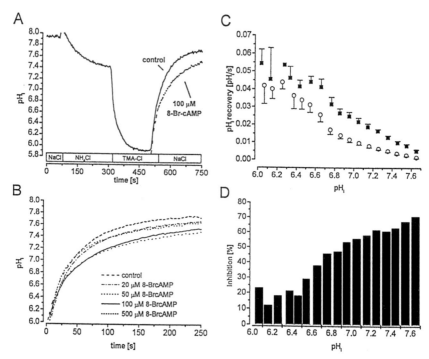

FIGURE 3B. 8-Br-cAMP inhibits Na^+/H^+ exchange in OK cells dose dependently through a change of the pH_i dependence of NHE3. **A:** Representative trace of cAMP-dependent inhibition of Na-dependent pH_i recovery. OK cells were acidified by ammonium prepulse (40 mM) and subsequent perfusion with TMA-Cl. pH_i recovery was facilitated by the readdition of NaCl (130 mM) (*solid line*). When indicated, 100 μM 8-Br-cAMP was present during the ammonium prepulse (*dotted line*). **B:** Concentration dependence of cAMP on NHE3. **D:** Dose response to 8-Br-cAMP. Different concentrations of 8-Br-cAMP were applied during the period of the ammonium prepulse, and the effect on Na-dependent pH_i recovery was recorded. **C:** pH_i dependency of the 8-Br-cAMP effect. pH_i recovery rates ($\Delta pH_i/\Delta t$) were calculated over short time intervals during the pH_i recovery under control conditions (■) and after treatment with 100 μM 8-Br-cAMP (O). The recovery rates ($\Delta pH_i/\Delta t$) are plotted at different intracellular pHs. **D:** From the data shown in panel **C**, the inhibition induced by 100 μM 8-Br-cAMP at different pH_is was calculated. (Reproduced with permission from Lamprecht *et al.*[19]).

REGULATION OF NHE3 IN CACO-2 CELLS INVOLVES CHANGES IN THE AMOUNT OF NHE3 ON THE PLASMA MEMBRANE

Changes in V_{max} can be achieved by altering the number of active plasma membrane transporters, the turnover number of each exchanger, or the inactive/active state of plasma membrane exchangers. Given that changes in V_{max} are involved in most protein kinase/growth factor regulation of NHE3, mechanistic studies were carried out to determine if NHE3 regulation was associated with changes in the number of exchangers in the plasma membrane.

Caco-2 Studies[15,16]

The boundaries of BB of Caco-2 were defined by confocal microscopy by marking the outside of the BB using the lectin phytohemagglutinin-E (PHA-E) and from the inside using actin. Both gave similar estimates of the size of the BB (3.8–4.0 μm). NHE3 localization in the BB was compared to that of PHA-E and actin. Under basal conditions, 81% of NHE3 was present in the BB, and ~19% was present in a compartment below the BB and above the nucleus, called the subapical compartments (19% estimated using PHA-E and 16% estimated from phalloidin-labeled actin). This most likely represents recycling endosomes plus the trafficking compartment from the golgi to the BB, which also is thought to traffic largely through the endosomes. Of note, this high percent of NHE3 present in the apical membrane of Caco-2 cells is very different from the surface expression of NHE3 in PS120 fibroblasts and AP-1 fibroblasts in which surface expression is 13–15% of total.[28,29] There are multiple differences in these studies of NHE3 reported in these different cell types, including not only the cell types but also overexpression versus endogenous expression. However, overexpression does not appear likely to be the major factor, given that NHE1 overexpressed in the same PS120 and AP-1 fibroblasts was nearly entirely expressed on the surface.[28,29] Of importance is that in renal proximal tubules it appears that ~50–80% NHE3 is on the apical membrane, similar to the Caco-2 studies, whereas a minority of NHE3 is on the BB in OK cells (~10%).[30,31] Because plasma membrane trafficking is important for regulation of NHE3 in both epithelial cells and fibroblasts (see below), the mechanisms of the differences in percent of NHE3 expressed on the surface will have to be understood to interpret mechanistic studies in these cell types.

Phorbol ester exposure within minutes was associated with inhibition of NHE3 transport activity on the apical surface of Caco-2 cells (28% inhibition analyzed using initial rates of Na^+/H^+ exchange), an effect entirely mediated by protein kinase C (FIG. 2).[15] As shown by confocal analysis, BB membrane NHE3 decreased by 13% in response to PMA, an effect also entirely inhibited by a protein kinase C inhibitor. Analyzed from the perspective of the amount of intracellular NHE3, PMA increased the amount of subapical compartment NHE3 from 19% to 29% of total, an ~50% increase. This PMA effect was not associated with a significant change in the size of the brush border.

Thus protein kinase inhibition of NHE3 is associated with less plasma membrane NHE3. Of note, the PMA inhibition of transport rate exceeds the magnitude of the change in amount of BB NHE3.[15] This strongly suggests that PMA also decreases the turnover number of each plasma membrane molecule of NHE3. There is no evidence that PMA inactivates plasma membrane NHE3 molecules, but this has not been studied adequately to eliminate the possibility.

Very similar results of the magnitude of change in the amount of NHE3 in the BB were obtained using reversible cell surface biotinylation.[15] Sulfo-NHS-SS-biotin was used to label surface proteins, including NHE3. This form of biotin can be removed from the surface by exposure to reducing agents such as glutathione. In control monolayers, 13% of initially identified BB NHE3 was present in the cytoplasm within 20 min at 37°C. This represents basal membrane recycling. Incubation of the cells with phorbol ester for 20 min led to 28% internalization of the initial plasma membrane NHE3. The increase represents more endocytosis. Note the similarity in

TABLE 1[a]

	Confocal Analysis		Na$^+$/H$^+$ Exchange	
	Relative Distribution of Apical NHE3 (SAC as percent of BB+SAC) $n = 40$ cells		H+ efflux rate (μM/sec) at pHi = 6.40 $n = 12$ monolayers	
Treatment	−H7	+H7	−H7	+H7
Control	18.7 ± 2.8	16.8 ± 3.5	360 ± 55	391 ± 41
PMA (1 μM)	$29.2 \pm 3.5^{b,c}$ $(12.3\%)^d$	17.2 ± 2.9	$259 \pm 29^{b,c}$ $(28\%)^e$	368 ± 37

[a]Comparison of the effect of PMA on the relative distribution of NHE3 between BB and SAC obtained by confocal analysis with the NHE3 *activity* evaluated by BCECF fluorometric method. Caco-2 cells (17 to 19 days postconfluency) were exposed to PMA (1 μM) for 20 min in the analysis (confocal analysis), and NHE3 activity was evaluated by the fluorometric method (Na+/H+ exchange). Averaged (mean ± SD) results from three separate experiments are shown for each method. (Modified from Ref. 15.)
[b]Significantly different $(p < 0.01)$ from respective control.
[c]Significantly different $(p < 0.0.1)$ from respective +H7 value.
[d]Percent of relative decrease in BB content of NHE3 in PMA-treated cells as compared to control.
[e]Percent of decrease of Na$^+$/H$^+$ exchange rate in PMA-treated cells as compared to control.

percent of initially identified NHE3, which increases intracellularly after PMA as identified by the confocal microscopic approach (10%) and as estimated by reversible cell surface biotinylation (15%).

EGF exposure to the basolateral surface of Caco-2 cells stimulated the rate of BB NHE3 (FIG. 1) by 19%. In addition, the distribution of BB NHE3 in cells exposed to EGF totally overlapped the distribution of the BB marked by PHA-E. In contrast to the PMA effect, this effect of EGF was associated with an increase in the size of the BB from 4.2 to 5.5 μm (30% increase). This indicates that in Caco-2 cells EGF stimulates exocytosis or inhibits endocytosis in a manner that leads to changes in the amount of BB, which is similar in magnitude to the change in rate of Na$^+$/H$^+$ exchange and amount of BB. A more detailed kinetic evaluation was carried out in PS120/NHE3 cells. In PS120 cells, FGF causes similar stimulation of NHE3 and increases the amount of plasma membrane NHE3 and also of total plasma membrane area.

We conclude that regulation of NHE3 by phorbol ester and EGF is due to changes in the amount of NHE3 on the plasma membrane and for phorbol ester partly due to changes in turnover number of the exchanger. Thus this regulation of NHE3 represents another example in which the amount of plasma membrane expression of a transporter represents at least part of the mechanism of its regulation. The intriguing issue raised by the EGF studies is that in understanding the mechanism of NHE3 regulation, effects that change the amount of plasma membrane rather than just changing the amount of the exchanger on the surface must be separated to determine whether these are specific mechanisms of regulation.

REGULATION OF NHE3 IN PS120 FIBROBLASTS ALSO INVOLVES CHANGES IN THE AMOUNT OF NHE3 ON THE PLASMA MEMBRANE

If fibroblasts that are devoid of endogenous Na^+/H^+ exchangers (PS120 cells and AP-1 cells) are to continue to be useful as models for understanding how NHE3 is regulated in epithelial cells, then each advance in understanding mechanisms of regulation that are shown to occur in epithelial cells must be shown to hold for the model fibroblasts. There are several indications that vesicle trafficking also is important in NHE3 regulation in these fibroblasts. These cells have both an intracellular as well as a plasma membrane pool of NHE3.[1,28,29] This has been identified by immunocytochemical approaches as well as by cell surface biotinylation. Grinstein *et al.* identified that NHE3 was present in a juxtanuclear pool in AP-1 cells.[29] Using established markers of intracellular organelles, they showed that this pool colocalizes with the endosomal markers, transferrin and cellubrevin, but not with golgi and ER markers. In addition, colchicine dispersed this pool of NHE3, and colchicine had previously been shown to disperse recycling endosomes. In addition, it has been suggested that a juxtanuclear vesicle population in fibroblasts was analogous to the subapical endosomal population in epithelial cells, and that fibroblasts mimicked some aspects of polarized cells by containing such a compartment as well as by having differences in organelles related to the attached and nonattached surfaces.

We also have observed such a juxtanuclear localization of NHE3 in PS120 fibroblasts and observed that this localization does not occur for NHE1.[1,28] Thus it appears that in fibroblasts, NHE3 is in an intracellular compartment that corresponds to the recycling endosome. Does NHE3 traffic to the surface as part of protein kinase/growth factor regulation?

To address this question, cell surface biotinylation was used to measure basal and serum-stimulated NHE3. An approach was developed in which PS120/NHE3V (epitope-tagged NHE3) cells were biotinylated at $0°C$, then total NHE3 determined by Western analysis and compared to avidin-agarose–precipitated NHE3 (plasma membrane NHE3) and NHE3 not precipitated by avidin (intracellular NHE3). These studies showed that ~15% of NHE3 was present on the plasma membrane. As a control for an effect of overexpression, similar studies were carried out with PS120/NHE1. PS120 cells stably expressing NHE1 produce both a mature glycosylated form of NHE1 and an immature unglycosylated form; 85% of the glycosylated form of NHE1 was on the plasma membrane, whereas all of the unglycosylated NHE1 was intracellular. That is, in PS120 cells having similar overexpression of NHE1 and NHE3, dramatically different amounts of these isoforms were present on the plasma membrane. Importantly, these results indicate that there is at least a biotinylation efficiency of 85% for NHEs in fibroblasts. Similar quantitative evaluation of the amount of NHE1 and NHE3 on the plasma membrane were found by Grinstein in AP-1 fibroblasts.[29]

The effect of serum on amount of NHE3 on the plasma membrane of PS120 cells was also determined. Serum pretreatment for 10 min increased the amount of NHE3 on the plasma membrane to 22%, an increase of approximately 40%. We previously have demonstrated that serum stimulates NHE3 expressed in PS120 cells by a V_{max} mechanism in which the V_{max} is increased by 50–120%.[13] Thus the serum stimulation of NHE3 in PS120 occurs by a mechanism in which there is an increase in the amount of NHE3 on the plasma membrane. Thus, at least for agonists that stimulate NHE3, similar mechanisms appear to occur in Caco-2 cells and fibroblasts inasmuch as trafficking appears to be involved in both. Note that serum stimulation of NHE1 is not associated with a change in amount of plasma membrane NHE1. NHE1 is stimulated by serum with an increase in affinity for intracellular H^+ ions without a change in amount on the plasma membrane NHE1.

BIOCHEMICAL MECHANISMS BY WHICH GROWTH FACTORS/PROTEIN KINASES REGULATE NHE3: ROLE OF PI 3–KINASE

PI 3–kinase has been shown to be involved in vesicular trafficking of transport proteins.[32,33] The best characterized is Glut4. PI 3–kinase colocalizes with Glut4 in transport vesicles, and inhibitors of PI 3–kinase and expression of dominant-negative forms of PI 3–kinase inhibit insulin stimulation of Glut4 amount/activity in the plasma membrane.

We have shown that PI 3–kinase is involved in EGF stimulation of NHE3 in intact ileum and Caco-2 cells.[34] In ileum, PI 3–kinase is present in both BB and basolateral membranes, but EGF stimulation of NHE3 is associated only with an increase in BB PI 3–kinase amount and activity, and tyrosine phosphorylation of the p85 subunit. These changes in PI 3–kinase occur quickly, within one minute of EGF exposure. Wortmannin, a PI 3–kinase inhibitor, prevents EGF stimulation of ileal NaCl absorption and BB Na^+/H^+ exchange. In Caco-2 cells, EGF stimulation of NHE3 is associated with an increase in PI 3–kinase activity. Also wortmannin inhibits both the increase in NHE3 transport rate and PI 3–kinase activity.

Grinstein et al.[27] reported studies of PI 3–kinase inhibitors on NHE3 in AP-1 cells under basal conditions. In AP-1 fibroblasts they demonstrated a role for PI 3–kinase in basal trafficking of NHE3 but not NHE1 expressed in the same cells. They demonstrated that wortmannin and LY294002, another PI 3–kinase inhibitor, inhibited basal NHE3 activity in a concentration- and time- (no effect of wortmannin was seen at 5 min, modest effects at 15 min, and peak inhibition at 30 min) dependent manner that maximally decreased the rate of NHE3 by 90% and led to a 72% decrease in amount of NHE3 on the plasma membrane in 30 minutes. The difference in effect on rate of Na^+/H^+ exchange and amount of plasma membrane NHE3 is consistent with a further effect of PI 3–kinase to regulate Na^+/H^+ exchange rate. This decrease in amount of plasma membrane NHE3 was accompanied by an increase in the amount of NHE3 in the juxtanuclear compartment. The major effect of the PI 3–kinase inhibitors was to inhibit exocytosis.

However, there is an additional complication that must be considered before assuming that PI 3–kinase/NHE3 studies in fibroblasts can be extrapolated to epithelial cells. Even though NHE1 in fibroblasts does not appear to traffic to and from the plasma membrane as part of protein kinase regulation, wortmannin did inhibit platelet-derived growth factor–induced stimulation of NHE1 in fibroblasts.[35]

Thus more detailed studies are required to understand the role of endo/exocytosis in the neurohumoral, protein kinase, and growth factor regulation of NHE3 that occurs as part of physiologic regulation in epithelial cells and fibroblasts and as part of the pathophysiology of diarrheal diseases.

ACKNOWLEDGMENTS

This work was supported in part by NIH Grants R01DK26523, R01DK5116, PO1DK44484, R29DK43778, K08DK02557, and T32DK0763205; the Meyerhoff Digestive Diseases Center; and the Hopkins Center for Epithelial Disorders.

REFERENCES

1. HOOGERWERF, W.A., S. TSAO, S.A. LEVINE *et al.* 1996. NHE2 and NHE3 are human and rabbit intestinal brush border proteins. Am. J. Physiol. **270:** G29–G41.
2. SILVIANI, V., M. GASTALDI, R. PLANELLS *et al.* 1997. NHE-3 isoform of the Na^+/H^+ exchanger in human gallbladder. Localization of specific mRNA by *in situ* hybridization. J. Hepatol. **26:**1281–1286.
3. KNICKELBEIN, R., P.S. ARONSON, C.M. SCHRON *et al.* 1985. Sodium and chloride transport across rabbit ileal brush border. II. Evidence for Cl-HCO3 exchange and mechanism of coupling. Am. J. Physiol. **249:** G236–G245.
4. MAHER, M.M., J.D. GONTAREK, R.E. JIMENEZ *et al.* 1996. Role of brush border Na^+/H^+ exchange in canine ileal absorption. Dig. Dis. Sci. **41:** 651–659.
5. MAHER, M.M., J.D. GONTAREK, R. BESS *et al.* 1997. The Na^+/H^+ isoform NHE3 regulates basal canine ileal Na^+ absorption in-vivo. Gastroenterology **112:**174–183.
6. DONOWITZ, M., C. DE LA HORRA, M.L. CALONGE *et al.* 1998. In birds NHE2 is the major brush border Na/H exchanger in the colon and is increased by a low NaCl diet. Am. J. Physiol. **274:** R1659–R1669.
7. WORMMEESTER, L., F. SANCHEZ DE MEDINA, F. KOKKE *et al.* 1998. Quantitative contribution of NHE2 and NHE3 to rabbit ileal brush border Na/H exchange. Am. J. Physiol. **274:** C1261–C1272.
8. BINDER, H.J., M. IKUMA & V.M. RAJENDRAN. 1998. Tissue-specific, differential and pretranslational regulation of Na-H exchange isoforms by aldosterone in rat colon. Gastroenterology **114:** A352.
9. DONOWITZ, M. & M.J. WELSH. 1987. Regulation of mammalian small intestinal electrolyte secretion. *In* Physiology of the Gastrointestinal Tract, 2nd edit. L.R. Johnson, Ed.: 1352–1388. Raven Press. New York.
10. CHANG, E.B. & M.C. RAO. 1994. Intestinal water and electrolyte transport: mechanisms of physiological and adaptive response. *In* Physiology of the Gastrointestinal Tract, 3rd edition. L.R. Johnson, Ed.: 2027–2082. Raven Press. New York.
11. HIRSCH, J.R., D.D.F. LOO & E.M. WRIGHT. 1996. Regulation of Na^+/glucose cotransporter expression by protein kinases in *Xenopus laevis* oocytes. J. Biol. Chem. **271:**14740–14746.
12. WRIGHT, E.M., J.R. HIRSCH, D.D.F. LOO & G.A. ZAMPIGHI. 1997. Regulation of Na^+/ glucose cotransporters. J. Exp. Biol. **200:** 287–293.
13. LEVINE, S.A., M.H. MONTROSE, C.M. TSE & M. DONOWITZ. 1993. Kinetics and regulation of three cloned mammalian Na^+/H^+ exchangers stably expressed in a fibroblast cell line. J. Biol. Chem. **268:** 25527–25535.
14. DONOWITZ, M., S.A. LEVINE, C.H. YUN *et al.* 1996. Molecular studies of members of the mammalian Na^+/H^+ exchanger gene family. *In* Molecular Biology of the Membrane Transport Disorders. S.G. Schultz, T.E. Andreoli, A.M. Brown *et al.*, Eds.: 259–275. Plenum Press. New York.
15. JANECKI, A.J., M.H. MONTROSE, P. ZIMNIAK *et al.* 1998. Subcellular redistribution is involved in acute regulation of the brush border Na^+/H^+ exchanger NHE3 in human colon adenocarcinoma cell line Caco-2: protein kinase C-mediated inhibition of the exchanger. J. Biol. Chem. **273:** 8790–8798.

16. JANECKI, A.J., M.H. MONTROSE, C.M. TSE et al. 1999. Development of an endogenous epithelial Na$^+$/H$^+$ exchanger (NHE3) in three clones of Caco-2 cells. Am. J. Physiol. 277(2 Pt 1): G292–305.

17. MCSWINE, R.L., M.W. MUSCH, C. BOOKSTEIN et al. 1998. Regulation of apical membrane Na$^+$/H$^+$ exchangers NHE2 and NHE3 in intestinal epithelial cell line C2/bbe. Am. J. Physiol. 275: C692–C701.

18. ZHAO, H., M.R. WIEDERKEHR, L. FAN et al. 1999. Acute inhibition of Na/H exchanger NHE3 by cAMP. J. Biol. Chem. 274: 3978–3987.

19. LAMPRECHT, G., E.J. WEINMAN & C.H.C. YUN. 1998. The role of NHERF and E3KARP in the cAMP-mediated inhibition of NHE3. J. Biol. Chem. 273(45): 29972–29978.

20. DONOWITZ, M., S. KHURANA, C. YUN & C.M. TSE. 1998. Asymmetry in plasma membrane signal transduction in intestinal epithelial cells: lessons from brush border Na$^+$/H$^+$ exchangers. Am. J. Physiol. 274: G971–G977.

21. YUN, C.H., C.M. TSE & M. DONOWITZ. 1995. Mammalian Na$^+$/H$^+$ exchanger gene family: structure and function studies. Am. J. Physiol. G1–G11.

22. NATH, S.K., C.Y. HANG, S.A. LEVINE et al. 1996. Hyperosmolarity inhibits the Na+/H+ exchanger isoforms NHE2 and NHE3: an effect opposite to that on NHE1. Am. J. Physiol. 270: G431–G441.

23. BIANCHINI, L, A. KAPUS, G. LUKACS et al. 1995. Responsiveness of mutants of NHE1 isoform of Na$^+$/H$^+$ antiport to osmotic stress. Am. J. Physiol. 269: C998–C1007.

24. YUN, C.H.C., S. OH, M. ZIZAK et al. 1997. Cyclic AMP mediated inhibition of the epithelial brush border Na$^+$/H$^+$ exchanger, NHE3, requires an associated regulatory protein. Proc. Natl. Acad. Sci. USA 94: 3010–3015.

25. YUN, C.H., G. LAMPRECHT, D.V. FORSTER & A. SIDOR. 1998. NHE3 kinase A regulatory protein E3KARP binds the epithelial brush border Na$^+$/H$^+$ exchanger NHE3 and the cytoskeletal protein ezrin. J. Biol. Chem. 273: 25856–25863.

26. AKHTER, S., S.K. NATH, C.M. TSE et al. 1999. Squalamine, a novel cationic steroid, specifically inhibits the brush border Na$^+$/H$^+$ exchanger isoform, NHE3. Am. J. Physiol. 276: C136–144.

27. KINSELLA, J.L. & J.P. FROEHLICH. 1998. Na$^+$/H$^+$ exchanger H$^+$ modifier site activation involves slow conformational changes in an oligomer. J. Am. Soc. Nephrology 9: 7A.

28. AKHTER, S., M. CAVET, C.M. TSE & M. DONOWITZ. 1998. Regulatory mechanisms of cloned Na$^+$/H$^+$ exchangers involve change in the amount of plasma membrane exchanger: use of cell surface biotinylation. Gastroenterology 114: A347.

29. KURASHIMA, K., E.Z. SZABO, G. LUKACS et al. 1998. Endosomal recycling of the Na$^+$/H$^+$ exchanger NHE3 isoform is regulated by the phosphatidylinositol 3-kinase pathway. J. Biol. Chem. 273: 20828–20836.

30. HENSLEY, C.B., M.C. BRADLEY & A.K. MIRCHEFF. 1989. Parathyroid hormone-induced translocation of Na-H antiporters in rat proximal tubules. Am. J. Physiol. 257: C637–C645.

31. ZHANG, Y., A.K. MIRCHEFF, C.B. HENSLEY et al. 1996. Rapid redistribution and inhibition of renal sodium transporters during acute pressure natriuresis. Am. J. Physiol. 270: F1004–F1014.

32. FREVERT, E.U., C. BJORBAEK, C.L. VENABLE et al. 1998. Targeting of constitutively active phosphoinositide 3-kinase to Glut4-containing vesicles in 3T3-L1 adipocytes. J. Biol. Chem. 273: 25480–25487.

33. CZECH, M.P. 1995. Molecular actions of insulin on glucose transport. Annu. Rev. Nutr. 15: 441–471.

34. KHURANA, S., S.K. NATH, S.A. LEVINE et al. 1996. Brush border phosphatidylinositol 3-kinase mediates epidermal growth factor stimulation of intestinal NaCl absorption and Na$^+$/H$^+$ exchange. J. Biol. Chem. 271: 9919–9921.

35. MA, Y.H., H.P. REUSCH, E. WILSON et al. 1994. Activation of Na$^+$/H$^+$ exchange by platelet-derived growth factor involves phosphatidylinositol 3-kinase and phospholipase C gamma. J. Biol. Chem. 269: 30734–30739.

Cl-Dependent Na-H Exchange

A Novel Colonic Crypt Transport Mechanism

HENRY J. BINDER,[a,c,d] VAZHAIKKURICHI M. RAJENDRAN,[a] AND JOHN P. GEIBEL[b,d]

Departments of [a]Internal Medicine, [b]Surgery, and [d]Cellular and Molecular Physiology, Yale University, New Haven, Connecticut 06520, USA

ABSTRACT: This communication summaries a series of observations of the transport function of the crypt of the rat distal colon. Development of methods to study both ^{22}Na uptake by apical membrane vesicles prepared from crypt cells and intracellular pH_i (pH_i), fluid movement (Jv), and bicarbonate secretion during microperfusion of the crypt has led to the identification of (1) a novel Cl-dependent Na-H exchange (Cl-NHE) that most likely represents the coupling of a Cl channel to a Na-H exchange isoform that has not as yet been identified and (2) bicarbonate secretion that appears to be most consistent with HCO_3 uptake across the basolateral membrane by a mechanism that is closely linked to Cl transport and its movement across the apical membrane via an anion channel. Na-dependent fluid absorption is the constitutive transport process in the crypt, while fluid secretion is regulated by one or more neurohumoral agonists. Cl-NHE is responsible for both the recovery/regulation of pHi in crypt cells to an acid load and fluid absorption.

The existing paradigm of fluid and electrolyte transport in the large and small intestine during the past 20 years has been that absorptive processes are located in surface cells in the colon (or villous cells in the small intestine), whereas secretory processes are crypt cell transport processes.[1] During this period there have been several indirect observations that indicate that absorptive processes may also be located in crypt cells and secretory ones in surface and/or villous cells.[2–4] Direct assessment of crypt cell function has not occurred as a consequence of the inability to determine fluid movement directly in these cells. During the past few years this laboratory has developed methods to study directly fluid movement in colonic crypts by adapting microperfusion techniques that had been initially developed and perfected for the study of renal tubules.[5–9]

Use of such methods has established that the colonic crypt of the rat distal colon has the capacity for both fluid absorption and secretion but that in the basal state the constitutive transport process is Na-dependent fluid absorption.[6] These observations have required reanalysis of the fundamental model of the spatial separation of absorptive and secretory processes to surface (and villous) epithelial and to crypt epi-

[c]Address for correspondence: Henry J. Binder, M.D., Department of Internal Medicine, Yale University, P.O. Box 208019, New Haven, CT 06520-8019. Fax: 203-737-1755.
Henry.Binder@Yale.edu

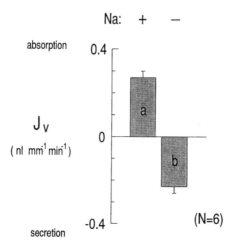

FIGURE 1. Net fluid movement in isolated, microperfused crypt from rat distal colon. Net fluid absorption was demonstrated during perfusion with a Na-Ringer solution (**a**). Substitution of lumen Na by choline resulted in net fluid secretion (**b**). It should be noted that when similar studies were performed in which NMG was substituted for Na, a decrease in net fluid absorption was observed. (Reproduced with permission from Singh et al.[6])

thelial cells. These studies revealed that microperfusion of colonic crypts with a Na-containing Ringer solution resulted in net fluid absorption.[6] Initial studies in which choline was used to substitute for lumen Na demonstrated net fluid secretion (FIG. 1). Subsequent experiments in which NMG was used as the Na substitute did not produce evidence of net fluid secretion but rather a *decrease* in net fluid absorption. As a consequence, it is likely that the observed net fluid secretion is due to choline and not to the absence of lumen Na. Although net fluid absorption was consistently demonstrated in the basal state, net fluid secretion was observed following the addition of either vasoactive intestinal peptide (VIP), acetylcholine (Ach) or dibutyryl cyclic AMP (DBcAMP) to the bath solution (FIG. 2). Such observations suggested that the underlying transport process in the resting state is a Na-dependent absorptive process, whereas secretory processes are regulated by one or more neurohumoral agonists released from lamina propria cells. This hypothesis was supported by the demonstration that the isolated perfused crypts were isolated without the underlying and attached myofibroblasts. These latter cells are present in the pericryptal sheath and are a potent source of eicosinoids, the synthesis of which is stimulated by neurohumoral agonists (e.g,. bradykinin) released from lamina propria cells.[10] It is possible that in intact tissue containing lamina propria cells and myofibroblasts the crypt may secrete fluid as a result of the tonic release of neurohumoral agonists. If the release of such agonists resulted in the stimulation of an active secretory process that is greater than the resting constitutive absorptive process, the result would be net fluid secretion.

Sodium absorption has been extensively studied for many years in both *in vivo* and *in vitro* studies with both physiological and molecular approaches. The primary mechanism of fluid absorption in the rat distal colon is electroneutral Na-Cl absorption that most likely represents the coupling of Na-H exchange and Cl-HCO$_3$ exchange via intracellular pH (pH$_i$).[11] Na-H exchange (NHE) is present in the apical membrane of surface epithelial cells and can be both up- and downregulated by agonists and second messengers.[12,13] (Apical membrane Cl-HCO$_3$ exchange in the rat distal colon is discussed elsewhere in this volume.[14]) Several NHE isoforms have been identified; the NHE isoform that is predominantly present in the apical mem-

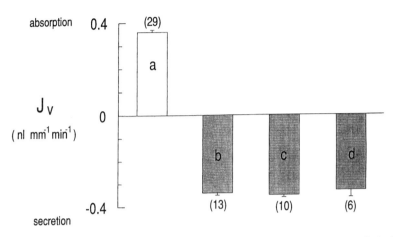

FIGURE 2. Effect of agonists and second messengers on net fluid movement in isolated, perfused, microperfused crypt from the rat distal colon. Net fluid absorption was present in the basal state during perfusion with a Na-Ringer solution (**a**). Addition of either vasoactive intestinal peptide (100 μM) (**b**), acetylcholine (60 μM) (**c**), or dibutyryl cyclic AMP (0.5 mM) (**d**) to the bath solution reversed net fluid absorption to net fluid secretion. (Reproduced with permission from Singh *et al.*[6])

brane of colonic surface epithelial cells is NHE-3 isoform, although NHE-2 isoform is also present.[12,15] NHE-3 isoform–specific mRNA is restricted to surface cells and not to crypt cells in *in situ* hybridization experiments, whereas immunocytochemical studies have localized NHE-3 isoform–specific protein to the apical membrane of surface, and not crypt, cells in the rat distal colon.[16,17] Although NHE-2 isoform may be present in crypt cells, NHE-2 isoform is not the primary NHE isoform in apical membranes of the rat distal colon.[15] Thus, this spatial distribution of NHE-3 isoform in the rat distal colon is not consistent with NHE-3 isoform being the Na transport process responsible for Na-dependent fluid absorption in the crypt of the rat distal colon. The possibility that an alternate NHE was present in the apical membrane of crypt epithelial cells was supported by preliminary studies that failed to establish the presence of [H] gradient–driven ^{22}Na uptake by apical membrane vesicles (AMV) isolated from crypt cells under conditions identical to those that resulted in [H] gradient–driven ^{22}Na uptake by AMV isolated from surface cells.

We subsequently reported that [H] gradient–driven ^{22}Na uptake by AMV isolated from crypt cells required the presence of Cl.[9] These studies provided compelling evidence that Na-H exchange (i.e., [H] gradient–driven ^{22}Na uptake) in crypt cell apical membrane was Cl dependent, whereas Na-H exchange in apical membrane of surface cells was Cl independent (FIG. 3). Further examination of Cl-dependent NHE revealed that other halides could stimulate Na-H exchange with Cl > Br > F > I. Most important were two additional observations that the characteristics of Cl-NHE were distinct from those of both NHE-2 and NHE-3 isoforms: (1) Cl-NHE was relatively resistant to amiloride, with an approximate K_i of more than 250 μM compared to the K_i for amiloride for the Cl-independent Na-H exchange in colonic apical membranes of 30 μM. All three of these NHEs are highly sensitive to the amiloride analogue,

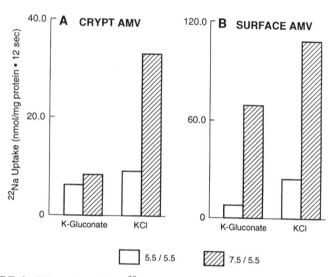

FIGURE 3. [H] gradient-driven ^{22}Na uptake by apical membrane vesicles from crypt and surface cells. Stimulation of ^{22}Na uptake by an outward-directed [H] gradient requires the presence of Cl in crypt membranes, but not in surface membranes. (Reproduced with permission from Rajendran et al.[9])

EIPA. (2) Aldosterone that inhibits electroneutral Na-Cl absorption and Cl-independent Na-H exchange in surface cells and substantially reduces both NHE-2 and NHE-3 isoform–specific mRNAs and proteins in the rat distal colon[15] did not alter Cl-dependent Na-H exchange in crypt apical membranes (unpublished observations). As a result, these observations suggested the presence of a novel NHE isoform in apical membranes of crypt epithelial cells of the rat distal colon. Additional studies were designed to assess (1) whether the Cl-NHE in crypt apical membranes participated in one or more cell functions and (2) the nature of the Cl dependence.

REGULATION OF pH$_i$

In addition to regulating transepithelial Na absorption, Na-H exchanges have important roles in the regulation of intracellular pH. NHE-1 isoform on the basolateral membrane of polarized cells and the plasma membrane of nonpolarized cells has an important role in pH$_i$ regulation. Studies were performed to assess the role of apical membrane Cl-dependent Na-H exchange in the regulation of pH$_i$ in microperfusion studies of rat colonic crypt cells.[9] Removal of Na from both the bath and lumen solutions resulted in a decrease in pH$_i$ from 7.03 ± 0.04 to 6.50 ± 0.16. The readdition of Na to the lumen solution did not alter pH$_i$ in the absence of lumen Cl. Further, the addition of Cl to the lumen solution in the continued absence of lumen Na also did not affect pH$_i$. By contrast, the readdition of both Na and Cl to the lumen solution resulted in an increase in pH$_i$ back to the initial resting value (FIG. 4). The presence of 1 mM amiloride in the luminal perfusion solution completely prevented the Na-/Cl-

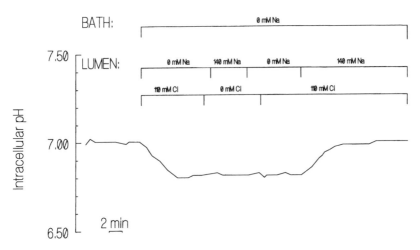

FIGURE 4. Role of Cl in Na-dependent recovery of pH_i to an acid load in isolated microperfused colonic crypts. Removal of Na from the lumen and bath solution resulted in a decrease in pH_i. Readdition of either lumen Na or lumen Cl alone did not alter pH_i. By contrast, the addition of both Na and Cl to the lumen solution resulted in a significant increase in pH_i to the resting level. (Reproduced with permission from Rajendran *et al.*[9])

dependent increase in pH_i. These observations provide supportive evidence of the presence of a Cl-dependent Na-H exchange in the apical membrane of the rat distal colon and suggest its role in the Na-dependent recovery/regulation of pH_i to an acid load.

MECHANISM OF Cl DEPENDENCE

The nature of the Cl dependence was not evident in these initial studies. As both Cl channels and Cl-anion exchanges are present in the apical membrane of the mammalian colon and the former are present in the crypt, studies were performed to address whether the Cl dependence of Na-H exchange represented its coupling to a Cl channel or to a Cl-anion exchange.[18] More than one type of Cl channel, including cystic fibrosis transmembrane conductance regulator (CFTR), exists in colonic apical membranes. To address this issue, NPPB, a relatively nonspecific Cl channel blocker, was used to determine its ability to alter the increase in pH_i in response to the luminal addition of both Na and Cl during crypt microperfusion studies. NPPB resulted in complete inhibition of Na-/Cl-dependent increase in pH_i, as shown in FIGURE 5. These studies provide evidence to suggest that a Cl channel was possibly coupled to a NHE in the crypt apical membrane and was responsible for the Cl dependence of Cl-NHE.

As a result of this observation, parallel studies were performed of the Cl dependence of [H] gradient–driven ^{22}Na uptake by crypt AMV.[18] In addition to the effect of potential inhibitors, studies also assessed the site of Cl dependence—that is, an intravesicular versus an extravesicular site of action. Surprisingly, a unilateral site for the preferential action of Cl could not be established. That is, both intravesicular

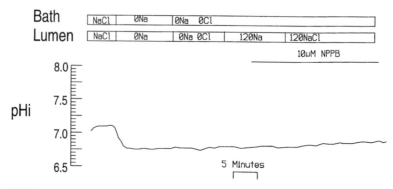

FIGURE 5. Effect of a Cl channel blocker, NPPB, on Na-/Cl-dependent recovery of pH$_i$ to an acid load. The protocol emloyed in this study is identical to that outlined in FIG. 4. Ten μM NPPB blocked Na-/Cl-dependent recovery of pH$_i$ to an acid load induced by the initial removal of Na from both lumen and bath solutions. (Reproduced with permission from Rajendran et al.[18])

and extravesicular sites (compared to the bilateral presence of Cl) appeared to be equally effective in stimulating [H] gradient–driven [22]Na uptake. This observation is most consistent with the presence of a Cl channel that resulted in the effective presence of Cl in both the extravesicular and intravesicular compartments regardless of the initial site to which Cl had been added to the vesicles.

To distinguish further between Cl channels and Cl-anion exchange, the effect of NPPB and DIDS on Cl-dependent Na-H exchange in studies of [22]Na uptake by AMV was determined.[18] The results shown in FIGURE 6 provide compelling evidence to support the possibility that a Cl channel is responsible for the Cl dependence of Cl-NHE. NPPB inhibited, in a dose-dependent manner, [H] gradient–driven [22]Na uptake in the presence of Cl. This observation is directly parallel to the results shown in FIGURE 5 that NPPB blocked the Na-/Cl-dependent recovery of pH$_i$ to an acid load.

DIDS, a stilbene inhibitor, inhibits Cl-anion exchanges at relatively low concentrations but acts as an inhibitor of Cl channels at higher concentrations. First, it should be noted that two distinct Cl-anion exchanges (Cl-HCO$_3$ and Cl-OH) are present in the apical membrane of the rat distal colon.[19] When the presence of these two exchanges was examined in surface and crypt apical membranes, Cl-OH exchange was present in apical membranes of both surface and crypt cells, but Cl-HCO$_3$ exchange was identified only in surface apical membranes.[20] Cl-HCO$_3$ exchange was not localized to apical membranes of crypt cells. In additon, Cl-HCO$_3$ and Cl-OH exchanges have quite different K$_i$s from DIDS (7.8 vs. 106.0 μM, respectively). Ten μM DIDS failed to inhibit [H] gradient–driven [22]Na uptake in the presence of Cl. By contrast, 500 μM DIDS substantially reduced Cl-dependent Na-H exchange (FIG. 6).

To confirm the relationship of Cl channels to Na-H exchange function, the effect of NPPB and DIDS was determined on Cl channel activity in these crypt AMV in parallel studies.[18] Cl channel activity was assessed by determining [36]Cl uptake in response to an electrical potential induced by K and valinomycin, a K ionophore. In

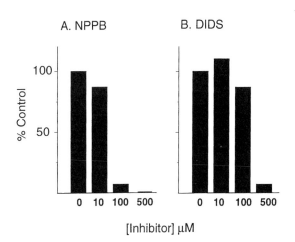

FIGURE 6. Effect of Cl channel and anion exchange inhibitors, NPPB and DIDS, on Cl-dependent H-gradient driven ^{22}Na uptake by apical membrane vesicles prepared from colonic crypt cells. **(A)** 100 μM NPPB, a Cl channel inhibitor, completely prevented Cl-dependent [H]-gradient stimulation of ^{22}Na uptake. **(B)** DIDS is an anion exchange inhibitor at low concentrations, but a Cl channel blocker at higher concentrations. Ten μM DIDS did not alter, whereas 500 μM DIDS almost completely inhibited, Cl-dependent [H]-gradient stimulation of ^{22}Na uptake. (Reproduced with permission from Rajendran *et al.*[18])

these experiments the presence of an intravesicular positive potential induced ^{36}Cl uptake, confirming the presence of Cl channels in the cyrpt apical membranes. NPPB inhibited potential-dependent ^{36}Cl uptake with a comparable dose dependency, as was observed for Cl-dependent [H] gradient–driven ^{22}Na uptake. In further studies, 500 μM, but not 10 μM, DIDS also inhibited potential-dependent ^{36}Cl uptake by crypt apical membranes. These several observations, when taken together with the results of Na-/Cl-dependent recovery of pH$_i$, are most consistent with the Cl dependence of Cl-NHE representing an apical membrane Cl channel, not an apical membrane Cl-anion exchange.

CFTR is one of several Cl channnels present in plasma membranes and has been identified in the apical membrane of colonic crypt cells. Therefore, Cl channel studies were performed with a polyclonal antibody to CFTR. The presence of both 1 μg and 2 μg CFTR polyclonal antibody resulted in an identical 40% inhibition of potential-dependent stimulation of ^{36}Cl uptake by crypt AMV.[18] As a result, this experimental finding raised the possibility that more than one Cl channel might be coupled to Na-H exchange in the crypt apical membrane and be responsible for the observed Cl dependence.

NET FLUID ABSORPTION

In view of the presence of Cl-dependent Na-H exchange in the crypt apical membrane, its involvement with pH$_i$ recovery, and the demonstration of Na-dependent fluid absorption in the crypt during microperfusion studies, we sought to establish whether Cl-dependent Na-H exchange was responsible for Na-dependent fluid absorption in the colonic crypt. Therefore, experiments of fluid movement were performed in which we assessed the role of lumen Cl and two different Cl transport inhibitors, DIDS and NPPB, on net fluid absorption. Removal of Cl resulted in a significant 66% inhibition of net fluid absorption. Consistent with the studies of Cl-dependent Na-H exchange in the [22]Na uptake studies with AMV[9] and the absence of Cl-HCO$_3$ exchange in crypt apical membranes,[18] 100 μM DIDS did not significantly inhibit net fluid absorption. DIDS, when studied at 500 μM, a concentration at which DIDS functions as a Cl channel blocker, and NPPB, a recognized Cl channel blocker, both inhibited net fluid absorption by approximately 65% (unpublished observations). These observations establish that net fluid absorption in the isolated microperfused colonic crypt is Cl dependent and is markedly inhibited by Cl channel blockers but not by an anion-exchange inhibitor. These present observations permit the speculation that net fluid absorption in the isolated colonic crypt is closely linked to Cl-dependent NHE, which is an NHE isoform that is distinct from other apical membrane NHE isoforms—for instance, the NHE-2 and NHE-3 isoforms.

BICARBONATE SECRETION

Most severe diarrheal illnesses—for instance, cholera and watery diarrheal syndrome—are associated with a metabolic acidosis that is a result of stool bicarbonate losses. These diarrheal disorders are often associated with fluid secretion that is plasmalike in composition but with a high bicarbonate concentration. Further, *in vivo* studies of experimental diarrheal models often yield evidence of elevated bicarbonate concentrations and stimulation of bicarbonate secretion. In general, this bicarbonate secretion has been thought to be secondary to the Cl-HCO$_3$ exchange that is present in the apical membrane of intestinal epithelial cells in the ileum and large intestine. By contrast, the extensive *in vitro* studies of active Cl secretion induced by one or more agonists implicated in diarrhea and related secretory processes in both native small and large intestinal epithelia and isolated intesinal cell culture lines consistently have not demonstrated bicarbonate stimulation.[21] In contrast to the investigation of active Cl secretion, only relatively limited studies of small and large intestinal bicarbonate transport have been reported.[22-25] Recently, there has been considerable interest in bicarbonate secretion in the proximal duodenum because of its potential importance in duodenal ulcer disease.[26-28]

Despite this paucity of information, three models of bicarbonate secretion have been advanced: (1) an apical membrane Cl-HCO$_3$ exchange; (2) an apical membrane Cl-HCO$_3$ exchange that is coupled to an apical membrane Cl channel; and (3) an apical membrane Cl channel (that is also permeable to HCO$_3$) or an apical membrane anion/HCO$_3$ channel. A general model of HCO$_3$ secretion with an understanding of both apical and basolateral events is not now available, nor is it known whether HCO$_3$ secretion is similar or differs (1) in the basal state versus following stimulation by agonists that induce active Cl secretion; (2) in varying segments of the small and large intestine; (3) following stimulation by agonists whose action is mediated by different second messengers; and (4) in surface versus crypt cells in view of the presence of Cl-HCO$_3$ exchange in surface cell apical membranes and of Cl channels in crypt cell apical membranes.

With this background of information we sought to determine whether the isolated microperfused colonic crypt might be a useful model to study bicarbonate secretion. Adapting methods that have been successfully used with isolated renal tubules,[5] [HCO$_3$] was determined using microcalorimetry that permits use of nanoliter samples.[29] Preliminary studies have recently established that the isolated microperfused colonic crypt is an excellent model for the investigation of colonic HCO$_3$ transport (unpublished observations). In the basal state with or without the presence of HCO$_3$ in either the lumen or bath perfusion solution, net HCO$_3$ *absorption* was consistently observed. By contrast, a substantial rate of HCO$_3$ secretion was induced by two agonists, VIP and Ach, that stimulate active Cl and fluid secretion. The actions of VIP are mediated by cyclic AMP, whereas those of Ach are mediated by intracellular Ca. In parallel studies the addition of DBcAMP to the bath solution reproduced the effects of both VIP and Ach. No studies have yet been performed with agonists that increase cyclic GMP. The experiments that have been completed to date that were designed to establish a cellular model of HCO$_3$ secretion have all been performed with DBcAMP.

Based on these recent experiments, several tentative conclusions can now be made. First, active HCO$_3$ secretion requires the presence of HCO$_3$ in the bath solution, but not in the lumen solution; and acetazolamide, an inhibitor of caronic anhydrase, does not inhibit HCO$_3$ secretion (unpublished observation). Thus, secreted HCO$_3$ is derived from the bath and is not generated from cellular methobolism. Because DBcAMP-stimulated fluid secretion is not substantially altered by the removal of bath HCO$_3$, it is possible that HCO$_3$ secretion is not a major driving force for fluid secretion. Second, active HCO$_3$ secretion is closely linked to active Cl secretion. That is, inhibition of active Cl secretion almost always resulted in inhibition of active HCO$_3$ secretion. For example, removal of bath Cl or addition of bumetanide, an inhibitor of Na-K-2Cl cotransport, to the bath solution inhibited active HCO$_3$ secretion. The exact nature of this relationship is not known but may be related to the movement of HCO$_3$ via an apical membrane Cl channel. This latter possibility is strongly supported by the observation that the presence of NPPB, a Cl channel blocker, in the lumen solution also inhibited active HCO$_3$ secretion.

This observation that active HCO$_3$ secretion could be eliminated by a Cl channel blocker in the lumen raised the important question of the relationship between a Cl channel and a Cl-HCO$_3$ exchange at an apical membrane locus in the genesis of HCO$_3$ secretion. Previous studies of HCO$_3$ secretion have observed a link between active Cl and active HCO$_3$ secretion and proposed that Cl is secreted across the apical membrane and then recycled back into the cell in exchange for HCO$_3$ via an apical membrane Cl-HCO$_3$ exchange.[26,30] Such a model of HCO$_3$ secretion would explain both the requirement for lumen Cl that has been often observed in the past[25] and the close link between active Cl and HCO$_3$ secretion in these present studies. Such coupling between active Cl and active HCO$_3$ secretion is not likely, as recent studies of the presence of Cl-anion exchanges in the apical membrane of surface and crypt cells have yielded perhaps unexpected findings. Although both Cl-HCO$_3$ and Cl-OH exchanges are present in the apical membrane of surface cells of rat distal colon, only Cl-OH exchange was identified in crypt apical membrane.[20] This observation is consistent with the association of Cl-HCO$_3$ exchange with transepithelial Cl movement, but not with HCO$_3$ secretion in crypt cells. The absence of Cl-HCO$_3$ exchange in the crypt apical membrane is consistent with absorptive processes being present in surface cells and with the presumed Cl absorption present in the crypt linked to Cl-dependent Na-H exchange, not to a Cl-anion exchange. Under these circumstances, Cl-OH exchange would be important in the regulation of one or more intracellular functons—for example, cell volume or pH_i.

Experiments of HCO_3 secretion were, therefore, performed in the isolated microfused colonic crypt in which Cl was removed from the lumen solution. DBcAMP-induced HCO_3 secretion was not affected by the removal of lumen Cl (unpublished observations). This observation is also consistent with the absence of a $Cl-HCO_3$ exchange from the apical membrane of the colonic crypt and permits the suggestion that HCO_3 movement across the crypt apical membrane is via a Cl channel. In view of the important role of Cl-dependent Na-H exchange in the apical membrane of the colonic crypt, it would be interesting to speculate that the Cl channel associated with Cl-dependent Na-H exchange was also linked to active HCO_3 secretion. At present this represents only speculation.

These studies do not provide insight into the nature of the association between an apical Cl channel and Na-H exchange. One possibility must include the coupling of more than one Cl channel (based on the partial inhibition of Cl-dependent Na-H exchange by a polyclonal CFTR antibody) with either an existing NHE isoform or a NHE isoform that has not as yet been cloned and identified. Alternatively, a novel transport protein may be present that has both Cl channel and Na-H exchange activities. In any event, it is now evident that a novel Cl-dependent Na-H exchange is present in the apical membrane of crypt cells of the rat distal colon that is important in the regulation of (1) Na-dependent recovery of pH_i to an acid load; (2) Na-dependent fluid absorption; and (3) possibly agonist-stimulated HCO_3 secretion.

ACKNOWLEDGMENT

This study was supported in part by USPHS Research Grant DK 14669, awarded by the National Institute of Diabetes and Digestive and Kidney Diseases.

REFERENCES

1. WELSH, M.J., P.L. SMITH, M. FROMM & R.A. FRIZELL. 1982. Crypts are the site of intestinal fluid and electrolyte secretion. Science **218:** 1219–1221.
2. KOCKERLING, A. & M. FROMM. 1993. Origin of cAMP-dependent Cl^- secretion from both crypts and surface epithelia of rat intestine. Am. J. Physiol. **264:** C1294–C1301.
3. STEWART, C.P. & L.A. TURNBERG. 1989. A microelectrode study of responses to secretagogues by epithelial cells on villus and crypt of rat small intestine. Am. J. Physiol. **257:** G334–G353.
4. PEDLEY, K.C. & R.J. NAFTALIN. 1993. Evidence from fluorescence microscopy and comparative studies that rat, ovine, and bovine colonic crypts are absorptive. J. Physiol. (Lond.) **460:** 525–547.
5. BURG, M., J. GRANTHAM, M. ABRAMOW & J. ORLOFF. 1966. Preparation and study of fragments of single rabbit nephrons. Am. J. Physiol. **210:** 1293–1298.
6. SINGH, S.K., H.J. BINDER, W.F. BORON & J. GEIBEL. 1995. Fluid absorption in isolated perfused colonic crypts. J. Clin. Invest. **96:** 2373–2379.
7. SINGH, S.K., H.J. BINDER, J.P. GEIBEL & W.F. BORON. 1995. An apical permeability barrier to NH_3/NH_4^+ in isolated perfused colonic crypts. Proc. Natl. Acad. Sci. USA **92:** 11573–11577.

8. RAJENDRAN, V.M., S.K. SINGH, J. GEIBEL & H.J. BINDER. 1998. Differential localization of colonic H^+-K^+-ATPase isoforms in surface and crypt cells. Am. J. Physiol. **274:** G424–G429.
9. RAJENDRAN, V.M., J. GEIBEL & H.J. BINDER. 1995. Chloride-dependent Na-H exchange. A novel mechanism of sodium transport in colonic crypts. J. Biol. Chem. **270:** 11051–11054.
10. BERSCHNEIDER, H.M. & D. W. POWELL. 1992. Fibroblasts modulate intestinal secretory responses to inflammatory mediators. J. Clin. Invest. **89:** 484–489.
11. BINDER, H.J. & G.I. SANDLE. 1987. Electrolyte absorption and secretion in the mammalian colon. *In* Physiology of the Gastrointestinal Tract, 2nd edit. L.R. Johnson, Ed.: 1389–1418. Raven Press. New York.
12. YUN, C.H.C., C.M. TSE, S.K. NATH *et al.* 1995. Mammalian Na^+/H^+ exchanger gene family structure and function studies. Am. J. Physiol. **269:** G1–G11.
13. RAJENDRAN, V.M., M. KASHGARIAN & H.J. BINDER. 1989. Aldosterone induction of electrogenic sodium transport in the apical membrane vesicles of rat distal colon. J. Biol. Chem. **264:** 18638–18644.
14. RAJENDRAN, V.M. & H.J. BINDER. 2000. Characterization and molecular localization of anion transporters in colonic epithelial cells. Ann. N.Y. Acad. Sci. **915:** this volume.
15. IKUMA, M., M. KASHGARIAN, H.J. BINDER & V.M. RAJENDRAN. 1999. Differential regulation of NHE isoforms by sodium depletion in proximal and distal segments or rat colon. Am. J. Physiol. **276:** G539–G549.
16. BOOKSTEIN, C., A.M. DEPAOLI, Y. XIE *et al.* CHANG. 1994. Na/H exchangers, NHE-1 and NHE-3, of rat intestine: expression and localization. J. Clin. Invest. **93:** 106–113.
17. CHO, J.H., M.W. MUSCH, A.M. DEPAOLI *et al.* 1994. Glucocorticoids regulate Na-H exchange expression and activity in region- and tissue-specific manner. Am. J. Physiol. **267:** C796–C803.
18. RAJENDRAN, V.M., J. GEIBEL & H.J. BINDER. 1999. Role of Cl channels in Cl-dependent Na-H exchange. Am. J. Physiol. **276:** G73–G78.
19. RAJENDRAN, V.M. & H.J. BINDER. 1993. Cl-HCO3 and Cl-OH exchanges mediate Cl uptake in apical membrane vesicles of rat distal colon. Am. J. Physiol. **264:** G874–G879.
20. RAJENDRAN, V.M. & H.J. BINDER. 1999. Distribution and regulation of apical Cl-anion exchanges in surface and crypt cells of rat distal colon. Am. J. Physiol. **276:** G132–G137.
21. CHANG, E.B. & M.C. RAO. 1994. Intestinal water and electrolyte transport. Mechanisms of physiological and adaptive responses. *In* Physiology of the Gastrointestinal Tract, 3rd Edit., Vol. 2. L.R. Johnson, Ed.: 2027–2081. Raven Press. New York.
22. SULLIVAN, S.K. & P.L. SMITH. 1986. Bicarbonate secretion by rabbit proximal colon. Am. J. Physiol. **251:** G436–G445.
23. SMITH, P.L., M.A. CASCAIRO & S.K. SULLIVAN. 1985. Sodium dependence of luminal alkalization by rabbit ileal mucosa. Am. J. Physiol. **249:** G358–G368.
24. FELDMAN, G.M. 1994. HCO3 secretion by rat distal colon: effects of inhibitors and extracellular Na^+. Gastroenterology **107:** 329–339.
25. HUBEL, K.A. 1968. The ins and outs of bicarbonate in the alimentary tract. Gastroenterology **54:** 647–651.
26. HOGAN, D.L., D.L. CROMBIE, J.I. ISENBERG *et al.* 1997. CFTR-mediated cAMP- and $Ca2^+$-activated duodenal epithelial HCO_3 secretion. Am. J. Physiol. **272:** G872–G878.
27. CLARKE, L.L. & M.C. HARLINE. 1998. Dual role of CFTR in cAMP-stimulated HCO_3 secretion across murine duodenum. Am. J. Physiol. **274:** G718–G276.
28. SEIDLER, U., I. BLUMENSTEIN, A. KRETZ *et al.* 1997. A functional CFTR protein is required for mouse intestinal cAMP-, cGMP- and Ca^{2+}-dependent HCO_3-secretion. J. Physiol. **505:** 411–423.
29. GEIBEL, J., G. GIEBISCH & W.F. BORON. 1989. Basolateral sodium-coupled acid-base transport mechanisms of the rabbit proximal tubule. Am. J. Physiol. **257:** F790–F797.
30. SUNDARAM, U., R.G. KNICKELBEIN & J.W. DOBBINS. 1991. Mechanism of intestinal secretion: effect of cyclic AMP on rabbit ileal crypt and villus cells. Proc. Natl. Acad. Sci. USA **88:** 6249–6523.

Coupling between Na$^+$, Sugar, and Water Transport across the Intestine

ERNEST M. WRIGHT[a] AND DONALD D. F. LOO

Department of Physiology, UCLA Medical Center,
Los Angeles, California 90095-1751, USA

ABSTRACT: Water is absorbed across the small intestine in the absence of external driving forces. However, it has been established that water transport is secondary to active sodium transport. In the upper intestine both sodium and water absorption are largely dependent on the presence of D-glucose. The link between active sodium transport and glucose is the coupled transport of sodium and glucose across the brush border membrane of enterocytes by the Na$^+$/glucose cotransporter (SGLT1). Na$^+$ that enters the cells with glucose is pumped out towards the blood by 3Na$^+$/2K$^+$ pumps on the basolateral membrane, and glucose passes out across the basolateral membrane by facilitated diffusion, the net result being that glucose and sodium are transported across the epithelium. The coupling between Na$^+$, glucose, and water transport is less well understood. It is commonly thought that Na$^+$ transport increases the local osmotic pressure in the lateral intercellular spaces, and that this in turn generates osmotic water flow across the epithelium. Recent work suggests a more direct link between Na$^+$, glucose, and water transport; that is, water is cotransported along with Na$^+$ and sugar through SGLT1. Here we review the evidence for Na$^+$/glucose/water cotransport.

INTRODUCTION

The small intestine is responsible for the uptake of nutrients, salts, and water into the body. The daily food intake of the average Western diet is 350 g of carbohydrate, 70 g of protein, 100 g of fat, and 1 liter of water. Food is digested in the stomach and upper small intestine to simple sugars (glucose, galactose, and fructose), small peptides and amino acids, and fatty acids. These products of digestion, the ingested salts and water, and the fluids secreted to aid digestion (saliva, gastric juice, pancreatic juice, bile, and intestinal secretions) are absorbed by the mature enterocytes lining the upper third of the intestinal villi. In the duodenum and jejunum these cells absorb about 1 mole of glucose and 8–9 liters of fluid a day.

The mechanism of intestinal fluid absorption has intrigued physiologists for over a century ever since Reid[1] demonstrated that the isolated small intestine was able to absorb fluid in the absence of any external driving forces (hydrostatic and osmotic pressures). It was subsequently observed that the rate of absorption required glucose in the intestinal lumen, and fluid absorption could occur even against an osmotic gradient. This is illustrated for the rat small intestine in FIGURE 1. Fluid absorption in-

[a]Address for correspondence: Ernest M. Wright, Department of Physiology, UCLA Medical Center, Los Angeles, California 90095-1751. Voice: 310-825-6905; fax: 310-206-5886.
ewright@mednet.ucla.edu

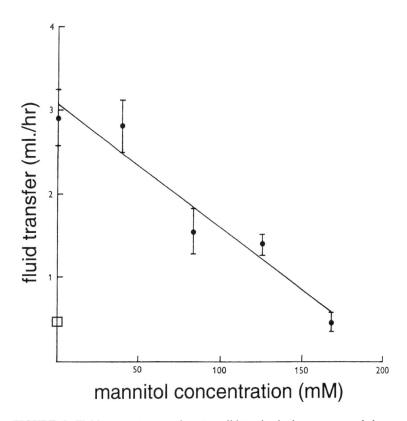

FIGURE 1. Fluid transport across the rat small intestine in the presence and absence of glucose. Fluid transport across everted sacs of rat jejunum was measured gravimetrically. Transport was measured in the presence (•) and absence (□) of 28 mM glucose and 0–160 mM mannitol in the mucosal fluid. (Modified from Smyth and Wright.[18])

creased sixfold in the presence of D-glucose (from 0.4 to 3 mL/h), and absorption continued even in the presence of large adverse osmotic gradients produced by the addition of an impermeant solute (mannitol) to the mucosal fluid.

The rate of water absorption was found to be linearly related to the rate of active salt (NaCl) absorption.[2,3] This is illustrated by a study where total solute (osmoles) and fluid transfers across the rat small intestine were measured simultaneously (FIG. 2). Net fluid absorption was proportional to the rate of solute transport (NaCl and sugar); the fluid transported was isotonic to saline bathing of the epithelium. Because fluid transport required the presence of Na^+, it was argued that water transport was secondary to active Na^+ transport across the epithelium. The central question that remained is how water movement is linked to active Na^+ transport, and this led to the hypothesis that active salt transport generated local osmotic gradients within the epithelial tissue.[4,5,21]

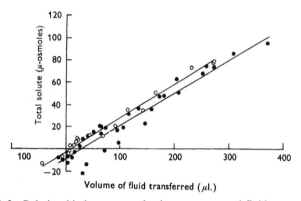

FIGURE 2. Relationship between total solute transport and fluid transport across the rat intestine. Transport data was obtained from everted sacs of rat intestine incubated at 37°C in the presence (○) and absence (•) of actively transported sugars. Fluid transport was estimated gravimetrically, and solute transport was estimated from the changes in osmolarity, and Na and sugar concentrations in the serosal compartment. (Reproduced with permission from Barry et al.[19])

MODEL FOR Na+ AND GLUCOSE TRANSPORT

A cartoon showing the overall characteristics of Na+ and sugar transport across enterocytes is presented in FIGURE 3. The columnar epithelial cells rest upon a basement membrane and are joined together at the apical surface by tight junctions. The apical surface area is expanded by the microvilli. The spaces between adjacent cells below the tight junctions are referred to as the *lateral intercellular spaces*. There are two potential routes for ions and molecules to cross the epithelium: one extracellular

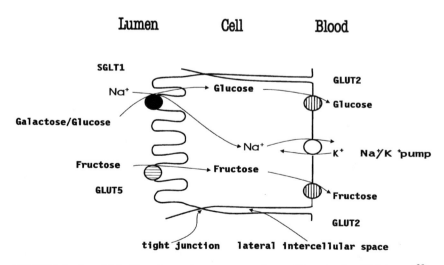

FIGURE 3. A model for Na and sugar transport across the small intestine (see Wright et al.[20]).

pathway through the tight junctions and lateral intercellular spaces, and the other though the cells where ions and molecules pass in turn through the brush border membrane, the cytoplasm, and the basolateral membrane. Although there is substantial evidence that ions may use the paracellular pathway in "leaky" epithelia, such as the small intestine, gall bladder, and proximal tubule, there is little direct evidence that molecules such as water use this path to any significant extent. Recently it was demonstrated that passive water flow across the proximal tubule occurs mostly through the cells.[6]

Sugars are transported across the brush border membrane by two different transport proteins: glucose and galactose are transported by the Na^+/glucose cotransporter (SGLT1), and fructose by the facilitated sugar transporter GLUT5. Glucose and galactose accumulated within the epithelium by SGLT1 are then transported out of the cell by another facilitated sugar transporter, GLUT2, in the basolateral membrane. Fructose also is transported across the basolateral membrane by GLUT2. Na^+ that enters the epithelium with sugar via the brush border Na^+/glucose cotransporter is then transported out of the cell across the basolateral membrane by the 3Na/2K pump. The net result is that sodium and glucose are absorbed across the epithelium in a fixed ratio (2 Na^+ and 1 glucose), and fructose diffuses across the cell down its concentration gradient. To maintain electroneutrality, two anions accompany Na^+ (chloride and bicarbonate). Furthermore, as discussed above, the net solute transport is followed by water transport. This link between Na^+, anions, and water transport provides the rationale for oral rehydration therapy (ORT) used to combat secretory diarrhea in infants and cholera patients.

As discussed earlier, active Na^+ absorption across the epithelium is thought to generate water absorption by creating a local osmotic gradient within the tissue. The most likely place for a local osmotic gradient is within the lateral intercellular spaces (FIG. 3). Arguably, the most elegant theory to account for isotonic water transport in the absence of external driving forces is the standing gradient hypothesis proposed 30 years ago by Diamond and Bossert.[7] In this model (FIG. 4), the basolateral $3Na^+$/ $2K^+$ pumps are concentrated near the blind end of the lateral spaces, and the tight junctions are assumed to be impermeable to Na^+ and water. Net solute pumping into the lateral spaces raises the solute concentration (osmotic pressure) in the spaces, largely because of the slow diffusion of salt down the long spaces. This increase in osmotic pressure pulls water out of the epithelium across the lateral plasma membranes into the lateral spaces. This in turn raises the hydrostatic pressure in the spaces and forces bulk flow of water towards the open end of the space. Given the rate of solute pumping, the osmotic water permeability of epithelial plasma membranes, and the dimensions of the lateral spaces, especially their length, the standing gradient model predicts that the fluid emerging from the mouth of the lateral space is isotonic to the external bathing solutions on each side of the epithelium.

There have been many attempts to prove, or disprove, the standing gradient hypothesis. Arguments against the hypothesis include (1) the $3Na^+$/$2K^+$ pumps are not concentrated at the dead end of the lateral spaces, but are uniformly distributed over the basal-lateral membrane; (2) the tight junctions are permeable to Na^+, and this would tend to dissipate the Na gradient by back leakage into the mucosal fluid; (3) the magnitude of the local osmotic gradients in the lateral spaces are, at best, a few milliosmoles; if indeed the osmotic gradients are small, this requires that the osmotic

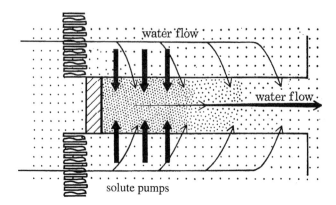

FIGURE 4. The standing gradient model for isotonic fluid absorption across epithelia.[7] Shown is the lateral intercellular space between two columnar epithelial cells. The solute pumps (3Na/2K pumps) are localized towards the blind end of the lateral spaces (i.e., towards the tight junctions). These pumps are proposed to generate an osmotic gradient in the spaces that result in the osmotic flow of water across the basolateral and brush border membranes of the epithelium. Water flows down the lateral spaces due to an increase in hydrostatic pressure, and osmotic equilibration over the long intercellular space results in an isotonic fluid emerging from the space. Simulations, using physiological parameters, predict that fluid transport across the epithelium is isotonic.

water permeability (Lp) of the lateral membranes of the enterocytes be high; (4) there is no direct evidence that water channels (aquaporins) are present on plasma membranes of enterocytes; and (5) neither standing gradient nor local osmosis theories account for water flow into the cell across the brush border membrane. It is implicit in the model shown in FIGURE 4 that osmotic flow out of the cell into the lateral spaces results in osmotic flow of water across the brush border to maintain cell volume and account for net water absorption. This would require a small increase in intracellular osmotic pressure and an even higher brush border membrane Lp. One possibility is that Na^+/glucose cotransport across the brush border membrane into the cell increases the intracellular osmolarity and osmotic water inflow across the brush border membrane, but his would require an even higher local osmotic gradient in the lateral intercellular spaces to pull water out of the cell.

COTRANSPORTERS AS WATER CHANNELS AND PUMPS

We now offer an alternative explanation of the link between Na^+, glucose, and water transport across the intestine; that is, cotransporters are water channels and molecular water pumps.[8–10,12,13,22] As a preamble to this alternative view of water transport across the intestine, we will first consider how water can cross biological membranes in general (FIG. 5).

The simplest mechanism for water transport across a membrane is the solubility and diffusion through the lipid bilayer. In this case water transport is proportional to the chemical potential gradient for water across the membrane (hydrostatic pressure,

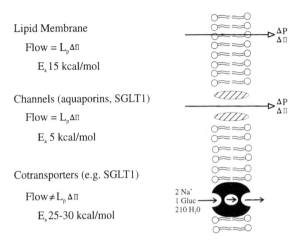

Lipid Membrane

Flow = $L_p \Delta \pi$

E_a 15 kcal/mol

Channels (aquaporins, SGLT1)

Flow = $L_p \Delta \pi$

E_a 5 kcal/mol

Cotransporters (e.g. SGLT1)

Flow ≠ $L_p \Delta \pi$

E_a 25-30 kcal/mol

FIGURE 5. Water transport across membranes. Shown are three mechanisms of water transport across biological membranes: solubility/diffusion through the lipid bilayer, water channels, and water cotransporters.

osmotic pressure, concentration gradient). Osmotic flow = Lp × osmotic gradient, where Lp is the hydraulic conductivity. We note that the Lp, expressed in the appropriate units ($P_f = Lp \times RT/V_m$, where R is the gas constant, T is the absolute temperature, and V_m the partial molar volume of water), is identical to the water permeability of the membrane (P_d) measured using radioactive tracers. The activation for water transport across the phospholipid bilayer is about 15 kcal/mole, consistent with the energy required to break the hydrogen bonds necessary to transfer water from the bulk aqueous solution into the hydophobic lipid bilayer.

Higher water permeabilities, Lp and P_d, are observed when water channels (aquaporins) are present in the membrane. Water transport through the channel is still proportional to the external driving forces, differences in the water chemical potential gradient (e.g., flow = Lp × osmotic gradient). The overall water permeability of a membrane containing water channels is much greater than that for the lipid bilayer alone, the actual value depending on the single-channel Lp and density of the channels in the membrane. Because diffusional water P_d is proportional to the area of the channels, and because Lp is proportional to the area of the channel and the square of the radius of each pore (according to Poiseuille's law), $P_f/P_d \neq 1$. For pores with radii of 0.4 nm, P_f/P_d is about 2. The activation energy for water permeation through pores is < 5 kcals/mole—that is, similar to that for self-diffusion of water (in water).

Cotransporters also behave as water channels.[10,12,13,22] This was established by overexpression of SGLT1 in a heterologous expression system, *Xenopus laevis* oocytes, and recording the change in cell volume after rapid changes in the osmotic pressure of the bathing solution. FIGURE 6 shows one such experiment on an oocyte expressing rabbit SGLT1. When the cell was bathed in an isotonic saline, there was no change in cell volume; but when the isotonic saline was replaced with a hypotonic one, the cell began to swell immediately. The subsequent addition of phlorizin, a potent and specific inhibitor of SGLT1, reduced the rate of increase in swelling by

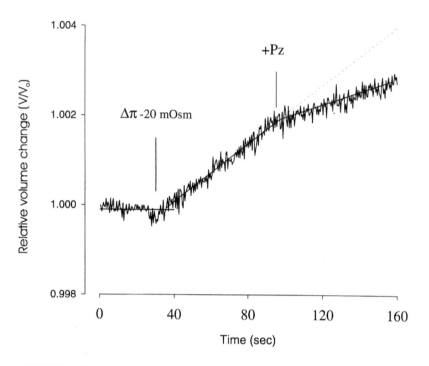

FIGURE 6. Passive water transport across the plasma membrane of an oocyte expressing rabbit SGLT1. The volume of the oocyte was measured optically and recorded with time.
At the point indicated the isotonic saline superfusing the oocyte was replaced with a hypotonic saline (−20 milliosmoles). The oocyte increased in volume, and this was partially
blocked by the addition of 100 µM phlorizin (from 3.3×10^{-3} to 1.5×10^{-3} %/s). (Reproduced with permission from Loo et al.[13]).

50%. The rate of swelling in phlorizin was identical to that in control oocytes not
expressing SGLT1. The phlorizin-sensitive rate of swelling (or shrinking) of SGLT1
oocytes was directly proportional to the osmotic gradient, and the slope was the
SGLT1 Lp. This Lp was independent of the presence or absence of sugars.

The number of SGLT1 molecules expressed in the oocyte plasma membrane can
be estimated by freeze-fracture electron microscopy or by carrier current (Q_{max})
measurements.[10,14] This enables us to estimate the Lp for a single SGLT1 molecule—4×10^{-16} cm^3/s, which is about 3% of that for AQP1. Although it is a low Lp,
the high number of SGLT1s per enterocyte, 250,000/cell, suggests a significant role
of these channels in water transport.

Similar Lp values to SGLT1 have been obtained for other cotransporters, such as
the Na$^+$/Cl$^-$/GABA, Na$^+$/citrate, and H$^+$/amino acid cotransporters. Additional evidence that passive water transport through cotransporters is through a channel-like
structure is that (1) the activation energy for passive water transport is 5 kcals/mole;
and (2) the magnitude of the water flow for a given osmotic gradient depends on the
reflection coefficient (sigma) of the solute used to generate the gradient. The water

flow produced by formamide was only 40% of that produced by impermeant mannitol.[12]

WATER COTRANSPORT

The third mechanism for water transport across membrane (FIG. 5) is cotransport with Na^+ and sugar through SGLT1. Again using the oocyte expression system, we recorded the volume of a cell before and after exposing the cell to glucose. The experiment shown in FIGURE 7 was conducted on an oocyte expressing rat SGLT1. In the absence of glucose in the bathing medium, there was no change in cell volume with time; but upon replacing the bathing medium with an isotonic solution containing glucose (10 mM glucose was used to replace 10 mM mannitol), the oocyte immediately swelled. The concurrently measured glucose-induced inward current recorded permitted us to relate the glucose-stimulated water transport to the rate of Na^+/glucose transport. The solid line in FIGURE 7 was fitted to the increase in volume, assuming that 300 water molecules are coupled to the inward transport of 1 glucose molecule and 2 Na^+ ions.[15]

What evidence is there that SGLT1 couples Na^+, glucose, and water transport across the membrane?

- There is a stoichiometric relationship between Na^+/glucose transport and water transport. The coupling is independent of the level of SGLT1 expression and the rate of Na^+/glucose cotransport. The rate of transport was varied by changing the membrane potential, the external Na^+ and sugar concentrations, and temperature (FIGS. 3 and 4, Loo et al.[22]; FIG. 4, Meinild et al.[12]).

- The activation energy for Na^+/glucose cotransport is identical (25–30 kcals/mole) to that for coupled water transport (FIG. 2, Loo et al.[22]; Meinild et al.[12]).

- Coupled water transport is independent of the osmotic gradient and even occurs against an osmotic gradient (FIG. 3, Meinild et al.[12]).

What is the evidence for direct rather than indirect coupling? One possible explanation for the link between Na^+/glucose cotransport and water transport is that the transport of Na^+ and glucose into the oocyte raises the intracellular osmotic pressure, and this in turn generates an osmotic flow into the cell. We think that this is unlikely for the following reasons:

- Water transport occurs immediately after the activation of Na^+/glucose transport by either addition of sugar to the external solution or changing the membrane potential (FIG. 7A, FIG. 4, Loo et al.;[22] FIG. 1, Meinild et al.[12]). Furthermore, the initial rate of water transport (0–5 seconds) was directly proportional to the rate of Na^+/glucose cotransport.

- No instantaneous water flow was observed when either ion channels—for example, Connexin 50—were expressed in oocytes, or when the oocytes were doped with ionophores such as gramicidin or nystatin (Meinild et al.[12]). Even when the rates of ion transport into the oocyte (Q_s) were comparable to the rates of Na^+/glucose transport (see FIG. 7B), there was no increase in cell volume for 30–40 seconds. Water uptake began after 40 seconds, and this continued for a time even after blocking ion uptake (by changing the membrane

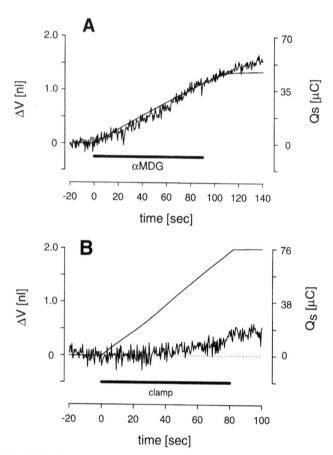

FIGURE 7. (A) The cotransport of water by rat SGLT1 expressed in oocytes. Water transport was measured optically and Na^+/glucose transport electrically. At the time indicated 10 mM sugar replaced 10 mM mannitol in the saline superfusing the oocyte (in mM: 90 NaCl, 20 mannitol, 2 KCl, 1 $MgCl_2$, 1 $CaCl_2$, 10 HEPES/pH 7.4). The *noisy record* is the oocyte volume and the *smooth line* the Na^+/glucose uptake reported as the charge influx (integral of the Na^+/glucose inward current). The charge uptake superimposes on the volume uptake, assuming that 300 water molecules accompany 2 Na^+ ions and 1 glucose molecule. **(B)** The relationship between Na^+ and water uptakes by an oocyte expressing the Connexin 50 (a Na^+-selective hemichannel). Inward Na^+ current through Connexin 50 was activated by voltage clamping the oocyte at –100 mV. Throughout the experiment the oocyte was incubated in a low calcium saline (10 μM). The integrated inward Na^+ current (Q_s microcoulombs) is plotted as a function of time. (Reproduced with permission from Wright *et al.*[15])

potential). This is the expected behavior for osmotic coupling between solute and water transport.

• Theoretical considerations (Loo *et al.*[22]) about "unstirred layer" effects predict an initial water flow that is only 3% of the observed flow.

- Finally, if the coupling were purely osmotic, the stoichiometry between solute and water transport should be invariant from transporter to transporter. In fact, the coupling coefficients varied considerably from cotransporter to cotransporter. For example, the coupling coefficients range from 50, for the plant H^+/amino acid cotransporter, through 176, for the renal Na^+/dicarboxylate cotransporter, through 210 for the human Na^+/glucose cotransporter, to as high as 390 for the rabbit Na^+/glucose cotransporter.[12,16,22]

MODEL FOR NA, GLUCOSE, AND WATER COTRANSPORT

A kinetic model for SGLT1 that incorporates water cotransport is given in FIGURE 8. This six-state model accounts for all the known characteristics of SGLT1.[11,14,16,17] In the model SGLT1 is shown as an integral membrane protein with a valence of –2 (for simplicity depicted by –). At a membrane potential of –50 mV in the absence of Na^+, 60% of the transporters are in the C1 form. Addition of saturating Na^+ to the external solution results in two Na^+ ions binding to the protein and a change in conformation to the C2 form (80%). In the absence of sugar, Na^+ transport occurs at a low rate (turnover 5/s, uniport mode). In the presence of external Na^+, sugar binds to C2 to form C3, and the transport cycle proceeds to C4, where sugar dissociates

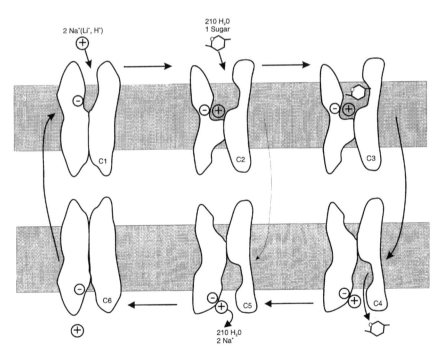

FIGURE 8. A six-state ordered model for the kinetics of Na^+, sugar, and water cotransport by SGLT1 (see Wright *et al.*[15]).

from SGLT1 at the interior surface, followed by Na^+ dissociation due to the low intracellular Na^+ concentration and negative membrane potential. The protein then returns to the original conformation C1 to complete the cycle. In the presence of saturating external Na and sugar, the turnover rate of transport cycle is about 25/second. This ordered reaction mechanism accounts for the strict coupling between Na and sugar transport and the kinetics of Na uniport in the absence of sugar. It is also indicated that both H^+ and Li^+ are able to substitute for Na in driving uniport and cotransport.

According to this scheme water is cotransported along with Na^+ and glucose simply as a consequence of the asymmetrical conformational changes that occur during the transport cycle. When Na^+ on the outside binds and produces the conformational change that permits high-affinity sugar binding (C2, C3), water "binds" to the open conformation C2 and passes to the cytoplasmic surface of the protein as a result of Na^+/glucose translocation (C3 to C4). As a consequence of the low intracellular Na^+ concentration, sugar and Na^+ dissociate from SGLT1, and the protein returns to the closed conformation (C6), extruding water to the cytoplasm. Finally, closed SGLT1 (C6) returns to the original state (C1) without any significant water movement—hence the asymmetry of the sodium concentrations across the membrane. The membrane potential provides the driving force for coupled inward Na^+, glucose, and water transport across the membrane. In theory, just as SGLT1 will transport Na and glucose in the reverse direction (out of the cell) if the sodium gradients and membrane potential are reversed, water cotransport will also be expected to reverse direction. If indeed SGLT1 is a perfect machine for harnessing sodium electrochemical potential gradients to drive sugar and water transport, coupling should be described by the Gibbs equation. Experiments are in progress to test this hypothesis.

If water can be cotransported by SGLT1 simply as a consequence of these asymmetrical conformation changes, why not other molecules? This has been tested by adding [^{14}C]urea to the solution bathing oocytes expressing SGLT1 and measuring radioactive urea uptake. Preliminary results show that urea is transported into the cell in strict proportion to the rate of Na^+/glucose transport.[23] To a first approximation, the ratio of urea to water transport is proportional to the ratio of molar urea and water concentrations. This suggests urea cotransport through SGLT1.

PHYSIOLOGICAL IMPORTANCE OF WATER COTRANSPORT

On the basis of the evidence summarized above, we conclude that the coupling between Na^+/glucose cotransport and water transport is direct and through SGLT1. The physiological importance of water transport by this mechanism can be assessed by considering the amounts of D-glucose and water absorption across the intestine each day. In humans the 350 grams of complex carbohydrate ingested each day yields about 1 mole of glucose after digestion. Taking a coupling coefficient of 210 for human SGLT1, the absorption of 1 mole of glucose across the brush border membrane would be accompanied by four liters of water—that is, about half of the total intestinal water absorption each day. Inasmuch as the osmolarity of the fluid transported by SGLT1 is considerably hypertonic (approximately 70 water molecules per solute molecules, as opposed to 275 water molecules per solute for an isotonic fluid), the cell should become hypertonic with time. This should, in turn, pull water osmotically through the SGLT1 water channel. The SGLT1 water channel is open in the presence and absence of sugar.[12] Therefore we conclude that water transport across the intestinal brush border membrane is the sum of two processes: the cotransport of $2 Na^+$, 1 glucose, and 210 water molecules; and the passive component through the SGLT1 water channel. This now explains the link between Na^+, glucose, and water absorption across the brush border of the intestine, and provides a molecular rationale for oral rehydration therapy.

How does Na^+, sugar, and water exit from the cell across the basolateral membrane? As described in FIGURE 3, Na^+ is pumped out by the basolateral $3Na^+/2K^+$ pump, and glucose exits via the facilitated transporter GLUT2. Given the geometry of the epithelium and the uniform density of pumps and transporters in the basolateral membrane, the bulk of the Na^+ and sugar will pass through the lateral intercellular spaces on their way across the tissue. Our hypothesis is that water is transported across the basolateral membrane by the K^+/Cl^- cotransporter. The choroid plexus K^+/Cl^- cotransporter transports water with a coupling coefficient of 1 K^+, 1 Cl^-, and 500 water molecules.[8,9] Thus K^+ accumulated within the epithelium by the basolateral $3Na^+/2K^+$ pump could drive the exit of water through the K^+/Cl^- water cotransporter.

In conclusion, we suggest that cotransport of water across the brush border and basolateral membranes of the enterocyte may account for the absorption of water by the intestine.

ACKNOWLEDGMENTS

Much of the recent work reported here was carried out in collaboration with Dr. Thomas Zeuthen and Anne-Kristine Meinild of the University of Copenhagen. Our research at UCLA was supported by grants from the USPHS (DK19567 and DK44602).

REFERENCES

1. REID, E.W. 1892. Preliminary report on experiments upon intestinal absorption without osmosis. Br. Med. J. **2:** 1133–1134.
2. MCHARDY, G.J.R. & D.S. PARSONS. 1957. The absorption of water and salt from the small intestine of the rat. Q. J. Exp. Physiol. **42:** 33–48.
3. CURRAN, P.F. & A.K. SOLOMON. 1957. Ion and water fluxes in the ileum of rats. J. Gen. Physiol. **41:** 143–168.
4. CURRAN, P.F. & G.F. SCHWARTZ. 1960. Na, Cl and water transport by the rat colon. J. Gen. Physiol. **43:** 555–571.
5. DIAMOND, J.M. 1962. The mechanism of water transport by the gall-bladder. J. Physiol. **161:** 503–527.
6. SCHERMANN, J., C.L. CHOU, T. MA et al. 1998. Defective proximal tubular fluid reabsorption in transgenic aquaporin-1 null mice. Proc. Natl. Acad. Sci. USA **95:** 9660–9664.
7. DIAMOND, J.M. & W.H. BOSSERT. 1967. Standing-gradient osmotic flow. A mechanism for coupling of water and solute transport in epithelia J. Gen. Physiol. **50:** 2061–2083.
8. ZEUTHEN, T. 1991. Secondary active transport of water across ventricular cell membrane of choroid plexus epithelium of *Necturus maculosus*. J. Physiol. (Lond.) **444:** 153–173.

9. ZEUTHEN, T. 1994. Cotransport of K^+, Cl^- and H_2O by membrane proteins from choroid plexus epithelium of *Necturus maculosus*. J. Physiol. (Lond.) **478:** 203–219.

10. ZAMPIGHI, G.A., M. KREMAN, K.J. BOORER *et al.* 1995. A method for determining the unitary functional capacity of cloned channels and transporters expressed in *Xenopus laevis* oocytes. J. Membr. Biol. **148:** 65–78.

11. LOO, D.D.F., B.A. HIRAYAMA, E.M. GALLARDO *et al.* 1998. Conformational changes couple Na^+ and glucose transport. Proc. Natl. Acad. Sci. USA **95:** 7789–7794.

12. MEINILD, A.-K., D. KLAERKE, D.D.F. LOO *et al.* 1998. The human Na^+/glucose cotransporter is a molecular water pump. J. Physiol. **508:** 15–21.

13. LOO, D.D.F., B.A. HIRAYAMA, A.-K. MEINILD *et al.* 1999. Passive water and ion transport by cotransporters. J. Physiol. **518:** 195–202.

14. LOO, D.D.F., A. HAZAMA, S. SUPPLISSON *et al.* 1993. Relaxation kinetics of the Na^+/glucose cotransporter. Proc. Natl. Acad. Sci. USA **90:** 5767–5771.

15. WRIGHT, E.M., D.D.F. LOO, M. PANAYOTOVA-HEIERMANN *et al.* 1998. Structure and function of the Na^+/glucose cotransporter. Acta Physiol. Scand. **163:** 257–264.

16. MEINILD, A.-K., D.D.F. LOO, A. PAJOR *et al.* 2000. Water transport by the renal Na^+/dicarboxylate cotransporter. Am. J. Physiol. **278:** F777–F783.

17. PARENT, L., S. SUPPLISSON, D.F. LOO & E.M. WRIGHT. 1992. Electrogenic properties of the cloned Na^+/glucose cotransporter: Part II. A transport model under non rapid equilibrium conditions. J. Membr. Biol. **125:** 63–79.

18. SMYTH, D.H. & E.M. WRIGHT. 1966. Streaming potentials in the rat small intestine. J. Physiol. **182:** 591-602.

19. BARRY, R.J.C., D.H. SMYTH & E.M. WRIGHT. 1965. Short circuit current and solute transfer by rat jejunum. J. Physiol. **181:** 410-431.

20. WRIGHT, E.M., B.A. HIRAYAMA, D.D.F. LOO *et al.* 1994. Intestinal sugar transport. *In* Physiology of Gastrointestinal Tract, 3rd edit. L.R. Johnson, Ed., Vol. II: 1751–1772. Raven Press. New York.

21. CURRAN, P.F. & J.R. MACINTOSH. 1962. A model system for biological water transport. Nature **193:** 347–348.

22. LOO, D.D.F., T. ZEUTHEN, G. CHANDY & E.M. WRIGHT. 1996. Cotransport of water by the Na^+/glucose cotransporter. Proc. Natl. Acad. Sci. USA **93:** 13367–13370.

23. LEUNG, D.W., D.D.F. LOO, B.A. HIRAYAMA *et al.* 2000. Urea transport by cotransporters. J. Physiol. (Lond.) **528.2:** 251–257.

Regulation of Chloride Secretion

Novel Pathways and Messengers

STEPHEN J. KEELY AND KIM E. BARRETT[a]

Department of Medicine, School of Medicine, University of California, San Diego, San Diego, California 92103, USA

ABSTRACT: The capacity for active chloride secretion, thereby driving the secretion of fluid, is an important property of the intestinal epithelium. Chloride secretion is stimulated by mechanisms involving increases in either cyclic nucleotide or cytoplasmic calcium concentrations. The calcium-dependent response is transient and limited in its magnitude, implying that negative signaling events may restrict the overall extent of this mode of chloride transport. We have uncovered a number of negative signaling mechanisms intrinsic to the epithelium that uncouple increases in calcium from the downstream response of chloride secretion. These involve various kinase cascades, the generation of messengers derived from membrane phospholipids, and interactions of G protein–coupled receptors with those for peptide growth factors such as epidermal growth factor. This chapter will review emerging information on the details of these negative signaling mechanisms, as well as points of convergence and divergence. The possible physiological and pathophysiological significance of such signaling will also be discussed.

INTRODUCTION

A major function of the epithelium of the mammalian intestine is to secrete fluid and electrolytes. Under normal circumstances, the small and large intestines, as well as the organs that drain into them, secrete approximately nine liters of fluid per day.[1] Much of this fluid secretion is driven osmotically by the active secretion of anions, especially chloride. It is evident that this process must be tightly regulated, inasmuch as both over- and underexpression of chloride secretion can lead to significant disease (secretory diarrheal states, such as cholera, and cystic fibrosis, respectively).[1] Moreover, epithelial chloride secretion underlies the normal function of a number of other organs in addition to the intestine, such as the lungs, pancreas, and biliary system. Thus, there are both basic and clinical reasons for wishing to develop a full understanding of how chloride secretion is accomplished and regulated.

In the intestine, chloride secretion is subject to two major levels of regulatory influences. Intercellular regulation is achieved via the actions of hormones, neurotransmitters, and other mediators that are released in the local environment of the epithelium and subsequently bind to epithelial receptors.[1] Intracellular regulation re-

[a]Address for correspondence: Dr. Kim E. Barrett, UCSD Medical Center 8414, 200 West Arbor Drive (for courier delivery, use CTF-A108, 210 Dickinson Street), San Diego, CA 92103. Voice: 619-543-3726; fax: 619-543-6969.

kbarrett@ucsd.edu

fers to the signaling pathways that link receptor occupancy to the downstream response of chloride secretion.[1] This article will focus on these latter intracellular mechanisms. Further, such intracellular regulatory pathways can be divided into those that promote chloride secretion and those that limit or terminate the secretory response. The pathways that stimulate chloride secretion have been quite well worked out at this point. In general, secretion is activated either by increases in the levels of cyclic nucleotides (cyclic AMP or cyclic GMP) or, in a mechanism that has distinct features, via elevations in cytoplasmic calcium concentrations.[1] Cyclic nucleotides evoke a sustained increase in chloride secretion, whereas the calcium-dependent response is transient, even in the continued presence of the stimulating agonist.[1-3]

The transient nature of the calcium-dependent response led us to hypothesize that there are negative signaling mechanisms, intrinsic to the epithelium, that modify the response of the chloride secretory machinery to ambient calcium concentrations.[3] Subsequent studies from our laboratory revealed that there are at least two such negative regulatory pathways. The first is activated by agents such as the muscarinic agonist carbachol, which acts initially as a stimulus of chloride secretion and subsequently renders cells refractory to further calcium-dependent chloride secretion without modifying calcium mobilization.[3] This inhibitory pathway is mediated by a soluble inositol phosphate messenger, inositol 3,4,5,6 tetrakisphosphate [Ins(3,4,5,6)P$_4$], and appears to target an apical chloride channel to block the effects of a rise in intracellular calcium.[4-6] Conversely, a second inhibitory pathway is activated by peptide growth factors, such as epidermal growth factor (EGF), which do not themselves act as agonists of the chloride secretory mechanism. This second inhibitory pathway is mediated by activation of the enzyme phosphatidylinositol 3-kinase and apparently blocks chloride secretion secondary to effects on a basolateral potassium channel.[6-8] However, more recent work from our laboratory has revealed that there may in fact be some cross talk between these negative signaling pathways.[9-12] It is the goal of the current article, therefore, to summarize the pathways that underlie this cross talk and also to propose how signaling specificity may be accomplished in the face of convergent signaling intermediates. The studies described derive from our work that has elucidated signaling pathways present in the T$_{84}$ human colonic epithelial cell line model of intestinal chloride secretion. However, preliminary evidence suggests that at least the broad outlines of these signaling pathways can likely be extrapolated to the setting of the native intestinal epithelium.

CROSS TALK BETWEEN MUSCARINIC AND EGF RECEPTORS AND IMPLICATIONS FOR CHLORIDE SECRETION

Our first indication that there might be a link between the negative signaling paradigms activated by carbachol and EGF was found in the observation that the inhibitory effects of carbachol on subsequent calcium-dependent chloride secretion could be reversed by the broad-spectrum tyrosine kinase inhibitor genistein. Similarly, the ability of carbachol to increase levels of Ins(3,4,5,6)P$_4$ was also genistein sensitive.[13,14] These findings implied that there was a tyrosine kinase–catalyzed step in the inhibitory signaling pathway. We went on to show that carbachol stimulates the tyrosine phosphorylation of a whole range of proteins in T$_{84}$ cells, including one

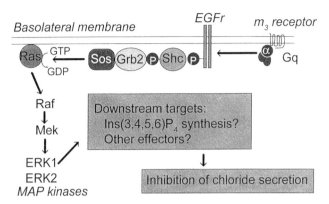

FIGURE 1. Scheme for the transactivation of the epidermal growth factor receptor (EGFr) by carbachol binding to a G_q-coupled m3 muscarinic receptor on the basolateral membrane of T_{84} colonic epithelial cells. EGFr transactivation results in receptor dimerization and tyrosine phosphorylation, recruitment of the adaptor proteins Shc and Grb2, activation of the guanine nucleotide exchange factor Sos, and subsequent activation of the mitogen-activated protein kinase (MAPK) cascade via the low-molecular-weight G protein, Ras. The ERK 1 and 2 isoforms of MAPK are then hypothesized to trigger inhibitory effectors that reduce chloride secretion, perhaps by inducing the synthesis of the inhibitory inositol phosphate $Ins(3,4,5,6)P_4$ or via other pathways that have yet to be identified. For further details, see text.

with a molecular mass of approximately 180 kDa, consistent with the molecular weight of the EGF receptor (EGFr).[9] In turn, carbachol-stimulated EGFr phosphorylation leads to the downstream recruitment of components of the mitogen-activated protein kinase (MAPK) signaling cascade via the adaptor proteins Shc and Grb-2, and consequent phosphorylation and activation of the ERK1 and ERK2 isoforms of MAPK.[9] The implications of this signaling cascade for the extent of chloride secretion were revealed by the observation that an EGFr kinase inhibitor, tyrphostin AG1478, reversed the effects of carbachol on EGFr phosphorylation and potentiated and prolonged chloride secretory responses to carbachol and other calcium-dependent secretagogues.[9] Similarly, inhibitors of MAPK activation, apigenin and PD 98059, also potentiated and prolonged chloride secretory responses evoked by carbachol.[9] These data can be interpreted in the context of the model shown in FIGURE 1, where transactivation of the EGFr in response to carbachol leads to downstream recruitment of MAPK activity and, via mechanisms that have yet to be defined [but perhaps involving $Ins(3,4,5,6)P_4$], the inhibition of chloride secretion.

The cross talk between the m_3 muscarinic receptor activated by carbachol in T_{84} cells and the EGFr was reminiscent of similar signaling paradigms that were being uncovered in other systems, where there is communication between G protein–coupled receptors, such as the lysophosphatidic acid receptor, and receptor tyrosine kinases, such as the platelet-derived growth factor receptor.[15–17] However, what remained unresolved was how the signaling events coupled to m_3 muscarinic receptor occupancy (i.e., activation of the G_q subtype of GTP-binding proteins and phospholipase C, generation of $Ins(1,4,5)P_3$, and mobilization of intracellular calcium) could be translated to a tyrosine kinase–dependent signaling event in this system. We therefore explored whether a nonreceptor tyrosine kinase, p[60] Src, and a novel calcium-dependent tyrosine kinase, Pyk-2, might be involved in this translation.[18–24]

Carbachol was shown to stimulate Pyk-2 tyrosine phosphorylation and the rapid association of Src with Pyk-2.[11,12] The agonist also induced the association of Src and the EGFr, but with slightly slower kinetics.[11,12] All of these events preceded carbachol-stimulated ERK activation. Moreover, PP2, an inhibitor of Src that is without effect on EGFr kinase activity, blocked the ability of carbachol to stimulate EGFr phosphorylation without altering EGFr phosphorylation in response to its cognate ligand, EGF.[11] Again, we were able to link these signaling events back to the control of chloride secretion with the observation that PP2, at doses that reduced the effect of carbachol on EGFr phosphorylation and MAPK activation, was also able to potentiate and prolong chloride secretory responses evoked by either carbachol itself or other calcium-dependent chloride secretagogues.[11] The ability of carbachol to activate Pyk-2, as the proximal event in this signaling cascade, is likely dependent upon the rise in intracellular calcium that is produced by this agonist and the secondary activation of a calmodulin-dependent protein kinase.[12] Thus, the calmodulin inhibitor fluphenazine inhibited the effect of carbachol on EGFr and ERK phosphorylation without altering the phosphorylation of these proteins that occurred in response to EGF.[12]

Thus, these data can be interpreted in the context of the model shown in FIGURE 2. Carbachol rapidly stimulates chloride secretion via a mechanism that involves an increase in cytoplasmic calcium concentrations and the activation of both basolateral potassium channels and apical chloride channels. However, subsequently, carbachol also recruits a negative signaling cascade that involves the sequential recruitment, in

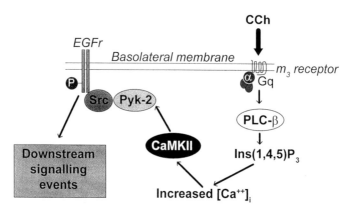

FIGURE 2. Pathway mediating transactivation of the EGFr in T_{84} cells in response to m_3 muscarinic receptor occupancy. Activation of G_q brings about the stimulation of phospholipase C-β (PLC-β), with resulting production of $Ins(1,4,5)P_3$ and mobilization of calcium from intracellular stores. This calcium, in addition to initiating chloride secretion (not shown), also activates calmodulin-dependent protein kinase type II (CAMKII), which, in turn, activates the tyrosine kinases Pyk-2 (calcium-dependent) and Src. These, in turn, mediate EGFr transactivation and the recruitment of downstream signaling events. For additional details, see text.

a calcium-dependent fashion, of Pyk-2, Src, EGFr, and components of the MAPK cascade. The net result is to terminate ongoing chloride secretion, hypothetically via an increase in $Ins(3,4,5,6)P_4$ and perhaps additional mechanism(s) yet to be defined. This also renders the cells refractory to further stimulation with calcium-dependent secretagogues, because, as we have shown, the inhibitory signal(s) generated by carbachol in T_{84} cells are long-lived.[4]

BASIS OF DIVERGENT SIGNALING VIA THE EGFr IN T_{84} CELLS

The foregoing discussion suggests that there is significant convergence of signaling events that lie downstream of carbachol and EGF binding to their respective receptors on the basolateral membrane of T_{84} cells. However, our earlier work had suggested that the inhibitory pathways that are recruited by carbachol versus EGF were in fact divergent. Thus, the effect of carbachol appears to be mediated, at least in part, by the effects of $Ins(3,4,5,6)P_4$ on an apical calcium-activated chloride channel; whereas the inhibitory effect of EGF was dependent on the ability of this growth factor to activate PI 3-kinase and apparently targeted a basolateral potassium channel, which in turn would reduce the driving force for apical chloride exit.[3,6] Moreover, when cells were treated with the combination of maximally inhibitory concentrations of carbachol and EGF, the extent to which the chloride secretory response to a subsequently applied calcium-dependent secretagogue was inhibited exceeded that seen with either carbachol or EGF alone.[7] This implies that the inhibitory mechanisms elicited by carbachol and EGF are at least partially independent. This therefore raised an intriguing question. If both carbachol and EGF induce inhibitory effects on chloride secretion via activation of the EGFr, what is the molecular mechanism whereby signaling is thereafter diversified to target the apical versus basolateral membrane? Put another way, we wished to define the mechanism whereby transactivation of the EGFr in response to carbachol, versus receptor activation in response to binding of the bona fide ligand EGF, is interpreted appropriately.

The EGFr is actually the prototypic member of a family of receptor tyrosine kinases known as the ErbB family of receptors (FIG. 3).[25] EGF and related ligands bind to ErbB1 (EGFr), whereas another family of growth factors, the heregulins, is specific for ErbB3 and ErbB4.[25,26] ErbB2 is notable for the fact that it is a so-called orphan receptor, with no known ligand (although, interestingly, it expresses the highest inherent catalytic activity of this family of receptors).[25] However, ErbB2 can be recruited to signaling cascades initiated by either EGF-related ligands or heregulins by heterodimerization with other receptor family members.[25] In fact, ligand binding to this receptor family has been shown to result in the formation of all possible combinations of homo- and heterodimers of ErbB receptors in cells that express multiple family members. We showed that T_{84} cells express not only ErbB1 (EGFr) but also other members of this receptor family. Thus, we wondered whether differential recruitment of other ErbB family members to the EGFr might underlie the divergent downstream consequences of EGFr phosphorylation in response to carbachol versus EGF in T_{84} cells.

In fact, immunoprecipitation and Western blotting studies suggest that this may indeed be the case. When cells are treated with carbachol or EGF, as expected, there is rapid phosphorylation of the EGFr.[9] However, only EGF, and not carbachol, also

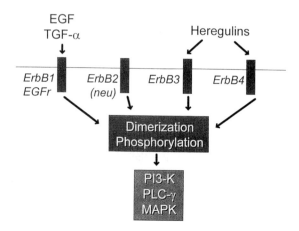

FIGURE 3. The ErbB family of receptor tyrosine kinases. EGF, transforming growth factor-alpha (TGF-α), and related growth factors are ligands for the prototypic member of this receptor family, EGFr (also known as ErbB1). Heregulins are ligands for ErbB3 and ErbB4. ErbB2, also known as *neu*, is an orphan receptor with no known ligand. Ligand binding induces receptor tyrosine phosphorylation, formation of homo- and heterodimers, and recruitment of downstream signaling components such as phosphatidylinositol 3-kinase (PI 3-K), PLC-γ, and the MAPK cascade.

stimulates phosphorylation of ErbB2 and coimmunoprecipitation of ErbB2 with EGFr.[10] Similarly, stimulation of T_{84} cells with EGF, but not carbachol, leads to the recruitment of the p85 regulatory subunit of PI 3-kinase to ErbB2.[10] Carbachol, by contrast, likely signals via the activity of EGFr homodimers inasmuch as it does not appear to cause association of EGFr with other ErbB family members.[10] Overall, these preliminary findings provide a basis for the divergent signaling and inhibitory mechanisms evoked by carbachol and EGF, despite the initial convergence of these signaling pathways on the EGFr (FIG. 4). It remains to be determined, however, why EGFr phosphorylation in response to carbachol fails to recruit ErbB2. We are currently exploring whether this may be related to differential patterns of EGFr tyrosine phosphorylation in response to carbachol versus EGF, perhaps reflecting substrate preferences of the EGFr kinase versus Src. The latter kinase is presumed to be the upstream kinase that mediates EGFr phosphorylation in response to the carbachol-stimulated increase in intracellular calcium, as discussed above (and depicted in FIG. 2).

SUMMARY AND IMPLICATIONS

In summary, we have defined a number of signaling pathways that are intrinsic to the epithelium, and at least partially interrelated, that serve to limit the overall extent of chloride secretion evoked by calcium-dependent secretagogues. The presence of such intrinsic negative signals implies that the chloride secretory responses occurring *in vivo* to substances such as acetylcholine are likely to be transient and small

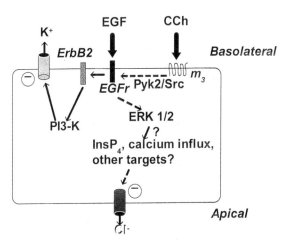

FIGURE 4. Hypothetical model for signaling diversification in the inhibition of chloride secretion mediated by either EGF binding to the EGFr or EGFr transactivation evoked by carbachol binding to the m_3 muscarinic receptor also located on the basolateral membrane of T_{84} cells. The EGF-dependent signaling pathway, depicted by the *solid arrows*, involves recruitment of ErbB2 and PI 3-K and targets a basolateral K^+ channel. The signaling cascade stimulated by carbachol, on the other hand, as depicted by *broken arrows*, involves Pyk-2/Src-dependent formation of EGFr homodimers, activation of the ERK isoforms of MAPK, and hypothetically an increase in Ins(3,4,5,6)P_4 (or other effectors, such as calcium influx, not discussed here, or pathways yet to be identified). This cascade ultimately targets an apical chloride channel to inhibit chloride secretion. For more details, see text.

in magnitude, unless coordinated with a simultaneous cyclic nucleotide–mediated response (indeed, we have shown previously that responses to combinations of agonists acting via calcium and cyclic nucleotides are synergistic[27]). In fact, the epithelium is the target of a plethora of signaling events that presumably serve to set the minute-to-minute, and perhaps longer-term, rate of chloride secretion.

It is tempting to speculate as to the significance of the signaling pathways that we have uncovered. Several scenarios can be envisaged, none of which are mutually exclusive. First, the calcium-dependent signaling mechanisms that lead to chloride secretion may represent the "fine-tuning" of the system. Thus, responses to calcium-dependent stimuli acting alone are normally limited but may be important under specific circumstances. Possible examples include fasting fluid secretion, which appears to be vagally mediated, relatively limited in extent, and significantly reduced by atropine.[28] Similarly, the negative signaling mechanisms that we have uncovered may be important to prevent inappropriately excessive secretory responses to luminal calcium–dependent chloride secretagogues (for example, bile acids, which are present at high concentrations in the postprandial period[29,30]). The temporally restricted chloride secretory response that is stimulated by calcium may also provide for an acute burst of secretion only when needed. For example, a reflex arc has been elucidated that provides for a burst of cholinergically mediated chloride secretion in response to mucosal stroking (presumed to model the physical stimulation induced by the passage of a food bolus).[31] This short secretory response may serve to lubricate the mucosa as the food passes.[31] A cyclic nucleotide–mediated response might be inappropriate in this setting, given that such secretory responses are sustained in the continued presence of agonist, and might then lead to excessive fluid losses.[1]

The negative signaling events we have described may also have pathophysiological significance. For example, in the setting of mucosal injury, where the release of growth factors such as EGF is enhanced (and EGF present in the intestinal lumen may gain access to basolateral receptors via the damaged epithelium), the ability of EGF to reduce chloride secretion may redirect the cell's energy resources from the energetically costly process of chloride secretion towards epithelial restitution and repair.[26] Indeed, exogenous administration of EGF and other growth factors in animal models of colitis leads to a resolution of diarrheal symptoms.[32,33] Conversely, negative signaling may compromise alternative pathways of chloride secretion in a disease state in which cAMP-mediated secretion is absent—namely, cystic fibrosis.[34] In fact, previous studies have revealed that, unlike the airways, the intestinal mucosa from patients with cystic fibrosis displays defects in both cyclic nucleotide– and calcium-mediated chloride secretion.[35–38] We propose that this may reflect the occurrence of negative signaling mechanisms, such as those described here, that limit the ability of the calcium-mediated secretory mechanism to serve as an alternate pathway for secretory function in cystic fibrosis. The extent to which a calcium-mediated secretory mechanism is expressed may be a determinant of disease severity in mouse models of cystic fibrosis.[39,40] Moreover, strategies to exploit the calcium-dependent secretory mechanism to bypass the defect in cAMP-mediated secretion are currently being explored in human patients with the disease.[41,42] This implies that a full understanding of the ways in which such responses may be limited, at least in the intestine, might ultimately lead to the optimization of such therapeutic approaches.

In conclusion, there are a number of complex and intertwining positive and negative signaling pathways that set the overall extent of chloride secretion in intestinal, and likely other, epithelial cells. In this article, we have summarized recent data from our laboratory that indicate the existence of at least two inhibitory signaling mechanisms intrinsic to the epithelium that serve to limit and/or terminate chloride secretory responses induced by an increase in levels of cytoplasmic calcium. Although the physiological significance of these mechanisms that limit calcium-dependent chloride secretion can only be the subject of speculation at present, a fuller understanding of such pathways may improve interventions for diseases of epithelial dysfunction, such as secretory diarrhea and cystic fibrosis.

ACKNOWLEDGMENTS

We are grateful to Ms. Glenda Wheeler for assistance with preparation of the manuscript and to the following colleagues for technical assistance, helpful suggestions, and other contributions to some of the studies described herein: Lone Bertelsen, Sean Calandrella, Jimmy Yip Chuen Chow, Silvia Resta-Lenert, Jane Smitham, and Jorge Uribe. Studies from the authors' laboratory have been supported by grants to KEB from the National Institutes of Health (DK28305 and DK53480), as well as by a Career Development Award to SJK from the Crohn's and Colitis Foundation of America.

REFERENCES

1. MONTROSE, M.H., S.J. KEELY & K.E. BARRETT. 1999. Secretion and absorption: small intestine and colon. *In* Textbook of Gastroenterology. T. Yamada, Ed.: 320–355. J.B. Lippincott. Philadelphia.
2. BARRETT, K.E. 1993. Positive and negative regulation of chloride secretion in T_{84} cells. Am. J. Physiol. **265:** C859–C868.
3. BARRETT, K.E. 1997. Integrated regulation of intestinal epithelial transport: intercellular and intracellular pathways. 1996 Bowditch Lecture. Am. J. Physiol. **272:** C1069–C1076.
4. VAJANAPHANICH, M., C. SCHULTZ, M.T. RUDOLF *et al.* 1994. Long-term uncoupling of chloride secretion from intracellular calcium levels by Ins(3,4,5,6)P_4. Nature **371:** 711–714.
5. KACHINTORN, U., M. VAJANAPHANICH, K.E. BARRETT & A.E. TRAYNOR-KAPLAN. 1993. Elevation of inositol tetrakisphosphate parallels inhibition of calcium-dependent chloride secretion in T_{84} colonic epithelial cells. Am. J. Physiol. **264:** C671–C676.
6. BARRETT, K.E., J. SMITHAM, A.E. TRAYNOR-KAPLAN & J.M. URIBE. 1998. Inhibition of Ca^{2+}-dependent Cl^- secretion in T_{84} cells: membrane target(s) of inhibition are agonist-specific. Am. J. Physiol. **274:** C958–C965.
7. URIBE, J.M., C.M. GELBMANN, A.E. TRAYNOR-KAPLAN & K.E. BARRETT. 1996. Epidermal growth factor inhibits calcium-dependent chloride secretion in T_{84} human colonic epithelial cells. Am. J. Physiol. **271:** C914–C922.
8. URIBE, J.M., S.J. KEELY, A.E. TRAYNOR-KAPLAN & K.E. BARRETT. 1996. Phosphatidylinositol 3-kinase mediates the inhibitory effect of epidermal growth factor on calcium-dependent chloride secretion. J. Biol. Chem. **271:** 26588–26595.
9. KEELY, S.J., J.M. URIBE & K.E. BARRETT. 1998. Carbachol stimulates transactivation of epidermal growth factor receptor and MAP kinase in T_{84} cells: implications for carbachol-stimulated chloride secretion. J. Biol. Chem. **273:** 27111–27117.
10. KEELY, S.J., M.J. HALVORSEN & K.E. BARRETT. 1998. Colonic epithelial ErbB2 receptors: a possible role in diversification of inhibitory signaling via the EGF receptor. Gastroenterology **114:** A385.
11. KEELY, S.J. & K.E. BARRETT. 1999. Carbachol stimulates p60[src] in T_{84} epithelial cells: implications for carbachol-stimulated chloride secretion. Gastroenterology **116:** A859.
12. KEELY, S.J., S.O. CALANDRELLA & K.E. BARRETT. 1999. Elevations in intracellular calcium are sufficient to evoke antisecretory signaling via the epidermal growth factor receptor and ERK in T_{84} cells. Gastroenterology **116:** A858.
13. VAJANAPHANICH, M., M. WASSERMAN, T. BURANAWUTI, K.E. BARRETT & A.E. TRAYNOR-KAPLAN. 1993. Carbachol uncouples chloride secretion from [Ca++]: association with tyrosine kinase activity. Gastroenterology **104:** A286.
14. WASSERMAN, M., M. VAJANAPHANICH, K.E. BARRETT & A.E. TRAYNOR-KAPLAN. 1993. $InsP_4$ elevation parallels inhibition of Cl^- secretion in T_{84} cells. FASEB J. **7:** A1259.
15. LUTTRELL, L.M., T. VAN BIESEN, B.E. HAWES *et al.* 1997. G-protein–coupled receptors and their regulation: activation of the MAP kinase signaling pathway by G-protein–coupled receptors. Adv. Second Messenger Phosphoprotein Res. **31:** 263–277.
16. LUTTRELL, L.M., Y. DAAKA & R.J. LEFKOWITZ. 1999. Regulation of tyrosine kinase cascades by G-protein–coupled receptors. Curr. Opin. Cell Biol. **11:** 177–183.
17. DAUB, H., C. WALLASCH, A. LANKENAU *et al.* 1997. Signal characteristics of G protein–transactivated EGF receptor. EMBO J. **16:** 7032–7044.
18. MURASAWA, S., Y. MORI, Y. NOZAWA *et al.* 1998. Role of calcium-sensitive tyrosine kinase Pyk2/CAKß/RAFTK in angiotensin II–induced Ras/ERK signaling. Hypertension **32:** 668–675.
19. BLAUKAT, A., I. IVANKOVIC-DIKIC, E. GRÖNROOS *et al.* 1999. Adaptor proteins Grb2 and Crk couple Pyk2 with activation of specific mitogen-activated protein kinase cascades. J. Biol. Chem. **274:** 14893–14901.
20. LEV, S., H. MORENO, R. MARTINEZ *et al.* 1995. Protein tyrosine kinase PYK2 involved in Ca^{2+}-induced regulation of ion channel and MAP kinase functions. Nature **376:** 737–745.
21. LUTTRELL, L.M., B.E. HAWES, T. VAN BIESEN *et al.* 1996. Role of c-SRC tyrosine kinase in G protein–coupled receptor- and Gß gamma subunit–mediated activation of mitogen-activated protein kinases. J. Biol. Chem. **271:** 19443–19450.

22. DIKIC, I., G. TOKIWA, S. LEV *et al.* 1996. A role for Pyk2 and Src in linking G-protein–coupled receptors with MAP kinase activation. Nature **383:** 547–550.
23. FELSCH, J.S., T.G. CACHERO & E.G. PERALTA. 1998. Activation of protein tyrosine kinase PYK2 by the m1 muscarinic acetylcholine receptor. Proc. Natl. Acad. Sci. USA **95:** 5051–5056.
24. DELLA ROCCA, G.J., T. VAN BIESEN, Y. DAAKA *et al.* 1997. Ras-dependent mitogen-activated protein kinase activation by G protein–coupled receptors. Convergence of G_i- and G_q-mediated pathways on calcium/calmodulin, Pyk2 and Src kinase. J. Biol. Chem. **272:** 19125–19132.
25. RIESE, D.J. II & D.F. STERN. 1998. Specificity within the EGF family/ErbB receptor family signaling network. Bioessays **20:** 41–48.
26. URIBE, J.M. & K.E. BARRETT. 1997. Non-mitogenic actions of growth factors: an integrated view of their role in intestinal physiology and pathophysiology. Gastroenterology **112:** 255–268.
27. VAJANAPHANICH, M., C. SCHULTZ, R.Y. TSIEN *et al.* 1995. Cross-talk between calcium and cAMP-dependent intracellular signaling pathways: implications for synergistic secretion in T_{84} colonic epithelial cells and rat pancreatic acinar cells. J. Clin. Invest. **96:** 386–393.
28. HOGENAUER, C., B.W. AICHBICHLER, J.L. PORTER & J.S. FORDTRAN. 1999. Effect of atropine and octreotide on normal active chloride secretion by the human jejunum *in vivo.* Gastroenterology **116:** A880.
29. DHARMSATHAPHORN, K., P.A. HUOTT, P. VONGKOVIT *et al.* 1989. Cl⁻ secretion induced by bile salts. A study of the mechanism of action based on a cultured colonic epithelial cell line. J. Clin. Invest. **84:** 945–953.
30. GELBMANN, C.M., C.D. SCHTEINGART, S.M. THOMPSON *et al.* 1995. Mast cells and histamine contribute to bile-acid–stimulated secretion in the mouse colon. J. Clin. Invest. **95:** 2831–2839.
31. SIDHU, M. & H.J. COOKE. 1995. Role for 5-HT and ACh in submucosal reflexes mediating colonic secretion. Am. J. Physiol. **269:** G346–G351.
32. GUGLIETTA, A. & P.B. SULLIVAN. 1995. Clinical applications of epidermal growth factor. Eur. J. Gastroenterol. Hepatol. **7:** 945–950.
33. PROCACCINO, F., M. REINSHAGEN, P. HOFFMAN *et al.* 1994. Protective effect of epidermal growth factor in an experimental model of colitis in rats. Gastroenterology **107:** 12–17.
34. FRIZZELL, R.A. 1993. The molecular physiology of cystic fibrosis. News Physiol. Sci. **8:** 117–120.
35. ANDERSON, M.P., D.N. SHEPPARD, H.A. BERGER & M.J. WELSH. 1992. Chloride channels in the apical membrane of normal and cystic fibrosis airway and intestinal epithelia. Am. J. Physiol. **263:** L1–L14.
36. GOLDSTEIN, J.L., A.B. SHAPIRO, M.C. RAO & T.J. LAYDEN. 1991. *In vivo* evidence of altered chloride but not potassium secretion in cystic fibrosis rectal mucosa. Gastroenterology **101:** 1012–1019.
37. BERSCHNEIDER, H.M., M.R. KNOWLES, R.G. AZIZKHAN *et al.* 1988. Altered intestinal chloride transport in cystic fibrosis. FASEB J. **2:** 2625–2629.
38. BOUCHER, R.C., E.H.C. CHENG, A.M. PARADISO *et al.* 1989. Chloride secretory response of cystic fibrosis human airway epithelia. Preservation of calcium but not protein kinase C– and A–dependent mechanisms. J. Clin. Invest. **84:** 1424–1431.
39. CLARKE, L.L., B.R. GRUBB, J.R. YANKASKAS *et al.* 1994. Relationship of a non-cystic fibrosis transmembrane conductance regulator-mediated chloride conductance to organ-level disease in *Cftr* (-/-) mice. Proc. Natl. Acad. Sci. USA **91:** 479–483.
40. GRUBB, B.R. 1995. Ion transport across the jejunum in normal and cystic fibrosis mice. Am. J. Physiol. **268:** G505–G513.
41. KNOWLES, M.R., L.L. CLARK & R.C. BOUCHER. 1991. Activation by extracellular nucleotides of chloride secretion in the airway epithelia of patients with cystic fibrosis. N. Engl. J. Med. **325:** 533–538.
42. CLARKE, L.L. & R.C. BOUCHER. 1993. Ion and water transport across airway epithelia. *In* Pharmacology of the Respiratory Tract. K.F. Chung & P.J. Barnes, Eds.: 505–550. Marcel Dekker. New York.

Neurotransmitters in Neuronal Reflexes Regulating Intestinal Secretion

HELEN J. COOKE[a]

Department of Neuroscience, The Ohio State University, Columbus, Ohio 43210, USA

ABSTRACT: The intestinal crypt cell secretes chloride into the lumen, resulting in accumulation of fluid that normally thins out mucus or, at higher secretory rates, flushes out the contents. The regulation of chloride secretion occurs by neural reflex pathways within the enteric nervous system. Mechanical stimulation releases 5-hydroxytryptamine (5-HT) from enterochromaffin cells with subsequent activation of intrinsic primary afferents that carry electrical signals to submucosal ganglia. After processing, interneurons activate cholinergic and vasoactive intestinal peptide (VIP) secretomotor neurons. Acetylcholine and VIP bind to epithelial receptors and stimulate sodium chloride and fluid secretion. Reflex-evoked secretory rates can be modulated by a variety of mediators at the level of the enterochromaffin cells, neurons within the reflex pathway, or epithelial cells. Understanding the complex regulatory mechanisms for chloride secretion is likely to provide mechanistic insights into constipation and diarrhea.

The intestinal epithelium serves as a barrier between the outside world and the host. This barrier consists of epithelial cells and associated tight junctions, sparsely dispersed enteroendocrine cells, and specialized epithelial cells called M cells. This barrier is not impermeant, as small solutes and water can penetrate via paracellular, transcellular, or M cell pathways. When the barrier is "leaky," macromolecules may gain access to the host's environment and trigger either tolerance or an inflammatory response.[1] One important aspect of this barrier is chloride secretion by epithelial cells. Chloride that is taken up by epithelial cells via a basolateral sodium/potassium/chloride cotransporter mechanism exits the cell via apical, cyclic AMP–activated cystic fibrosis transmembrane regulator chloride channels (CFTR) or through a distinct class of calcium-activated chloride channels whose existence is somewhat controversial.[2,3] Subsequent opening of basolateral potassium channels sustains the appropriate electrochemical driving forces for continued chloride secretion. Whereas elevation of intracellular cyclic AMP causes sustained chloride secretion, elevation of intracellular calcium causes transient chloride secretion due to generation of the negative regulatory messenger, 1,3,4,5-tetrakisphosphate (IP4), which limits the response.[4] Chloride secretion into the lumen provides a driving force for sodium to

[a]Address for correspondence: Helen J. Cooke, Department of Neuroscience, The Ohio State University, 333 W. 10th Avenue, Columbus, OH 43210. Voice: 614-292-5660; fax: 614-688-8742.

cooke.1@osu.edu

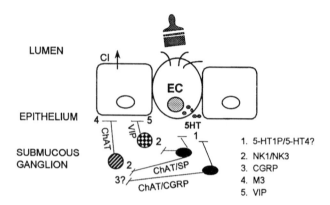

FIGURE 1. Working model of mechanically evoked neural reflex in the submucosal plexus regulating chloride secretion by crypt epithelial cells in the guinea pig colon. A mechanical stimulus (brush stroke) releases 5-hydroxytryptamine (5-HT) from enterochromaffin cells (EC). 5-HT acts at 5-HT$_{1P}$/5-HT4? receptors (**1**) on intrinsic primary afferents containing acetylcholine (ChAT)/substance P (**2**) and ChAT/calcitonin gene–related peptide (CGRP) (**3**). Substance P is released in the submucosal ganglion to activate NK$_1$/NK$_3$ receptors on cholinergic or vasoactive intestinal peptide (VIP) secretomotor neurons. Binding of acetylcholine or VIP to epithelial muscarinic M$_3$ (**4**) or VIP receptors (**5**), respectively, causes chloride secretion. A mechanical stimulus also releases prostaglandins (not shown). No cholinergic nicotinic transmission is involved. In addition, the mechanical stimulus may release an unknown transmitter from submucosal primary afferent collaterals associated with the epithelium in an axon reflex.

follow passively along with water. Normally secretion maintains the fluidity of mucus, and excess secretion flushes the intestine of its contents and any noxious agents.

Two neurotransmitter receptors that signal through cyclic AMP or calcium are vasoactive intestinal peptide (VIP) and acetylcholine, respectively. The presence of these receptors on epithelial cells suggests that the enteric nervous system is one of the systems that regulates chloride secretion. Extracellular messengers, in the form not only of neurotransmitters, but also of paracrine/endocrine mediators and chemicals released from cells of the immune system, can regulate chloride secretion as well. This paper will focus on the role of the enteric nervous system in reflex regulation of chloride secretion.

The enteric nervous system consists of two ganglionated plexuses, the myenteric plexus sandwiched between the longitudinal and circular muscles and the submucous plexus lying in the submucosa. Both participate in coordinating secretion and muscle contraction that lead to segmenting or propulsive contractile patterns. Neurons within the submucous plexus are arranged in reflex circuits that consist of primary afferent neurons carrying signals to the ganglia from the periphery, interneurons, and secretomotor neurons that project to the epithelium.[5] Sensory cells that transduce mechanical stimuli include enterochromaffin cells that release 5-hydroxytryptamine (5-HT) and cells that release prostaglandins.[6–9] In addition, mast cells from sensitized animals or enteroendocrine cells are also candidates for sensory cells.

In the intestine of the guinea pig, an animal model that has been used extensively to identify the types of neurons, their chemical codes, and their functional characteristics, mucosal stroking causes release of 5-HT from enterochromaffin cells.[6,8] Its binding to $5-HT_{1P}/5-HT_4$ receptors on intrinsic primary afferents in the colon activates a neural reflex within the submucous plexus that causes chloride secretion. Associated with chloride secretion is muscle contraction necessary to propel secretions and residual contents along the intestine's length. In the small intestine, there appear to be two types of primary afferents that are cholinergic, and they can be distinguished by different combinations of neurotransmitters: (1) Choline acetyltransferase (ChAT)/substance P with a possible subset containing glutamate, and (2) ChAT/calcitonin gene–related peptide (CGRP).[10,12–14] Both of these are activated by 5-HT. Whether this is also true for the colon is unknown. In the small intestine, ChAT/substance P primary afferents may synapse directly with cholinergic secretomotor neurons whose cell bodies express NK_1 receptors.[15] This implies that ChAT/ substance P neurons release substance P that activates cholinergic secretomotor neurons and not VIP secretomotor neurons that lack an NK_1 receptor. That substance P is involved in stroking-evoked secretion is evident in the colon, because NK_1 receptor antagonists inhibit the response.[7] In the colon, both ChAT and VIP secretomotor neurons responded to NK_1 and NK_3 receptor agonists, and this suggests that ChAT/ substance P primary afferents may synapse directly with both cholinergic and VIP secretomotor neurons.[16,17]

In the guinea pig colon, of the total number of ChAT- and VIP-positive neurons, the mucosa is innervated by a few ascending pathways that are primarily ChAT (9% total, with 90% ChAT and 8% VIP positive), by many descending pathways with a majority containing VIP (60% total, with 88% VIP and 4% ChAT positive), and by circumferential projections (32% total, with 71% ChAT and 28% VIP positive).[16,17]

Mechanical stimulation evokes neural reflexes that regulate the rate of chloride secretion. This reflex-activated chloride secretion can be modified at several sites within the reflex. Regulation can occur at the level of the enterochromaffin cell, neurons within the reflex pathway, and epithelial cells. With respect to enterochromaffin cells, 5-HT is synthesized and stored in granules. 5-HT is released by influx of calcium through voltage-sensitive ion channels and by receptor-mediated release of intracellular calcium. There are multiple cell surface receptors that are inhibitory or stimulatory.[18–20] Inhibitory receptors include adrenergic α_2, histamine H_3, $GABA_A$, $GABA_B$, adenosine A_1, purinergic P_{2y}, $5-HT_4$, neurokinin NK_1, VIP, PACAP, and somatostatin. Stimulatory receptors include adrenergic β, muscarinic M_3, and nicotinic and $5-HT_3$.

These same receptors and many more are found on neurons and some of them on epithelial cells. Only by studying integrated responses as well as the effects of each of these receptors individually will we begin to understand all the ramifications of regulatory mechanisms for chloride secretion. The challenge for the future is understanding how integration of cellular function occurs to produce chloride secretion and important host defense mechanisms.

REFERENCES

1. BERIN, M.C., A.J. KILIAAN, P.C. YANG *et al.* 1997. Rapid transepithelial antigen transport in rat jejunum: impact of sensitization and the hypersensitivity reaction. Gastroenterology **113**: 856–864.

2. JI, H.L, M.D. DUVALL, H.K. PATTON *et al.* 1998. Functional expression of a truncated Ca(2+)-activated Cl- channel and activation by phorbol ester. Am. J. Physiol. **274:** C455–464.
3. CARTWRIGHT, C.A., J.A. MCROBERTS, K.G. MANDEL *et al.* 1984. Electrolyte transport in a human colonic cell line: regulation by cAMP and calcium. Kroc Found. Ser. **17:** 65–75.
4. KACHINTORN, U., M. VAJANAPHANICH, A.E. TRAYNOR-KAPLAN *et al.* 1993. Activation by calcium alone of chloride secretion in T84 epithelial cells. Br. J. Pharmacol. **109:** 510–517.
5. COOKE, H.J. 1998. Enteric tears: chloride secretion and its neural regulation. News Physiol. Sci. **13:** 275–280.
6. SIDHU, M. & H.J. COOKE. 1995. Role for 5-HT and ACh in submucosal reflexes mediating colonic secretion. Am. J. Physiol. **269:** G346–G351.
7. COOKE, H.J., M. SIDHU, P. FOX *et al.* 1997. Substance P as a mediator of colonic secretory reflexes. Am. J. Physiol. **273:** G238–G245.
8. COOKE, H.J., M. SIDHU & Y.-Z. WANG. 1997. 5-HT activates neural reflexes regulating secretion in guinea-pig colon. Neurogastroenterol. Motil. **9:** 181–186.
9. COOKE, H.J., M. SIDHU & Y.-Z. WANG. 1997. Activation of 5-HT$_{1P}$ receptors on submucosal afferents subsequently triggers VIP neurons and chloride secretion in guinea-pig colon. J. Auton. Nerv. Syst. **66:** 105–110.
10. KIRCHGESSNER, A.L., H. TAMIR & M.D. GERSHON. 1992. Identification and stimulation by serotonin of intrinsic sensory neurons of the submucosal plexus of the guinea pig gut: activity induced expression of fos immunoreactivity. J. Neurosci. **12:** 235–248.
11. GRIDER, J.R., J.F. KUEMMERLE & J.G. JIN. 1996. 5-HT released by mucosal stimuli initiates peristalsis by activating 5-HT4/5-HT1P receptors on sensory CGRP neurons. Am. J. Physiol. **270:** G778–G782.
12. FURNESS, J.B., W.A. KUNZE, P.P. BERTRAND *et al.* 1998. Intrinsic primary afferent neurons of the intestine. Prog. Neurobiol. **54:** 1–18.
13. PAN, H. & M.D. GERSHON. 1998. Identification and characterization of intrinsic sensory neurons of the submucosal plexus. Gastroenterology **114:** A818.
14. LIU, M.-T., J.D. ROTHSTEIN, M.D. GERSHON & A.L. KIRCHGESSNER. 1997. Glutamatergic enteric neurons. J. Neurosci. **17:** 4764–4784.
15. MOORE, B.A., S. VANNER, N.W. BUNNETT & K.A. SHARKEY. 1997. Characterization of neurokinin-1 receptors in the submucosal plexus of guinea pig ileum. Am. J. Physiol. **273:** G670–G678.
16. FRIELING, T., G. DOBREVA, E. WEBER *et al.* 1999. Different tachykinin receptors mediate chloride secretion in the distal colon through activation of submucosal neurons. Naunyn-Schmiedebergs Arch. Pharmakol. **359:** 71–19.
17. NEUNLIST, M. & M. SCHEMANN. 1998. Polarised innervation pattern of the mucosa of the guinea pig distal colon. Neurosci. Lett. **246:** 161–164.
18. RACKE, K., A. REIMANN, H. SCHWORER & H. KILBINGER. 1996. Regulation of 5-HT release from the enterochromaffin cells. Behav. Brain Res. **73:** 83–87.
19. GINAP, T. & H. KILBINGER. 1997. NK1- and NK3-receptor mediated inhibition of 5-hydroxytryptamine release from the vascularly perfused small intestine of the guinea pig. Naunyn-Schmiedebergs Arch. Pharmakol. **356:** 689–693.
20. GEBAUER, A., M. MERGER & H. KILBINGER. 1993. Modulation of 5-HT3 and 5-HT4 receptors of the release of 5-hydroxytryptamine from the guinea-pig small intestine. Naunyn-Schmiedebergs Arch. Pharmakol. **347:** 137–140.

Three 5′-Variant mRNAs of Anion Exchanger AE2 in Stomach and Intestine of Mouse, Rabbit, and Rat

HEIDI ROSSMANN, [a,b] SETH L. ALPER,[c] MANUELA NADER,[d] ZHUO WANG,[e] MICHAEL GREGOR,[a] AND URSULA SEIDLER[a]

[a]I. Department of Medicine, Eberhard-Karls University, Tübingen, Germany

[c]Molecular Medicine and Renal Units, Beth Israel Deaconess Medical Center, Departments of Medicine and Cell Biology, Harvard Medical School, Boston, Massachusetts 02215, USA

[d]II. Department of Medicine, Technical University Munich, Munich, Germany

[e]Department of Molecular Genetics, Biochemistry and Microbiology, University of Cincinnati College of Medicine, Cincinnati, Ohio 45267, USA

ABSTRACT: AE2 is one of three known isoforms of the anion exchanger (AE) gene family. The use of alternative promoters, resulting in a tissue-specific transcript pattern, was reported for all AE genes. Three N-terminal variant AE2 subtypes are described: AE2a, AE2b, and AE2c. Although the basolaterally located parietal cell anion exchanger is known to be an AE2, the molecular identity of the basolateral and apical anion exchangers throughout the gut are still unknown. This article summarizes functional, immunohistochemical, and Western blot data demonstrating the basolateral localization of the gastric and intestinal AE2 in rabbit, mouse, and rat, and showing the AE2 subtype mRNA expression pattern in the stomach and along the intestine of rabbit and mouse: AE2a is expressed in all studied tissues, but most strongly in the colon; AE2b is expressed mainly in the stomach; and AE2c is detected nearly exclusively in the stomach. Further investigation is necessary to characterize the apical anion transport protein involved in NaCl absorption and HCO_3^- secretion in the gut.

THE ANION EXCHANGER GENE FAMILY

Cl^-/HCO_3^- exchangers contribute to the regulation of intracellular pH, cell volume, intracellular chloride concentration, and transepithelial acid/base transport in a wide variety of organs and cell types. Under physiological extra- and intracellular Cl^- and HCO_3^- concentrations, the sodium-independent anion exchangers (AE) normally mediate electroneutral, stilbene-sensitive one-for-one exchange of extracellular Cl^- for intracellular HCO_3^- across the cell membrane. This activity is acid loading and, if uncompensated, results in intracellular acidification.

[b]Address for correspondence: Dr. Heidi Rossmann, Abteilung Innere Medizin I, Universitätsklinikum auf dem Schnarrenberg, Eberhard-Karls-Universität Tübingen, Otfried-Müller-Str. 10, 72076 Tübingen, Germany. Fax: +49-7071-295692.
heidi.rossmann@uni-tuebingen.de

The AE proteins are encoded by the anion exchanger gene family. The molecular identity of three members of this gene family has been elucidated, and there is evidence for the presence of further isoforms (reviewed in Refs. 1–3). AE1, also called "band 3," was the first isoform to be cloned.[4] AE1 has been found in erythrocytes, where it contributes to CO_2 release and membrane stability, and in the kidney.[5–7] In contrast to the restricted expression of AE1, AE2 is a widely expressed protein[8,9] that is thought to play a "housekeeping" role on the one hand and is involved in transepithelial secretion and absorption—for example, in the stomach, the intestine, the choroid plexus, and the renal medulla—on the other.[10–14] AE3 is expressed mainly in excitable tissue—for example, in heart and brain[9,15]—but in gut, as well.[9]

All three anion exchanger proteins are composed of two distinct domains. The C-terminal, membrane-associated domain is highly conserved among the isoforms and mediates the Cl^-/HCO_3^- exchange. The N-terminal cytoplasmic domain shows much lower homology among the AE proteins.

INTRON/EXON ORGANIZATION AND PROMOTER USAGE OF THE AE GENE FAMILY WITH FOCUS ON AE2

Due to alternative promoters, different mRNAs are transcribed from all three AE genes in a tissue-specific manner. Intron/exon organization, promoter usage, and the splicing pattern of the rat,[4,7,16] mouse,[14,15] and rabbit (Wang & Shull, unpublished) anion exchanger gene family are quite similar. The rat AE1 form expressed in the kidney (kAE1) is a truncated form of the erythrocyte AE1 (eAE1, "band 3") and is transcribed from a different promoter, located in intron 3 of the AE1 gene. In the rodent, kAE1 lacks the N-terminal 79 amino acids (65 amino acids in the human). The mouse AE3 detected in cardiac myocytes (cAE3) is shorter than brain AE3 (bAE3). The cAE3 mRNA is transcribed from an alternative promoter, located in intron 6 of the AE3 gene. The resulting cAE3 protein contains 73 N-terminal amino acids in place of the 270 N-terminal amino acids of bAE3. After the detection of multiple AE2 transcripts of different sizes,[9] Wang et al. searched for and defined three rat mRNAs (AE2a, AE2b, and AE2c; FIG. 1A), transcribed from alternative promoters.[16] AE2a is a 1234–amino acid protein, encoded by a 4.4-kb mRNA. AE2b, a 1220–amino acid protein, encoded by a 4.2-kb mRNA, is transcribed from a promoter, localized in intron 2 of the AE2 gene. The N-terminal 17 amino acids of AE2a are replaced by 3 alternative amino acids in the AE2b protein. The amino acid sequence of AE2c is identical to that of AE2a, except for the absence of the first 199 amino acids of AE2a. AE2c is encoded by two mRNAs of 3.8 and 4.1 kb, and both are transcribed from a common promoter located in intron 5. The 4.1-kb splice variant retains intronic sequence between exon 1c and exon 6, whereas this sequence is spliced out from the 3.8-kb transcript. Although the AE2a transcript was detected in all examined tissues, the expression of AE2b and AE2c is more restricted: AE2b is expressed in the lung, liver, kidney, and gastrointestinal tract, for example, although high levels have been detected only in the stomach. AE2c is expressed almost exclusively in the stomach.[16,17]

The complete coding sequence of rabbit AE2a[11] along with sequence from exons 1b and 1c and intron 5 of the rabbit AE2 gene (Wang & Shull, unpublished data) suggests that the AE2 gene of rabbit resembles that of the rat and mouse[14] and predicts rabbit AE2 transcripts of about 4.2 kb (AE2a), 4.0 kb (AE2b), 3.5 kb (AE2c1), and 3.8 kb (AE2c2).

A parallel splicing pattern has been found for murine AE2a and AE2b.[14] Amplification and sequencing of murine intron 5[14] resulted in the prediction of a novel AE2c2 protein that contains 32 additional amino acids, compared to the rat AE2c polypeptide. The first 27 of the predicted 32 amino acids are not present in AE2a and AE2b. Low-abundance AE2c2 transcript has been found in murine kidney[14] in addition to AE2a and minor levels of AE2b mRNAs.

The biological significance of the N-terminal variant AE2 mRNAs (summarized in FIG. 1a), transcribed from alternative promoters in a tissue-specific pattern, has not yet been established. The predicted differences among the N-terminal amino acid sequences of the AE2 encourage the speculation that the various subtypes may be differentially regulated by protein kinases and/or sorted into various membrane domains in the various organs.

EXPRESSION OF AE2 ANION EXCHANGER IN THE GASTROINTESTINAL TRACT

A highly active anion exchanger is located in the basolateral membrane of the parietal cell of the gastric mucosa (FIG. 1b), which is responsible for the import of Cl^- that is secreted and for the export of HCO_3^- that is produced intracellularly during acid secretion.[18–20] Rossmann *et al.* cloned an AE2 from highly purified rabbit parietal cells[21] by RT-PCR; AE2 is highly expressed in rabbit parietal cells (FIG. 2b). The functional properties of the rabbit basolateral parietal cell anion exchanger that affect its regulation by intracellular pH and osmolarity—its stilbene sensitivity and anion selectivity[22,23]—resemble those of recombinant AE2 expressed in *Xenopus* oocytes[24,25] and in transfected mammalian cells.[26,27] Northern hybridization of rabbit parietal cell poly (A^+) RNA shows three strong AE2 hybridization products (FIG. 2B) at 4.2, 4.0, and 3.5 kb, corresponding to the theoretical sizes of rabbit AE2a, AE2b, and AE2c1, described above. RT-PCR analysis, using specific primers for the different AE2 subtypes, confirms this assumption (data not shown): from rabbit kidney only AE2a (35 cycles) was amplified, and Northern analysis of kidney RNA shows only the 4.2-kb transcript (FIG. 2C). Thus, it represents AE2a. AE2a, AE2b, and AE2c1, but not AE2c2, were amplified (35 cycles) from rabbit gastric mucosa, and three transcripts were discriminated by Northern analysis. Therefore, at least in rabbit stomach, the 4.0-kb transcript likely represents AE2b and the 3.5-kb transcript AE2c1. Several antibodies against various AE2 peptide antigens detect AE2 in the gastric mucosa (FIG. 2C) and delineate the parietal cell basolateral membrane.[12,28–30] Thus, the parietal cell anion exchanger involved in acid secretion is very likely an AE2 (FIG. 1C).

Anion exchange activity has also been reported throughout the gut, from duodenum to colon, basolateral as well as apical, in villus and crypt enterocytes.[22,31–34] An apical anion exchange process has been functionally characterized in the brush border membrane of small and large intestine more than a decade ago.[22,33,35] The coupled activity of Na^+/H^+ exchangers (NHE2 and NHE3[36]) and Cl^-/HCO_3^- or Cl^-

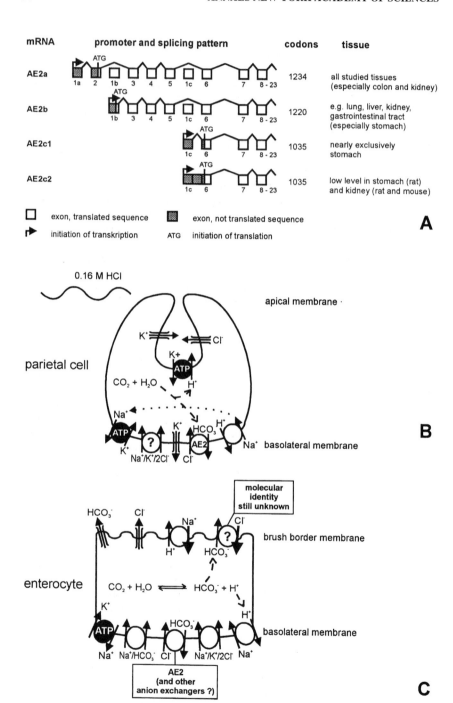

FIGURE 1. *See following page for caption.*

/OH$^-$ exchangers is thought to mediate intestinal electroneutral NaCl absorption across the apical membrane of the enterocyte.[35,37] In addition, apical Cl$^-$/HCO$_3^-$ exchange is thought to be involved in duodenal HCO$_3^-$ secretion. The molecular identity of the anion exchange protein(s), mediating these functions, remained unknown (FIG. 1C), but it was early suggested to be an AE2. Chow *et al.* cloned the rabbit AE2 cDNA from the ileum and localized the AE2 protein in the microvillar membrane fraction by Western analysis.[11] Indeed, we were able to detect the 4.2-kb and a 4.0-kb AE2 transcript in rabbit intestine (probably representing AE2a and AE2b), but AE2 mRNA expression levels were low (FIG. 2A), and we were not able to detect AE2 in the brush border membrane fractions of rabbit ileum using an antibody directed against the C-terminus of AE2. However, there was a small signal in the basolateral membrane fraction (FIG. 2C). By contrast, anion exchange rates, measured by DIDS-inhibitable ^{36}Cl$^-$ uptake into Cl$^-$-loaded rabbit ileal vesicles, were at least twice as high in brush border membrane vesicles (BBM, FIG. 3A) than in basolateral membrane vesicles (BLM, FIG. 3b). Furthermore, BBM and BLM anion exchanger displayed different kinetics (FIG. 4), so they are probably different entities. In many other epithelial cell types, among them gastric epithelium,[12] choroid plexus epithelium,[38] and kidney tubular cells,[13,14] a basolateral localization of the AE2 protein was reported. Moreover, the anion exchanger of the brush border membrane vesicles from the small and large intestine of several species displays crucial differences from that reported for AE2.[22,31,39,40]

Inasmuch as AE2a and AE2b have differences in their N-terminal amino acid composition, it is possible that they are sorted to different membranes. Thus, it has been speculated that AE2a represents the brush border anion exchanger in the intestine and AE2b is sorted to the basolateral membrane. To get further insight into the distribution of the different AE2 subtypes, we studied AE2 mRNA and protein expression in mouse intestine.[17] Using a semiquantitative PCR technique, we studied AE2 mRNA abundance of AE2a, AE2b, and AE2c in mucosal tissue of selected parts of the gastrointestinal tract: stomach, duodenum, jejunum, ileum, and colon. AE2 expression was highest in the gastric mucosa, substantial in the colonic mucosa, and low in the other parts of the intestine. AE2a mRNA abundance was higher in distal colon than in more proximal segments. Interestingly, high levels of AE2a mRNA were also found in the muscularis of the colon. AE2b mRNA was expressed at high levels in the stomach and at much lower levels in the other tested tissues. AE2c mRNA was nearly exclusively expressed in the stomach, with intestinal levels at the threshold of detection. The abundance of AE2 protein as detected by immunoblot qualitatively paralleled that of mRNA, whereas AE2 immunostaining exhibited a more continuous decrease in intensity from colon to duodenum. AE2 polypeptide was more abundant in colonic surface cells than in crypts, whereas ileal crypts and villi exhibited similar AE2 abundance. Localization of AE2 epitopes was restricted to the basolateral membranes of epithelial cells throughout the intestine with only few exceptions. This basolateral localization corresponds to the results of localization studies for AE2 in other epithelial tissues. An AE2, basolaterally located in the

---

FIGURE 1. (A) Intron/exon organization, promoter usage, and splicing patterns of the rat AE2 gene (according to Ref. 16). Mouse and rabbit genes appear similar in alternative 5'-exon usage. Ion transport mechanisms in the parietal cell **(B)** and the enterocyte **(C)**.

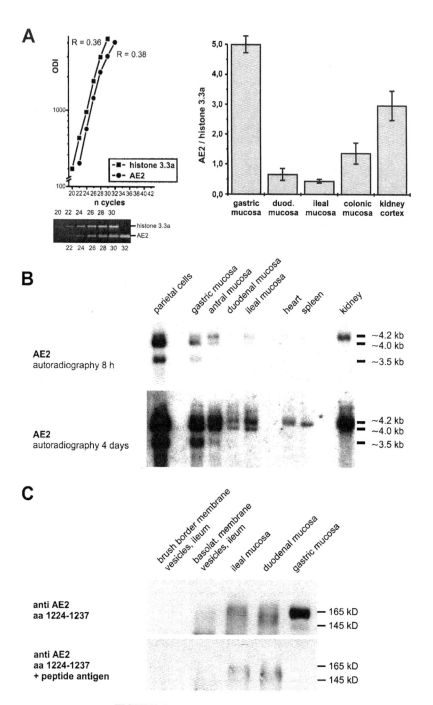

FIGURE 2. *See following page for caption.*

enterocyte, could contribute to intracellular pH and to cell volume regulation during transepithelial secretion and absorption events. Rajendran *et al.* have functionally characterized Cl^-/HCO_3^- and Cl^-/OH^- exchangers in basolateral membrane vesicles of rat distal colon,[41] and Tosco *et al.* have done so in the basolateral membrane of the rat jejunum.[42] It is likely that at least part of this basolateral anion exchange activity represents AE2 function.

Whereas the basolateral parietal cell anion exchanger is an AE2 (FIG. 1b), the molecular identity of the apical anion exchanger proteins, contributing to HCO_3^- secretion and NaCl absorption in the intestine, remains unclear (FIG. 1c). Transport proteins such as AE3,[17,22] DRA (downregulated in adenoma[17,43,44]), AE1, or yet-uncloned proteins have been suggested as possible candidates.

The biological reason for the evolutionary development of alternative promoters within the AE2 gene remains unclear, but the tissue-specific expression patterns of AE2a, AE2b, and AE2c suggest that the alternative promoters may serve to confer tissue specificity to AE2 expression. At least in part, AE2b and AE2c, lacking potential N-terminal regulative sequences of AE2a, could have been developed to make a genetic mechanism available, facilitating the high anion exchange rates required in the stomach. AE2a, which is detected in all tissues studied, could play the "housekeeping" role. Different N-terminal proteins could be differentially regulated—for example, by the loss or addition of phosphorylation sites.[16] However, to this date no

---

FIGURE 2. (A) Semiquantitative RT-PCR analysis of AE2 in rabbit stomach and intestine. The *left diagram* shows an example of the parallel amplification curves of AE2 and histone 3.3a from rabbit gastric mucosa (0.8 μg total RNA); the *gel picture below* shows the bands that were analyzed (ODI: integrated optical density). The numbering represents the cell cycles. The slope of both curves (R) was calculated and the ratio AE2/histone 3.3a was determined during the exponential phase of both reactions. For the *bar graph on the right*, the ratio of AE2/histone 3.3a (representing the relative expression level of AE2[47]) was plotted for the stomach and different parts of the gut ($n = 3$). To ascertain the identity of the PCR fragments, they were both cloned and sequenced. **(B)** High-stringency Northern analysis of 5 μg poly (A$^+$) RNA (1x-oligo-*dT*-enriched) from rabbit parietal cells; rabbit gastric, antral, duodenal, and ileal mucosa; and heart, spleen, and kidney, probed with a ^{32}P-labeled homologous AE2 cDNA fragment (bp 2484 to 2742 of sequence S45791 [GenBank accession number]) and exposed for 8 hours and 4 days. Three AE2 transcripts of different size (4.2 kb, 4.0 kb, and 3.5 kb) were detected: The 4.2-kb transcript was found in all tested organs, the 4.0-kb transcript only in the gastrointestinal organs, and the 3.5-kb transcript exclusively in the stomach. **(C)** Western analysis of AE2 protein in rabbit gastric, duodenal, and ileal mucosa. Brush border membrane vesicles (BBM, enrichment 14.7 ± 1.9–fold for alkaline phosphatase and 0.4 ± 0.06–fold for Na$^+$/K$^+$-ATPase relative to total membranes) and basolateral membrane vesicles (BLM, enrichment 10.8 ± 3.8–fold for Na$^+$/K$^+$-ATPase and 1.6 ± 0.1–fold for alkaline phosphatase) were isolated from ileal mucosa according to Knickelbein *et al.*[48] and Nader *et al.*[22] Two hundred and fifty μg protein from BBM and BLM, ileal and duodenal total mucosa, and 30 μg protein from gastric mucosa were electrophoresed, blotted, and incubated with an antibody directed against aa 1224 to 1237 of mouse AE2 *(upper panel)* or with the antibody in the presence of peptide antigen *(lower panel;* negative control). The *strong band* at 165 kDa, found in gastric mucosa, shows that the antibody recognizes rabbit AE2. *Fuzzy bands* are detected in duodenal and ileal mucosa, as well as in the ileal BLM fraction, whereas there was no band in the BBM fraction. Although the antibody binding in gastric mucosa is specifically blocked by addition of the peptide antigen, the small bands in duodenum and ileum do not completely disappear.

A

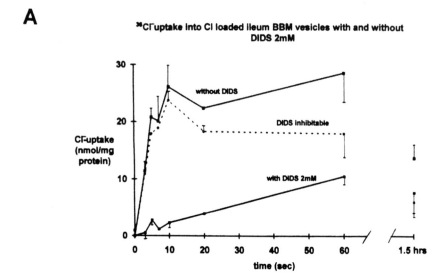

B

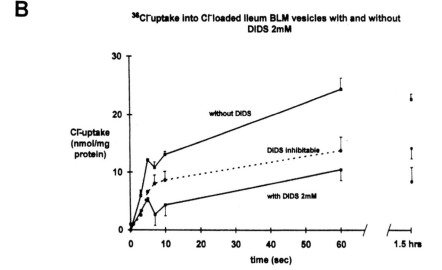

FIGURE 3. Time course of $^{36}Cl^-$ uptake into Cl^--loaded ileal brush border membrane vesicles (BBM, enrichment 10.4 ± 0.92–fold for alkaline phosphatase and 1.5 ± 0.55–fold for Na^+/K^+-ATPase) (**A**) and ileal basolateral membrane vesicles (BLM, enrichment 16.2 ± 1.53–fold for Na^+/K^+-ATPase and 1.6 ± 0.55–fold for alkaline phosphatase) (**B**) prepared and studied in parallel. Anion exchange rates were measured as DIDS-inhibitable $^{36}Cl^-$ uptake into Cl^--loaded vesicles (anion uptake against a concentration gradient) in the presence of a valinomycin voltage clamp, using the rapid filtration technique.[22] DIDS-sensitive $^{36}Cl^-/Cl^-$ exchange was present in both ileal BBM and ileal BLM vesicles. The initial rate of DIDS-sensitive uptake and the initial plateau accumulation were each approximately two times higher in BBM than in BLM vesicles, despite the smaller vesicular volume of the BBM vesicles.

requirement for the N-terminal–most 199 amino acids of AE2 has been detected for acute regulation of anion transport.[45] (Also, Chernova & Alper, in preparation) Alternatively, the multiple N termini of AE2 polypeptides may encode membrane sorting or retention signals.[46] Further investigation will be necessary to elucidate the physiological significance of the subtypes of the AE isoforms and may be forthcoming from analysis of mice genetically modified in AE gene expression.

REFERENCES

1. KOPITO, R.R. 1990. Molecular biology of the anion exchanger gene family. Int. Rev. Cytol. **123:** 177–199.
2. ALPER, S.L. 1994. The band 3-related AE anion exchanger gene family. Cell Physiol. Biochem. **4:** 265–281.
3. TANNER, M.J. 1997. The structure and function of band 3 (AE1): recent developments (review). Mol. Membr. Biol. **14**(4): 155–165.
4. KOPITO, R.R. & H.F. LODISH. 1985. Primary structure and transmembrane orientation of the murine anion exchange protein. Nature **316**(6025): 234–238.
5. BROSIUS, F.C., S.L. ALPER, A.M. GARCIA & H.F. LODISH. 1989. The major kidney band 3 gene transcript predicts an amino-terminal truncated band 3 polypeptide. J. Biol. Chem. **264**(14): 7784–7787.
6. KUDRYCKI, K.E. & G.E. SHULL. 1989. Primary structure of the rat kidney band 3 anion exchange protein deduced from a cDNA. J. Biol. Chem. **264**(14): 8185–8192.
7. KUDRYCKI, K.E. & G.E. SHULL. 1993. Rat kidney band 3 Cl⁻/HCO₃⁻ exchanger mRNA is transcribed from an alternative promoter. Am. J. Physiol. **264**(3 Pt. 2): F540–547.
8. ALPER, S.L., R.R. KOPITO, S.M. LIBRESCO & H.F. LODISH. 1988. Cloning and characterization of a murine band 3–related cDNA from kidney and from a lymphoid cell line. J. Biol. Chem. **263**(32): 17092–17099.
9. KUDRYCKI, K.E. P.R. NEWMAN & G.E. SHULL. 1990. cDNA cloning and tissue distribution of mRNAs for two proteins that are related to the band 3 Cl⁻/HCO₃⁻ exchanger. J. Biol. Chem. **265**(1): 462–471.
10. LINDSEY, A.E., K. SCHNEIDER, D.M. SIMMONS *et al.* 1990. Functional expression and subcellular localization of an anion exchanger cloned from choroid plexus. Proc. Natl. Acad. Sci. USA **87**(14): 5278–5282.
11. CHOW, A., J.W. DOBBINS, P.S. ARONSON & P. IGARASHI. 1992. cDNA cloning and localization of a band 3–related protein from ileum. Am. J. Physiol. **263**(3 Pt. 1): G345–352.
12. STUART-TILLEY, A., C. SARDET, J. POUYSSEGUR *et al.* 1994. Immunolocalization of anion exchanger AE2 and cation exchanger NHE-1 in distinct adjacent cells of gastric mucosa. Am. J. Physiol. **266**(2 Pt. 1): C559–568.
13. ALPER, S.L., A. STUART-TILLEY, D. BIEMESDERFER *et al.* 1997. Immunolocalization of AE2 anion exchanger in rat kidney. Am. J. Physiol. **273**(4 Pt. 2): F601–614.
14. STUART-TILLEY, A., B.E. SHMUKLER, D. BROWN & S.L. ALPER. 1998. Immunolocalization and tissue-specific splicing of AE2 anion exchanger in mouse kidney. J. Am. Soc. Nephrol. **9**(6): 946–959.
15. LINN, S.C., K.E. KUDRYCKI & G.E. SHULL. 1992. The predicted translation product of a cardiac AE3 mRNA contains an N terminus distinct from that of the brain AE3 Cl⁻/HCO₃⁻ exchanger. Cloning of a cardiac AE3 cDNA, organization of the AE3 gene, and identification of an alternative transcription initiation site. J. Biol. Chem. **267**(11): 7927–7935.
16. WANG, Z., P.J. SCHULTHEIS & G.E. SHULL. 1996. Three N-terminal variants of the AE2 Cl⁻/HCO₃⁻ exchanger are encoded by mRNAs transcribed from alternative promoters. J. Biol. Chem. **271**(13): 7835–7843.
17. ALPER, S.L., H. ROSSMANN, S. WILHELM *et al.* 1999. Expression of AE2 anion exchanger in mouse intestine. Am. J. Physiol. Accepted for publication.
18. MUALLEM, S., D. BLISSARD, E.J. CRAGOE, JR. & G. SACHS. 1988. Activation of the Na⁺/H⁺ and Cl⁻/HCO₃⁻ exchange by stimulation of acid secretion in the parietal cell. J. Biol. Chem. **263**(29): 14703–14711.

19. PARADISO, A.M., M.C. TOWNSLEY, E. WENZL & T.E. MACHEN. 1989. Regulation of intracellular pH in resting and in stimulated parietal cells. Am. J. Physiol. 257(3 Pt. 1): C554–561.
20. SEIDLER, U., S. ROITHMAIER, M. CLASSEN & W. SILEN. 1992. Influence of acid secretory state on Cl^-/base and Na^+/H^+ exchange and pH_i in isolated rabbit parietal cells. Am. J. Physiol. 262(1 Pt. 1): G81–91.
21. ROSSMANN, H., U. SEIDLER, B. OBERMAIER et al. 1993. Expression of the AE2 isoform of the anion exchanger gene family in rabbit parietal cells. Gastroenterology 104: A717.
22. NADER, M., G. LAMPRECHT, M. CLASSEN & U. SEIDLER. 1994. Different regulation by pH_i and osmolarity of the rabbit ileum brush-border and parietal cell basolateral anion exchanger. J. Physiol. (Lond.) 481(Pt. 3): 605–615.
23. SEIDLER, U., M. HUBNER, S. ROITHMAIER & M. CLASSEN. 1994. pH^i and HCO_3^- dependence of proton extrusion and Cl^-/base exchange rates in isolated rabbit parietal cells. Am. J. Physiol. 266(5 Pt. 1): G759–766.
24. HUMPHREYS, B.D., L. JIANG, M.N. CHERNOVA & S.L. ALPER. 1994. Functional characterization and regulation by pH of murine AE2 anion exchanger expressed in Xenopus oocytes. Am. J. Physiol. 267(5 Pt. 1): C1295–1307.
25. HUMPHREYS, B.D., L. JIANG, M.N. CHERNOVA & S.L. ALPER. 1995. Hypertonic activation of AE2 anion exchanger in Xenopus oocytes via NHE-mediated intracellular alkalinization. Am. J. Physiol. 268(1 Pt. 1): C201–209.
26. LEE, B.S., R.B. GUNN & R.R. KOPITO. 1991. Functional differences among nonerythroid anion exchangers expressed in a transfected human cell line. J. Biol. Chem. 266(18): 11448–11454.
27. JIANG, L., A. STUART-TILLEY, J. PARKASH & S.L. ALPER. 1994. pHi and serum regulate AE2-mediated Cl^-/HCO_3^- exchange in CHOP cells of defined transient transfection status. Am. J. Physiol. 267(3 Pt. 1): C845–856.
28. JONS, T., B. WARRINGS, A. JONS & D. DRENCKHAHN. 1994. Basolateral localization of anion exchanger 2 (AE2) and actin in acid-secreting (parietal) cells of the human stomach. Histochemistry 102(4): 255–263.
29. ZOLOTAREV, A.S., M.N. CHERNOVA, D. YANNOUKAKOS & S.L. ALPER. 1996. Proteolytic cleavage sites of native AE2 anion exchanger in gastric mucosal membranes. Biochemistry 35(32): 10367–10376.
30. JONS, T. & D. DRENCKHAHN. 1998. Anion exchanger 2 (AE2) binds to erythrocyte ankyrin and is colocalized with ankyrin along the basolateral plasma membrane of human gastric parietal cells. Eur. J. Cell Biol. 75(3): 232–236.
31. VAANDRAGER, A.B. & H.R. DE JONGE. 1988. A sensitive technique for the determination of anion exchange activities in brush-border membrane vesicles. Evidence for two exchangers with different affinities for HCO_3^- and SITS in rat intestinal epithelium. Biochim. Biophys. Acta 939(2): 305–314.
32. ORSENIGO, M.N., M. TOSCO & A. FAELLI. 1991. Cl^-/HCO_3^- exchange in the basolateral membrane domain of rat jejunal enterocyte. J. Membr. Biol. 124(1): 13–19.
33. SUNDARAM, U., R.G. KNICKELBEIN & J.W. DOBBINS. 1991. pH regulation in ileum: Na^+/H^+ and Cl^-/HCO_3^- exchange in isolated crypt and villus cells. Am. J. Physiol. 260(3 Pt. 1): G440–449.
34. ROSSMANN, H., M. NADER, U. SEIDLER et al. 1995. Basolateral membrane localization of the AE2 isoform of the anion exchanger family in both stomach and ileum. Gastroenterology 108: A319.
35. KNICKELBEIN, R., P.S. ARONSON, C.M. SCHRON et al. 1985. Sodium and chloride transport across rabbit ileal brush border. II. Evidence for Cl^-/HCO_3^- exchange and mechanism of coupling. Am. J. Physiol. 249(2 Pt .1): G236–245.
36. WORMMEESTER, L., F. SANCHEZ DE MEDINA, F. KOKKE et al. 1998. Quantitative contribution of NHE2 and NHE3 to rabbit ileal brush-border Na^+/H^+ exchange. Am. J. Physiol. 274(5 Pt. 1): C1261–1272.
37. TURNBERG, L.A., J.S. FORDTRAN, N.W. CARTER & F.C. RECTOR, JR. 1970. Mechanism of bicarbonate absorption and its relationship to sodium transport in the human jejunum. J. Clin. Invest. 49(3): 548–556.

38. ALPER, S.L., A. STUART-TILLEY, C.F. SIMMONS *et al.* 1994. The fodrin-ankyrin cytoskeleton of choroid plexus preferentially colocalizes with apical Na^+/K^+-ATPase rather than with basolateral anion exchanger AE2. J. Clin. Invest. **93**(4): 1430–1438.

39. KRETZ, A., M. GREGOR & U. SEIDLER. 1998. Identification of two functionally different anion exchangers in rat proximal duodenum. Gastroenterology **114**: G1589.

40. JACOB, P., G. LAMPRECHT, H. ROSSMANN *et al.* 1998. Functional characterization of apical and basolateral anion transporters in rabbit proximal duodenum. Gastroenterology **114**: G1555.

41. RAJENDRAN, V., P. SANGAN, M, MANN *et al.* 1998. Tissue-specific expression of basolateral anion exchange (AE2) isoform in rat colon. Gastroenterology **114**: A408.

42. TOSCO, M., M.N. ORSENIGO, G. GASTALDI & A. FAELLI. 1998. pH dependence of Cl^-/ HCO_3^- exchanger in the rat jejunal enterocyte. Biochim. Biophys. Acta **1372**(2): 323–330.

43. HOGLUND, P., S. HAILA, J. SOCHA *et al.* 1996. Mutations of the down-regulated in adenoma (DRA) gene cause congenital chloride diarrhoea. Nat. Genet. **14**(3): 316–319.

44. SILBERG, D.G., W. WANG, R.H. MOSELEY & P.G. TRABER. 1995. The down regulated in adenoma (dra) gene encodes an intestine-specific membrane sulfate transport protein. J. Biol. Chem. **270**(20): 11897–11902.

45. ZHANG, Y., M.N. CHERNOVA, A.K. STUART-TILLEY *et al.* 1996. The cytoplasmic and transmembrane domains of AE2 both contribute to regulation of anion exchange by pH. J. Biol. Chem. **271**(10): 5741–5749.

46. COX, K.H., T.L. ADAIR-KIRK & J.V. COX. 1995. Four variant chicken erythroid AE1 anion exchangers. Role of the alternative N-terminal sequences in intracellular targeting in transfected human erythroleukemia cells. J. Biol. Chem. **270**(34): 19752–19760.

47. ROSSMANN, H., O. BACHMANN, D. VIEILLARD-BARON *et al.* 1999. Na^+/HCO_3^- cotransport and expression of NBC1 and NBC2 in rabbit gastric parietal and mucous cells. Gastroenterology **116**: 1389–1398.

48. KNICKELBEIN, R., P. ARONSON & J.W. DOBBINS. 1990. Characterization of Na^+/H^+-exchangers on villus cells in rabbit ileum. Am. J. Physiol. **259**: G802–806.

Differential Regulation of ENaC by Aldosterone in Rat Early and Late Distal Colon

S. AMASHEH,[a] H. J. EPPLE,[b] J. MANKERTZ,[b] K. DETJEN,[a] M. GOLTZ,[c]
J. D. SCHULZKE,[b] AND M. FROMM[a,d]

[a]Institut für Klinische Physiologie, Freie Universität Berlin, 12200 Berlin, Germany

[b]Medizinische Klinik I Gastroenterologie und Infektiologie, Universitätsklinikum Benjamin Franklin, Freie Universität Berlin, 12200 Berlin, Germany

[c]Xenotransplantation, Robert Koch-Institut, 13535 Berlin, Germany

The epithelial sodium channel (ENaC) is the limiting factor for sodium absorption in different mammalian epithelia. In kidney and distal colon, the channel is regulated primarily by the corticosteroid hormone aldosterone. In distal colon, activation of channel proteins can be elicited by nanomolar concentrations of aldosterone *in vitro*.[1] This part of the gut, furthermore, displays a strong segmental heterogeneity.[1] In order to determine the effect of nanomolar concentrations of aldosterone on ENaC in different segments of rat distal colonic epithelia, we combined electrophysiological experiments with biochemical techniques.[2] Namely, the four-electrode Ussing chamber technique was combined with subsequent Northern blotting analysis. A differential regulation both on the functional and on the transcriptional level was observed.

Male Wistar rats (250–300 g) with free access to tap water and a standard rat diet (Altromin 1320, Altromin, Lage, Germany) were killed with CO_2. Rat "early distal colon" and rat "late distal colon 2" were obtained following experimental protocols, described in more detail previously.[1] These tissues were 6–7 cm (early) and 1–2 cm (late) apart from rat anus, respectively. A so-called "total strip" of the colonic epithelium was performed leaving only epithelium, lamina propria, and the outer layer of the muscularis mucosae. The conventional four-electrode Ussing chamber technique was employed to report the electrogenic Na^+ absorption (J_{Na}) induced by aldosterone. Both half chambers were oxygenated with 95% O_2 and 5% CO_2. J_{Na} was determined after an 8-h incubation of the tissues in solution containing aldosterone (10^{-10} to $3 \cdot 10^{-7}$ M) by the drop in short circuit current caused by mucosal addition of amiloride (10^{-4} M). α-, β- and γ-subunit mRNA were detected by Northern blot analysis, subsequently, using DIG-dUTP–labeled DNA probes.

Maximum aldosterone-induced, amiloride-sensitive short circuit currents representing J_{Na} were observed after an 8-h incubation of the tissues in $3 \cdot 10^{-9}$ M aldosterone. FIGURE 1 shows a comparison of these data obtained from early and late distal colon. In early distal colon, aldosterone had only little effect (J_{Na}= 1.1 ± 0.4 µmol·h^{-1}·cm^{-2}, n = 9). In late distal colon, a distinct increase of J_{Na} (15.3 ± 1.1 µmol·h^{-1}·cm^{-2}, n = 9) was

[d]Address for correspondence: Prof. Dr. Michael Fromm, Institut für Klinische Physiologie, Universitätsklinikum Benjamin Franklin, Freie Universität Berlin, Hindenburgdamm 30, 12200 Berlin, Germany. Voice: +49 30 8445 2787; fax: +49 30 8445 4239.
m.fromm@medizin.fu-berlin.de

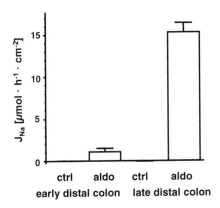

FIGURE 1. Maximum aldosterone effects on rat early and late distal colon *in vitro*. Maximum J_{Na} induced by an 8-h incubation with aldosterone ($3 \cdot 10^{-9}$ M) in early and late distal colon ($n = 9$).

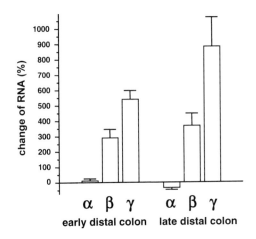

FIGURE 2. Histogram of relative alterations of α-, β-, and γENaC mRNA quantities induced by $3 \cdot 10^{-9}$ M aldosterone. Controls were set as 100%. Whereas in early distal colon aldosterone-induced alteration of αENaC mRNA was not observed significantly, in late distal colon a minimum of αENaC mRNA was reached at $3 \cdot 10^{-9}$ M aldosterone ($n = 6$–9 /column).

observed. Controls without aldosterone addition did not display significant J_{Na}. Application of higher aldosterone concentrations ($3 \cdot 10^{-7}$ M) did not further enhance J_{Na}.

After Ussing chamber experiments, tissues were removed and used for Northern blot analysis detecting the effect of aldosterone on the quantity of α-, β-, and γENaC mRNA. A distinct pattern of transcriptional regulation was observed. In early distal colon, the α-subunit RNA remained constant. The β- and γENaC mRNA was upregulated dose dependently, reaching a maximum at $3 \cdot 10^{-9}$ M aldosterone. Surprisingly, in late distal colon the α-subunit RNA decreased with increasing aldosterone concentrations, reaching a minimum of $63 \pm 3.5\%$ at a concentration of $3 \cdot 10^{-9}$ M aldosterone. At this concentration, maximum increase of β- and γENaC mRNA was detected in both early and late distal colon. A histogram depicting the results is shown in FIGURE 2.

This study confirms and extends data on the heterogeneity of regulation of the ENaC by aldosterone in distal colon of the rat.[1] The finding that in early distal colon the α subunit is constant is in accordance with the literature, indicating a constitutive role of the α subunit in that segment.[3] However, in late distal colon, which is the segment of highest mineralocorticoid sensitivity, the αENaC mRNA is downregulated with increasing aldosterone levels. This is the first report of a negative regulation of an ENaC subunit mRNA by aldosterone.

It was discovered recently that the ENaC bears a constant subunit stoichiometry consisting of two α subunits, one β subunit, and one γ subunit per channel protein.[4,5] Downregulation of the α subunit on the RNA level may provide a counter-regulatory mechanism for the α subunit in order to obtain an optimized subunit proportion, increasing velocity of channel assembling mechanisms.

REFERENCES

1. FROMM, M., J.D. SCHULZKE & U. HEGEL. 1993. Control of electrogenic Na$^+$ absorption in rat late distal colon by nanomolar aldosterone added *in vitro*. Am. J. Physiol. **264:** E68–E73.
2. EPPLE, H.J., S. AMASHEH, J. MANKERTZ *et al.* 2000. Early aldosterone effect in distal colon by transcriptional regulation of ENaC subunits. Am. J. Physiol. **278:** G718–G724.
3. ASHER, C., H. WALD, B.C. ROSSIER & H. GARTY. 1996. Aldosterone-induced increase in the abundance of Na$^+$ channel subunits. Am. J. Physiol. **271:** C605–C611.
4. KOSARI, F., S. SHENG, J. LI *et al.* 1998. Subunit stoichiometry of the epithelial sodium channel. J. Biol. Chem. **273:** 13469–13474.
5. FIRSOV, D., I. GAUTSCHI, A.-M. MERILLAT *et al.* 1998. The heterotetrameric architecture of the epithelial sodium channel (ENaC). EMBO J. **17:** 344–352.

Epithelial Phosphate Transporters in Small Ruminants

K. HUBER,[a,b] C. WALTER,[a] B. SCHRÖDER,[a] J. BIBER,[c] H. MURER,[c] AND G. BREVES[a]

[a]Department of Physiology, School of Veterinary Medicine, Hannover, Germany

[b]Institute of Physiology, University of Zürich, Switzerland

INTRODUCTION

Functional properties of renal and intestinal phosphate (P_i) transporters are clearly different in monogastric animals and ruminants. During adaptation to an alimentary phosphorus (P) depletion, the activity of renal and intestinal P_i transporters of monogastric animals are stimulated by calcitriol.[1,2] In ruminants alimentary P depletion modulates intestinal but not renal P_i absorption. The intestinal enhancement of P_i transport is not correlated with higher calcitriol concentrations in plasma.[3,4] In kidneys, maximal P_i transport capacity is reached within physiological plasma P_i levels.[5] This could be based on variations in the structure of the P_i transporters as well as in the regulation of gene expression and in the modulation of transporter function. The molecular structures of renal and intestinal P_i transporters are well characterized for monogastric animals;[1,2] for ruminants, no structural information is available. It was the aim of our study to find out the molecular identity of renal and intestinal P_i transporters in ruminants in order to clarify the mechanisms that are responsible for modulating barrier function of renal and intestinal epithelia.

MATERIAL AND METHODS

Tissue was obtained from sheep and goats immediately after slaughter. The animals had been either kept under control conditions with adequate P supply or were P depleted according to Schröder et al.[4] RT-PCR was performed in goat and sheep kidneys using primers specific for murine renal Na^+/P_i cotransporter type II (primer 1: 5′CGGAGTGATGGCTGAGGTGA3′; primer 2: 5′ATGGTCTCCTCTGGCTT-GGTTG3′). Derived PCR fragments were cloned into pGEMT vector (Promega) and were sequenced by a commercial lab (MWG-Biotech). For Northern analysis poly (A)+ RNA was prepared from goat jejunum of control and P-depleted animals. After separating mRNA in a 1.2% formamide/agarose gel and transferring it to a nylon membrane, hybridization was performed in $6 \times$ SSC, $5 \times$ Denhardt's, and 0.5% SDS (and 100 µg/mL herring sperm DNA) with a 3.2-kb radioactive-labeled probe de-

[b]Address for correspondence: Dr. Korinna Huber, Department of Physiology, School of Veterinary Medicine, Hannover, Bischofsholer Damm 15/102, 30173 Hannover, Germany. Voice: +49 511 856-7331; fax: +49 511 856-7687.

khuber@physiology.tiho-hannover.de

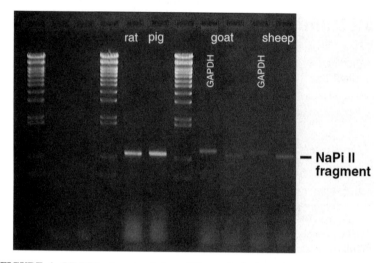

FIGURE 1. RT-PCR of goat and sheep kidneys. In both, goat and sheep NaP$_i$ IIb–specific fragments were isolated. After separating the DNA in a 1% agarose gel, ethidium bromide–stained bands were located in the same range as in control reactions of rat and pig kidneys.

rived from mouse intestinal Na$^+$/P$_i$ cotransporter type IIb clone.[2] The washing procedure was 5´ in 2 × SSC/0.1% SDS at room temperature; 10´ in 1 × SSC/0.1% SDS at 40°C; and 10´ in 0.5 × SSC/0.1% SDS at 50°C. The membrane was exposed to X-ray film for 4 hours. Simultaneously hybridization of L28 was performed to ensure mRNA integrity and to quantitate intensity of gene expression dependent on alimentary P supply. For immunodetection, proteins of brush border membranes of goat jejunum (control and P-depleted animals) were separated by SDS-PAGE and transferred onto a nitrocellulose membrane. Immunodetection was performed by the ECL system (Amersham) after incubation with mouse Na$^+$/P$_i$ cotransporter type IIb–specific antibodies. Specificity of immunoreaction was determined by peptide protection (50 µL antigenic peptide/mL).

RESULTS

Using RT-PCR in goat and sheep kidney, Na$^+$/P$_i$ cotransporter type IIb–like fragments were isolated (FIG. 1). After cloning and sequencing these PCR fragments, 89% homology of amino acid sequences between goat and sheep with respect to rat was found. In Northern analysis of small intestines of goats (control and P-depleted animals), Na$^+$/P$_i$ cotransporter type IIb–specific mRNA could be observed (FIG. 2). Alimentary P depletion stimulates specific gene expression twofold. Western analysis in goat jejunal brush border membranes resulted in a strong Na$^+$/P$_i$ cotransporter type IIb–specific immunoreaction using antibodies raised against N terminal–specific oligopeptide of mouse Na$^+$/P$_i$ cotransporter type IIb (data not shown).

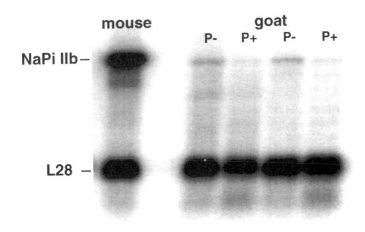

FIGURE 2. Northern analysis of mRNA of mouse duodenum and goat jejunum. Clear hybridization signals appeared in goat intestinal mRNA (P+: goats with adequate supply; P–: P-depleted goats) located in the same range as the mouse-specific NaP_i IIb band. With respect to the L28 band, the specific signal was twofold stronger in P-depleted animals.

DISCUSSION

Hybridization studies on nucleic acid and protein levels demonstrated that renal and intestinal P_i transporters in small ruminants were structurally related to Na^+/P_i cotransporter type II and IIb as described for rat and mouse despite variations in their function. An alimentary P depletion seemed to stimulate specific mRNA expression. At present it is not clear whether this stimulation is also present at the protein level. To clarify the overall homology, structural identification of renal and intestinal P_i transporters is in progress. Inasmuch as Na^+/P_i transporter cDNA and specific antibodies of goat are available, the effect of an alimentary P depletion on specific gene expression will be determined in more detail.

REFERENCES

1. MURER, H. & J. BIBER. 1997. A molecular view of proximal tubular inorganic phosphate reabsorption and of its regulation. Pflügers Arch. Eur. J. Physiol. **433:** 379–389.
2. HILFIKER, H. *et al.* 1998. Characterization of a murine type II sodium-phosphate cotransporter expressed in mammalian small intestine. Proc. Natl. Acad. Sci. USA **95**(24): 14564–14569.
3. SCHRÖDER, B. & G. BREVES. 1996. Mechanisms of phosphate uptake into brush border membrane vesicles from goat jejunum. J. Comp. Physiol. B **166:** 230–240.
4. SCHRÖDER, B. *et al.* 1995. Mechanisms of intestinal phosphate transport in small ruminants. Br. J. Nutr. **74:** 635–648.
5. WIDIYONO, I. *et al.* 1998. Renal phosphate excretion in goats. J. Vet. Med. A **45:** 145–153.

Inflammatory Mediators Influencing Submucosal Secretory Reflexes

THOMAS FRIELING,[a,b] ECKHARD WEBER,[a] AND MICHAEL SCHEMANN[c]

[a]Department of Gastroenterology, Hepatology, and Infection,
Heinrich-Heine-University of Düsseldorf, 40225 Düsseldorf, Germany

[c]Department of Physiology, School of Veterinary Medicine, Hannover, Germany

The enteric nervous system is an integrative network of neurons that regulates gastrointestinal functions relatively independently of the central nervous system.[1] Although the neuronal components of the reflexes are hardwired, they may be modulated and fine-tuned by various stimuli arising from various sources. One of these is the immune system in the gut, which is able to respond to noxious stimuli by releasing a variety of mediators. Although inflammatory mediators might influence epithelial function directly, there is increasing evidence that parts of inflammatory responses involve modulation of nervous activity.[2,3] This has been shown by studies indicating that a significant part of the secretory response to bacterial toxins, and during anaphylaxis, is due to the activation of submucosal secretomotor neurons.[3–8]

We have investigated neuroimmune interactions in the guinea pig colon. The rationale for this was the finding that colonic type 1 hypersensitivity reactions in guinea pigs sensitized to cow's milk or the parasite *Trichinella spiralis* are associated with a significant increase in submucosal neuronal activity.[7,8] Analysis of the effects of the inflammatory mediators serotonin, histamine, prostaglandins, and leukotrienes have indicated that neuronal excitation is mediated by a direct tetrodotoxin-sensitive action at the postsynaptic membrane of submucosal neurons mediated via specific postsynaptic receptors[9–13] (FIG.1). However, indirect effects also appear to account for the enhanced excitability of submucosal neurons indicated by the frequent occurrence of nicotinic fast excitatory postsynaptic potentials (fEPSP). By contrast, some inflammatory mediators (serotonin, histamine) may also modulate signal processing between neurons by presynaptic inhibition of fEPSP[9,10] (FIG.1).

Characterization of the functional secretory correlate for the inflammatory mediator-evoked changes of neural activity in Ussing chamber studies revealed that the direct epithelial secretory responses to serotonin, to the prostaglandins PGD_2, PGE_2, PGI_2, and PGF_{2a} and to the leukotrienes LTC_4, LTD_4, and LTE_4 are potentiated by the activation of submucosal neurons and the release of acetylcholine and noncholinergic neurotransmitters at the neuroepithelial junction.[9,11–13,15] In addition to the neuronal potentiation of mediator-evoked secretory responses, histamine, LTC_4, and LTD_4 evoke coordinated activation of submucosal neurons by the induction of long-

[b]Address for correspondence: Thomas Frieling, M.D., Dept. of Gastroenterology, Hepatology and Infectiology, Heinrich-Heine-University of Düsseldorf, Moorenstr. 5, 40225 Düsseldorf, Germany. Voice: +49-211-811-7833; fax: +49-211-811-8752.
frielint@uni-duesseldorf.de

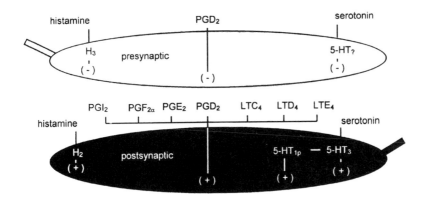

FIGURE 1. Pre- and postsynaptic inhibition/activation by inflammatory mediators.

lasting recurrent discharge of bursts of action potentials leading to ongoing cyclical chloride secretion and coordinated circular muscle contractions.[10,14,15]

In addition to the effects of inflammatory mediators on basal neuronal activity, reflex-evoked activation of submucosal neurons may be modulated as well. This has been shown in modified Ussing flux chamber studies where distending stimuli induce secretory responses mediated via polarized projecting descending VIPergic and ascending cholinergic secretomotor neurons.[16–18] Distension evokes a volume-dependent increase in short-circuit current that is significantly reduced by tetrodotoxin, atropine, VIP-receptor antagonist (VIP6-28), neurokinin-receptor antagonists (CP99,994-1, SR142801), and capsaicin. By contrast, hexamethonium, cimetidine, renzapride, tropisetron, piroxicam, and the lipoxygenase inhibitor MK886 are without effect. Distension-evoked short-circuit current is significantly enhanced by submaximal concentrations of dimaprit, PGE_2, and LTC_4 (FIG. 2). Subthreshold concentrations of the mediators modulate distension-evoked secretion differently. Dimaprit is without effect, whereas PGE_2 significantly reduces, and LTC_4 significantly increases, short-circuit current (FIG. 2). In addition, antigen challenge with β-lactoglobulin in cow's milk–sensitized tissues significantly increases short-circuit current compared to sensitized and nonsensitized control tissues.

In conclusion, inflammatory mediators exhibit differentiated effects within the submucosal plexus that consist of postsynaptic excitation, presynaptic inhibition, and coordinated activation of a large number of neurons. Based on the findings that inflammatory mediators predominantly activate submucosal neurons, the release of excitatory neurotransmitters at the neuroepithelial junction may amplify direct epithelial effects of inflammatory mediators. Neuronal potentiation of direct epithelial, as well as reflex-evoked secretory effects of inflammatory, mediators may be interpreted in two ways. First, a full response to inflammatory mediators requires intact innervation. Second, the overall response may be modulated and fine-tuned by the enteric nervous system. The net result will be an activation of effector systems that is responsible for the dilution of harmful materials. The enteric nervous system

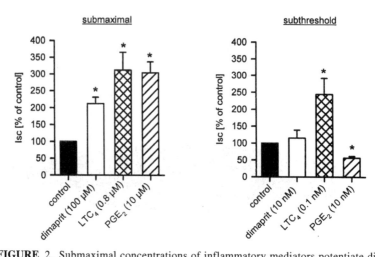

FIGURE 2. Submaximal concentrations of inflammatory mediators potentiate distension-evoked increase in short-circuit current (I_{sc}). Subthreshold concentrations of inflammatory mediators modulate distension-evoked I_{sc} differently. *, significant difference compared to control; $p < 0.05$, ANOVA, Dunnett's post hoc; distension volume: 300 µL.

might act as a defense system that will be switched on by antigenic and noxious stimulation, resulting in a sustained activation of secretory cells.

ACKNOWLEDGMENTS

This work was supported by Grants DFG Fr 733/3-3, Fr 733/13-1, Sche 267/4-2, and SFB 280.

REFERENCES

1. WOOD, J.D. 1994. Physiology of the enteric nervous system. *In* Physiology of the Gastrointestinal tract. L.R. Johnson, Ed.: 423–481. Raven Press. New York.
2. FRIELING, T. & G. STROHMEYER. 1995. Neuroimmune Interaktionen im Verdauungstrakt. Z. Gastroenterol. **33:** 219–224.
3. COOKE, H.J. 1994. Neuroimmune signaling in regulation of intestinal ion transport. Am. J. Physiol. **266:** G167–G178.
4. CASTAGLIUOLO, I., J.T. LAMONT, R. LETOURNEAU *et al.* 1994. Neuronal involvement in the intestinal effects of *Clostridium difficile* toxin A and *Vibrio cholerae* enterotoxin in rat ileum. Gastroenterology **107:** 657–665.
5. FRIELING, T., H.J. COOKE & J.D. WOOD. 1991. Electrophysiological properties of neurons in submucous ganglia of the guinea-pig distal colon. Am. J. Physiol. **260:** G835–G841.
6. FRIELING, T., H.J. COOKE & J.D. WOOD. 1991. Synaptic transmission in submucous ganglia of the guinea-pig distal colon. Am. J. Physiol. **260:** G842–G849.
7. FRIELING, T., J.M. PALMER, H.J. COOKE *et al.* 1994. Neuroimmune communications in the submucous plexus of guinea-pig colon after infection with *Trichinella spiralis*. Gastroenterology **107:** 1602–1609.

8. FRIELING, T., H.J. COOKE & J.D. WOOD. 1994. Neuroimmune communication in the submucous plexus of guinea pig colon after sensitization to milk antigen. Am. J. Physiol. **30:** G1087–G1093.
9. FRIELING, T., H.J. COOKE & J.D. WOOD. 1991. Serotonin receptors on submucous neurons in the guinea-pig colon. Am. J. Physiol. **261:** G1017–G1023.
10. FRIELING, T., H.J. COOKE & J.D. WOOD. 1993. Histamine receptors on submucous neurons in guinea-pig colon. Am. J. Physiol. **264:** G74–G80.
11. FRIELING, T., C. RUPPRECHT, A.B.A. KROESE *et al.* 1994. Effects of the inflammatory mediator prostaglandin D_2 (PGD_2) on submucosal neurons and secretion in guinea pig colon. Am. J. Physiol. **266:** G132–G139.
12. FRIELING, T., C. RUPPRECHT, G. DOBREVA *et al.* 1994. Prostaglandin E_2 (PGE_2)-evoked chloride secretion in guinea-pig colon is mediated by nerve-dependent and nerve-independent mechanisms. Neurogastroenterol. Motil. **6:** 95–102.
13. FRIELING, T., C. RUPPRECHT, G. DOBREVA *et al.* 1995. Effect of prostaglandin F_{2a} (PGF_{2a}) and prostaglandin I_2 (PGI_2) on nerve-mediated secretion in guinea-pig colon. Pfluegers Arch. **431:** 212–220.
14. COOKE, H.J., Y.Z. WANG & R. ROGERS. 1993. Coordination of Cl secretion and contraction by histamine H_2-receptor agonist in guinea pig distal colon. Am. J. Physiol. **265:** G973–G978.
15. FRIELING, T., K. BECKER, C. RUPPRECHT *et al.* 1997. Leukotriene-evoked cyclic chloride secretion is mediated by enteric neuronal modulation in guinea-pig colon. Naunyn-Schmiederbergs Arch. Pharmakol. **355:** 625–630.
16. FRIELING, T., J.D. WOOD & H.J. COOKE. 1992. Submucosal reflexes: distension-evoked ion transport in the guinea pig distal colon. Am. J. Physiol. **263:** G91–G96.
17. NEUNLIST, M., T. FRIELING, C. RUPPRECHT *et al.* 1998. Polarized enteric submucosal circuits involved in secretory responses of the guinea-pig proximal colon. J. Physiol. (Lond.) **506:** 539–550.
18. NEUNLIST, M. & M. SCHEMANN. 1998. Polarised innervation pattern of the mucosa of the guinea pig distal colon. Neurosci. Lett. **246:** 161–164.

Tumor Necrosis Factor–α Potentiates Ion Secretion Induced by Muscarinic Receptor Activation in the Human Intestinal Epithelial Cell Line HT29cl.19A

JUDITH C. J. OPRINS,[a] HELEN P. MEIJER, AND JACK A. GROOT

Institute for Neurobiology, University of Amsterdam,
1098 SM Amsterdam, the Netherlands

INTRODUCTION

Patients suffering from inflammatory bowel disease (IBD) have severe diarrhea due to excessive water secretion into the intestinal lumen. Mucosal biopsies of these patients show enhanced levels of cytokines, secreted by inflammatory cells.[1,2] The contribution of these cytokines to the severe diarrhea remains unknown. Antibodies against tumor necrosis factor (TNF)–α were revealed to be a good therapy against the diarrhea in these patients, suggesting that TNF-α plays an important role in diarrhea.[3,4] In human distal colon TNF-α is thought to increase ion secretion indirectly by activating subepithelial prostaglandin release.[5] However, not much is known about the direct effects of TNF-α on ion secretion in intestinal epithelia. In this study we investigated the effects of TNF-α on the ion secretion in the human intestinal epithelial cell line HT29cl.19A.

METHODS

HT29cl.19A cells were cultured as monolayers on filters and electrophysiological parameters were measured as described previously.[6] In brief: the filters were mounted in a horizontal Ussing chamber, and the transepithelial potential (V_t) across the monolayer and the transepithelial resistance (R_t) were measured. Concomitantly a microelectrode was impaled in the cells to measure the apical membrane potential (V_a) and fractional resistance ($fR_a = R_a/[R_a+R_b]$). The I_{sc} was calculated from V_t and R_t and is a measure of the chloride secretion across the epithelium.

[a]Address for correspondence: Dr. Judith C.J. Oprins, Institute for Neurobiology, Biological Faculty, University of Amsterdam, Kruislaan 320, 1098 SM Amsterdam, the Netherlands. Voice: +31-20-5257650; fax: +31-20-5257709.
oprins@bio.uva.nl

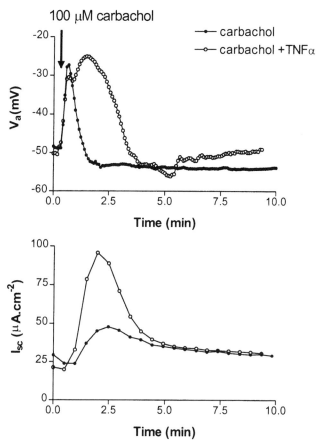

FIGURE 1. Changes in V_a and I_{sc} after addition of 100 μM carbachol to the basolateral side of the cells with (*open dots*) or without (*solid dots*) exposure to 10 ng/ml TNF-α for 48 hours. V_a, intracellular potential; I_{sc}, short circuit current.

RESULTS AND DISCUSSION

Incubation for 48 hours with TNF-α (10 ng/ml bilateral) did not affect the basal electrophysiological characteristics of the monolayer. However, the response to the muscarinic receptor agonist carbachol was largely changed. FIGURE 1 shows a typical recording of an intracellular measurement and of I_{sc} after addition of carbachol (shown by the solid dotted lines). A fast depolarization and a small serosa negative change in I_{sc} occur during the first phase of the response. This phase is followed by a hyperpolarization of V_a, together with an increase in I_{sc}. Previous studies have shown that activation of phospholipase C by carbachol results in formation of IP_3, which triggers calcium release from intracellular pools.[7] The rise in intracellular calcium levels results in fast activation of calcium-dependent chloride channels in both

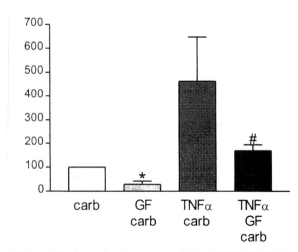

Effect of 30 minutes' preincubation with 1 μM GF109203X (bilaterally) on
)y carbachol with or without exposure to TNF-α. Data are presented as per-
; compared to pairewise control experiments without the inhibitor. *Bars* rep-
values ± SE from four to five monolayers. * indicates $p < 0.05$ compared to
licates $p < 0.05$ compared to carbachol + TNF-α.

the cell (reflected in the first depolarization), followed by an activa-
;olateral calcium-dependent potassium conductance (reflected as hy-
1 of V_a). The latter increases the driving force for chloride efflux
chloride channels, which are regulated by PKC-α, activated by the re-
and DAG from PLC action. In FIGURE 1 a typical recording of V_a and
irs' exposure to 10 ng/ml TNF-α is shown by the open dotted line. The
t depolarization has not been changed after incubation with TNF-α.
second phase shows a prolonged depolarization of V_a together with a
in I_{sc} and decrease of fR_a. Experiments in which the mucosal chloride
eplaced with gluconate revealed that TNF-α potentiates the carbachol-
tion by an increase of apical chloride conductance.
e first phase depolarization was not affected by TNF-α, we exluded the
it TNF-α exposure resulted in an increased number of muscarinic

shows that incubation with the PKC inhibitor GF109203X (1 μM, bi-
:ed the secretory response to carbachol alone as well as after exposure
is indicates an important role for PKC in the carbachol response and
n by TNF-α.
ition can occur via formation of DAG, which in turn can be formed by
'LC and PLD. PLD-dependent DAG formation can be blocked by pro-
dition of 100 μM propranolol to the apical side of the cells inhibited
to carbachol by 70%, suggesting an important role for PLD-derived
)n in the carbachol response. In FIGURE 3 a typical recording of V_a after
rbachol after exposure to TNF-α is shown by the open dotted line. Ad-
ranolol, prior to carbachol, completely abolished the prolonged depo-

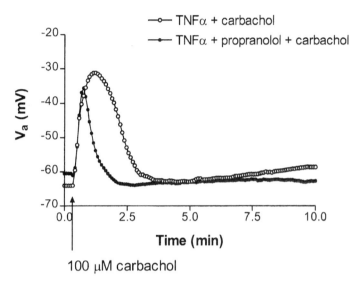

FIGURE 3. Effect of 10 minutes' preincubation with 100 μM propranolol (apical) on changes in V_a induced by addition of carbachol after exposure to TNF-α. *Tracings* represent six experiments.

larization, and now the recording of V_a is fully comparable with control responses to carbachol. Thus, inhibition of PLD-dependent DAG formation abolishes the potentiating effect of TNF-α.

These results indicate that activation of PLD is involved in ion secretion induced by carbachol and that TNF-α can potentiate this pathway. These results could give an explanation for the important role TNF-α plays in the diarrhea in patients suffering from IBD.

REFERENCES

1. RADEMA, S.A. *et al.* 1995. *In situ* detection of interleukin-1 beta and interleukin-8 in biopsy specimens from patients with ulcerative colitis. Adv. Exp. Med. Biol. **371B:** 1297–1299.
2. BREESE, E.J. *et al.* 1994. Tumor necrosis factor alpha–producing cells in the intestinal mucosa of children with inflammatory bowel disease. Gastroenterology **106**: 1455–1466.
3. STACK, W.A. *et al.* 1997. Randomised controlled trial of CDP571 antibody to tumour necrosis factor–alpha in Crohn's disease. Lancet **349**: 521–524.
4. WATKINS, P.E. *et al.* 1997. Treatment of ulcerative colitis in the cottontop tamarin using antibody to tumour necrosis factor alpha. Gut **40**: 628–633.
5. SCHMITZ, H. *et al.* 1996. Tumor necrosis factor-alpha induces Cl– and K+ secretion in human distal colon driven by prostaglandin E2. Am. J. Physiol. **271**: G669–G674.
6. BAJNATH, R.B. *et al.* 1991. Electrophysiological studies of forskolin-induced changes in ion transport in the human colon carcinoma cell line HT-29 cl.19A: lack of evidence for a cAMP-activated basolateral K+ conductance. J. Membr. Biol. **122**: 239–250.
7. BAJNATH, R.B. *et al.*1992. Biphasic increase of apical Cl– conductance by muscarinic stimulation of HT-29cl.19A human colon carcinoma cell line: evidence for activation of different Cl– conductances by carbachol and forskolin. J. Membr. Biol. **127**: 81–94.

8. TOULLEC, D. *et al.* 1991. The bisindolylmaleimide GF109203X is a potent and selective inhibitor of protein kinase C. J. Biol. Chem. **266:** 15771–15781.
9. THOMPSON, N.T. *et al.* 1991. Receptor-coupled phospholipase D and its inhibition. Trends. Pharmacol. Sci. **12:** 404–408.

Investigation of Motility and Secretion in Perfused Guinea Pig Colon *ex Situ*

DETLEF WERMELSKIRCHEN[a,b] AND KERSTIN SCHNEIDER[c]

[a]*Janssen Research Foundation, 41470 Neuss, Germany*

[c]*Justus-Liebig-University, Medizinische Klinik III und Poliklinik, Rodthohl 6, 35392 Gießen, Germany*

INTRODUCTION

Intestinal function is the result of a delicate balance between intestinal motility and secretion/absorption. Due to different requirements in the intestine, the concrete regulation of intestinal function varies with the intestinal segment. Moreover, intestinal fluid handling depends on intestinal motility and vice versa. However, most studies addressing intestinal function allow the investigation of just one aspect of intestinal function and, moreover, only in one segment of the intestine. Hence, the present study was conducted to develop a method allowing a simultaneous investigation of colon motility and secretion. In this respect a guinea pig colon was isolated and luminally and vascularly perfused *ex situ*. Motility was investigated at the distal and proximal colon by force transducers (contraction of the circular muscle) and pressure transducers (luminal pressure). Intestinal fluid movements were determined by using the nonpermeating ^{14}C-polyethylene glycol 4000 (^{14}C-PEG 4000) as a marker. Arterial application of the carbachol and loperamide was used to influence colon function and to prove reliability of this method.

METHODS

Male Dunkin-Hartley guinea pigs (body weight 301 ± 14 g) were fasted 16 h before surgery but had free access to water. Animals were anesthetized with chloral hydrate (500 mg/kg body weight, i.p.). The abdomen was opened, and the colon was isolated. A catheter was inserted into the lumen of the proximal colon. The distal colon was opened, the lumen was carefully rinsed, and a catheter was introduced into the distal colon. The caudal mesenteric artery was ligated and cut off, and the mesenteric artery and the portal vein were cannulated. Vascularly perfusion of the organ (1 mL/min) was provided by a peristaltic pump (Reglo 8, Ismatec, Switzerland). Perfusion pressure was continuously monitored by a pressure transducer (Statham P10EZ, Ohmeda, Germany). One force transducer (F-04IS, 3×4 mm, Hugo Sachs Electronics, Germany) were sutured 7 cm from the proximal end of the colon, and the other force transducer was placed 5 cm to the distal end. The colon was placed

[b]Address for correspondence: Detlef Wermelskirchen, Ph.D., Janssen Research Foundation, Raiffeisenstrasse 8, 41470 Neuss, Germany. Voice: +49 2137 955519; fax: +49 2137 955256.
dwermels@jacde.jnj.com

TABLE 1. Effect of carbachol and loperamide on motility of the guinea pig colon[a]

	Signal	Dose	n	Vehicle	n	Compound
Carbachol	distal luminal pressure	10^{-8}M	4	-0.25 ± 0.62	3	0.33 ± 1.32
		10^{-7}M		-0.83 ± 1.11		2.00 ± 0.87^b
		10^{-6}M		-0.75 ± 1.06		2.22 ± 0.97^b
	proximal luminal pressure	10^{-8}M	5	1.00 ± 1.07	5	-0.13 ± 1.19
		10^{-7}M		0.40 ± 1.64		1.07 ± 1.39
		10^{-6}M		0.80 ± 2.11		1.40 ± 2.20
	contraction of proximal circular muscles	10^{-8}M	5	0.20 ± 1.01	5	-0.13 ± 1.30
		10^{-7}M		-0.13 ± 1.64		1.20 ± 1.32
		10^{-6}M		-0.07 ± 1.75		2.13 ± 1.25^b
	contraction of distal circular muscles	10^{-8}M	5	0.00 ± 0.38	5	0.20 ± 1.32
		10^{-7}M		-0.47 ± 0.92		0.53 ± 0.83^b
		10^{-6}M		-0.33 ± 0.82		1.87 ± 1.13^b
Loperamide	distal luminal pressure	10^{-5}M	7	-0.71 ± 0.90	9	-1.58 ± 1.25^b
	proximal luminal pressure	10^{-5}M	7	-1.33 ± 0.73	8	-1.44 ± 1.42
	contraction of proximal circular muscles	10^{-5}M	7	-1.44 ± 0.57	8	-0.79 ± 1.22^b
	contraction of distal circular muscles	10^{-5}M	7	-0.67 ± 0.86	5	-1.53 ± 1.13^b

[a] Data are based on individual scoring and are shown as mean ± standard deviation.
[b] Significantly different from corresponding vehicle, $p < 0.05$.

into an organ bath (37°C), and the arterial flow rate was set to 3 mL/min, resulting in an intraarterial pressure of 60–80 mm Hg. Luminal perfusate was administered and recirculated (peristaltic pump, Reglo 8, Ismatec, Switzerland). Flow rate for luminal perfusion was set to 200 μL/minute.

Solutions

Modified Krebs-Henseleit buffer: NaCl (116 mM), KCl (5.5 mM), $CaCl_2$ (1.2 mM), NaH_2PO_4 (3.5 mM), and $MgSO_4 \times 7 H_2O$ (1.1 mM).

Vascular perfusate was composed of 20–25% bovine erythrocytes (Froschek, Mühlheim, Germany) in modified Krebs-Henseleit buffer (gassed with 95% O_2/5% CO_2), 3% bovine serum albumin, initial fraction (Sigma, Deisenhofen, Germany), 3% aminoplasmal (Braun, Melsungen), 20 IU/mL heparin (Liquemin, Hoffman-La Roche, Grenzach Wyhlen, Germany), and 8 mmol glucose.

Luminal perfusate consisted of NaCl (30.0 mM), $MgCl_2$ (1.1 mM), HEPES (15 mM), NaH_2PO_4 (10 mM), KCl (4 mM), $NaHCO_3$ (10 mM), $CaCl_2$ (1.2 mM), sodium butyrate (10 mM), sodium acetate (60 mM), sodium propionate (10 mM), 5 g/L polyethylene glycol 4000 (Fluka, Germany), and ^{14}C-PEG 4000 (Amersham, USA) (360 Bq/mL). Osmolarity was 287 ± 11 mosmol (mean ± standard deviation).

Calculation of Fluid Absorption

The total luminal volume at each measure point was calculated via the specific activity of ^{14}C-PEG 4000. Luminal volume after equilibration time was regarded as 100%. The content at each measurement was related to the luminal volume after the equilibration time.

Statistics

All data were shown as the mean ± standard deviation. Statistical analysis was performed by Wilcoxon rank sum test, with significance accepted at the 5% level.

RESULTS AND DISCUSSION

At an arterial flow rate of 3 mL/min, the viability of the colon was maintained, as confirmed by histological examination (data not shown). The guinea pig colon was in an absorptive state (FIG. 1), although fluid absorption was lower than absorption in rat colon.[1,2] To characterize the function of the isolated perfused colon, the cholinergic agonist carbachol was used to stimulate colon muscle contraction and to provoke colonic secretion.

As in previous studies, carbachol dose-dependently increased colonic motility (TABLE 1). The effect of loperamide (10^{-5} mol/L) was found to be segment dependent. Distal luminal pressure and circular muscle contraction were decreased, whereas proximal circular muscle contraction was increased by loperamide (TABLE 1). Despite the well-documented inhibitory effect of loperamide,[3] comparable excitatory effects of loperamide have been reported in human and pig small intestine.[4,5]

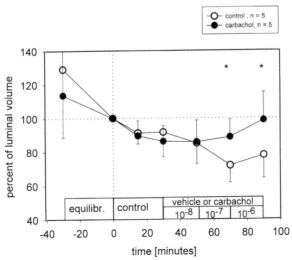

FIGURE 1. Effect of increasing concentrations of carbachol on luminal fluid absorption. Each carbachol concentration was given for 20 minutes. *, indicates significantly different, $p < 0.05$.

In agreement with previous studies,[6,7] carbachol dose-dependently increased net secretion (FIG. 1). Although loperamide has been reported to depress intestinal secretion in rat colon,[8] the present study showed no significant effect of loperamide (data not shown).

In conclusion, the vasculary and luminally perfused colon allows the simultaneous investigation of secretion and motility. The effects of carbachol and loperamide confirmed the suitability of this model. Moreover, the simultaneous investigation of both intestinal motility and secretion offers unique possibilities to gain further insight into colonic regulation and function as well as the effects of compounds thereon.

REFERENCES

1. BANDI, J.C., G. ROSEMBECK, A. DE PAULA et al. 1993. Action of cisapride on rat colonic secretion. Naunyn-Schmiedeberg's Arch. Pharmacol. **348:** 319–324.
2. KRUGLIAK, P., D. HOLLANDER, T.Y. MA et al. 1989. Mechanisms of polyethylene glycol 400 permeability of perfused rat intestine. Gastroenterology **97:** 1164–1170.
3. VAN NUETEN, J.M. & J.A.J. SCHUURKES. 1981. Effect of loperamide on intestinal motility. Clin. Res. Rev. **1**(Suppl. 1): 175–185.
4. REMINGTON, M., J.R. MALAGELADA, A. ZINSMEISTER & C.R. FLEMING. 1983. Abnormalities in gastrointestinal motor activity in patients with short bowels: effect of a synthetic opiate. Gastroenterology **85:** 629–636.
5. THEODOROU, V., J. FIORAMONTI, T. HACHET & L. BUENO. 1991. Absorptive and motor components of the antidiarrhoeal action of loperamide: an in vivo study in pigs. Gut **32:** 1355–1359.
6. MORISSET, J., L. GEOFFRION, L. LAROLSE et al. 1981. Distribution of muscarinic receptors in the digestive tract organs. Pharmacology **22:** 189–195.
7. KILBINGER, H., H. SCHWORER & K.D. SUSS. 1989. Muscarinic modulation of acetylcholine release: receptor subtypes and possible mechanisms. EXS **57:** 197–203.
8. FARACK, U.M. & K. LOESCHKE. 1984. Inhibition of deoxycholic acid induced intestinal secretion. Naunyn-Schmiedeberg's Arch. Pharmacol. **325:** 286–289.

Stimulation by Portal Insulin of Intestinal Glucose Absorption via Hepatoenteral Nerves and Prostaglandin E_2 in the Isolated, Jointly Perfused Small Intestine and Liver of the Rat

FRANK STÜMPEL,[a] BETTINA SCHOLTKA, AND KURT JUNGERMANN

Institute for Biochemistry and Molecular Cell Biology, Georg-August-Universität, Humboldtallee 23, 37073 Göttingen, Germany

ABSTRACT: Insulin infused into the portal vein acutely enhanced intestinal glucose and galactose absorption via the sodium-dependent glucose cotransporter–1 in the isolated, jointly perfused small intestine and liver of the rat. Atropine and tetrodotoxin infused into the superior mesenteric artery completely prevented the portal insulin-dependent increase in intestinal glucose absorption, and carbachol caused an increase similar to that of portal insulin. Thus, a signal was transmitted against the bloodstream in a retrograde direction from the portal vein to the small intestine via hepatoenteral cholinergic nerves. The intracellular messenger in the enterocytes was cAMP, and the link between the muscarinic receptors, which do not increase cAMP concentrations, and adenylate cyclase was found to be prostaglandin E_2.

ADAPTIVE CHANGES IN INTESTINAL CARBOHYDRATE ABSORPTION

The monosaccharides glucose, galactose, and fructose are absorbed by the intestine via specific translocators following the digestion of dietary carbohydrates. Glucose and galactose are taken up via the sodium-dependent glucose cotransporter-1 (SGLT1)[1] and fructose via the sodium-independent glucose transporter-5 (GLUT5),[2] both located in the apical (luminal) plasma membrane of mature enterocytes. On the basolateral membrane of the enterocytes all three carbohydrates share the same translocator and enter the circulation via the sodium-independent glucose transporter-2 (GLUT2).[2]

The absorptive capacity for carbohydrates in small intestine was found not to be static, but to respond tightly to several physiological and pathophysiological long-term alterations. The classical observations have been made with weaning lambs. The switch from carbohydrate-rich maternal milk to carbohydrate-poor grass resulted in a marked decrease in the intestinal capacity to absorb glucose; however, refeeding adult sheep a glucose-rich diet increased the glucose absorptive capacity to the

[a]Address for correspondence: Dr. Frank Stümpel, Institute for Biochemistry and Molecular Cell Biology, Humboldtallee 23, 37073 Göttingen, Germany. Voice: +49-551-395946; fax: +49-551-395960.

fstuemp@gwdg.de

preweaning level.[3] In rats pregnancy and lactation,[4] semistarvation,[5] or streptozotocin-induced diabetes mellitus[6] were shown to increase the intestinal capacity for glucose absorption. The common underlying mechanism of these alterations in intestinal glucose transport was a long-term adaptation in the number of the SGLT1 or the GLUT2 by *de novo* synthesis.

ACUTE INCREASE IN INTESTINAL CARBOHYDRATE ABSORPTION

Contrary to these long-term adaptations, recently short-term alterations in intestinal glucose absorption have been described with an isolated perfused organ system,[7] the joint perfusion of small intestine and liver of the rat. This experimental system allows the perfusion of the isolated small intestine and liver of the rat (FIG. 1)[7, 8] in a nonrecirculating manner via the celiac trunk (CT) and superior mesenteric artery (SMA). The organs are completely dissected from the body and transferred into an organ bath. Samples of the vascular perfusion medium can be obtained from the por-

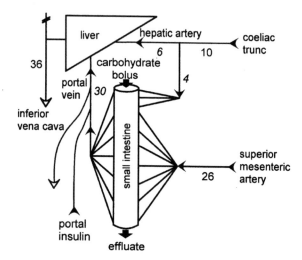

FIGURE 1. Scheme of the isolated, nonrecirculating, joint perfusion of small intestine and liver of the rat. This perfusion model allows the combined, vascular perfusion of small intestine and liver of rats via the celiac trunk (CT) and superior mesenteric artery (SMA). *Arrows* indicate the flow direction. The perfusion medium infused into the CT supplied the liver via the hepatic artery and the proximal part of the duodenum via the gastroduodenal artery; the medium infused into the SMA supplied the remainder of the small intestine. The flow in the SMA was measured by means of an ultrasonic flow probe. The perfusion medium draining from the small intestine (flow through gastroduodenal artery plus SMA) reached the liver through the portal vein (PV). Outflow from the inferior vena cava (IVC) was collected in calibrated tubes for the measurement of total flow. Flow in the celiac trunk was the difference between total and SMA flow. Insulin was infused into the portal vein via a flexible catheter. Flow rates are given in $ml \cdot min^{-1}$. The carbohydrate bolus was applied via the pylorus into the lumen of the duodenum. Perfusion medium samples were obtained from the portal vein via a second flexible catheter for measurements of glucose and prostaglandin E_2.

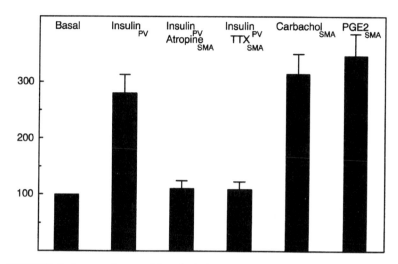

FIGURE 2. Increase by portal insulin in intestinal glucose absorption in the isolated, jointly perfused small intestine and liver of the rat. Inhibition by arterial atropine and tetrodotoxin and mimicking by arterial carbachol and prostaglandin E_2. Intestine and liver were perfused as described in FIGURE 1 with a Krebs-Henseleit bicarbonate buffer containing 5 mM glucose, 2 mM lactate, 0.2 mM pyruvate, 3% dextran, and 1% bovine serum albumin equilibrated with 95% O_2 and 5% CO_2. Glucose (5.5 mmol = 1 g within 1 min) was applied as an intraluminal bolus. Insulin (100 nM sinusoidal concentration) was infused into the portal vein (PV). Atropine (10 nM), tetrodotoxin (TTX; 1 µM), carbachol (10 µM), and prostaglandin E_2 (10 µM) were infused into the superior mesenteric artery (SMA). Glucose absorption is the area under the curve of glucose concentration in PV vs. time following a glucose bolus.[7] Basal (unstimulated) glucose absorption was taken as 100%. Values are means ± SEM of 4–6 experiments each.

tal vein for determination of portal glucose concentrations. Since the portal vein is the draining vessel of the entire small intestine, elevations in portal glucose concentration reflect intestinal glucose absorption.

Acute Stimulation by Portal Insulin of Intestinal Glucose Absorption via the SGLT1

In this experimental system glucose absorption following an intestinal glucose bolus was acutely enhanced by simultaneous infusion of insulin (100 nM) into the portal vein when compared to basal (unstimulated) glucose absorption without infusion of insulin (FIG. 2). The onset of the stimulatory action of portal insulin was rapid and dose dependent with a half-maximally effective concentration at a physiological level of 2–5 nM (data not shown).[7] Glucagon, infused into the portal vein like insulin, had no effect on intestinal glucose absorption (data not shown).[9] Thus, the increase in intestinal glucose absorption appeared to be a specific effect of insulin. Because a stimulatory effect of portal insulin could be observed only on absorption of glucose and galactose (both transported via the SGLT1), but not of fructose (transported via the GLUT5), the involved translocator was the SGLT1 (data not shown).[7]

Transmission of the Stimulatory Action of Portal Insulin to the Small Intestine by Hepatoenteral Nerves

As the isolated small intestine and liver were perfused in a nonrecirculating mode, insulin infused into the PV could not reach the small intestine. Thus, the stimulatory signal of portal insulin must have been transmitted against the blood stream in a retrograde manner and could therefore be expected to be of a nervous rather than hormonal nature.

In previous experiments with the perfused small intestine and liver, the stimulation by portal insulin of intestinal glucose absorption could be completely blocked by atropine infused into the superior mesenteric artery and mimicked by carbachol.[7] Therefore it was concluded that the signal pathway from the portal vein to the small intestine involved hepatoenteral cholinergic nerves. To confirm this conclusion, an additional series of experiments was performed using the neurotoxin tetrodotoxin (TTX), which blocks sodium channels of axons and other excitable membranes.[10] In the isolated, jointly perfused small intestine and liver of the rat, TTX, infused into the SMA, entirely prevented the portal insulin-stimulated increase in intestinal glucose absorption (FIG. 2). These data support the previous results, which indicated the involvement of hepatoenteral nerves.

Possible Involvement of Prostaglandin E_2 in the Increase in Intestinal Glucose Absorption by Portal Insulin

In the isolated, jointly perfused rat small intestine and liver, dibutyryl-cAMP infused into the SMA acutely increased intestinal glucose absorption almost threefold (data not shown). Therefore, cAMP might be involved in the stimulation by portal insulin of intestinal glucose absorption. However, none of the so-far-described muscarinic receptor subtypes has been shown to increase intracellular cAMP concentrations; rather, subtypes M1, M3, and M5 increase inositol trisphosphate (IP3); and subtypes M2 and M4 decrease cAMP.[11] Thus, if portal insulin stimulated intestinal glucose absorption by increasing intracellular cAMP concentrations via muscarinic hepatoenteral nerves, acetylcholine must trigger the release of a signal substance from cells other than enterocytes that would increase cAMP in enterocytes. Infusion of prostaglandin E_2 (PGE$_2$), which can elevate intracellular cAMP concentrations via EP$_2$ and EP$_4$ receptors, into the SMA increased intestinal glucose absorption acutely by threefold when compared with basal glucose absorption (FIG. 2).[12] Thus, PGE$_2$ might be the link between the cholinergic nerves and the absorptive cells, the enterocytes. To prove this hypothesis in an additional series of experiments, the overflow of prostanoids into the portal vein was measured, which reflects their synthesis in the small intestine. Following a luminal bolus of glucose without additional infusion of an effector, the overflow of PGE$_2$ increased only slightly. However, with an infusion of insulin into the portal vein, the overflow of PGE$_2$ was markedly elevated (data not shown). This increase must not necessarily be due to the stimulatory signal of portal insulin, but could also result from the increased glucose absorption. Therefore, the overflow was measured also with portal insulin infusion, but without a glucose bolus. Again, a marked increase in the portal PGE$_2$ concentrations could be detected (data not shown). Thus, portal insulin increased PGE$_2$ formation in small

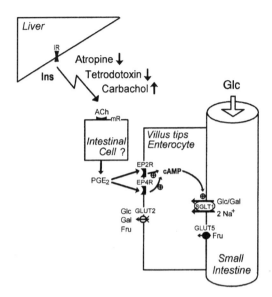

FIGURE 3. Hepatoenteral nervous signal chain regulating intestinal glucose absorption (hypothesis). Portal insulin stimulates intestinal glucose absorption in a retrograde manner via hepatoenteral nerves leading to a muscarinic receptor, which then causes an increase in cAMP in enterocytes (ACH = acetylcholine; mR = muscarinic receptor). The intraenterocytic messenger is cAMP and the involved translocator the sodium-dependent glucose transporter SGLT1. Since muscarinic receptors are not known to increase intracellular cAMP concentrations, the cholinergic hepatoenteral nerves induce prostaglandin E_2 release (PGE_2) from a so-far-unknown intestinal cell, which leads to an increase in cAMP concentrations in the enterocytes via the PGE_2 receptor subtypes EP2R and EP4R (GLUT2 = glucose transporter-2; GLUT5 = glucose transporter-5; Glc = glucose; Gal = galactose; Fru = fructose).

intestine, which could increase intracellular cAMP concentrations in the enterocytes and thus stimulate glucose absorption.

Hypothetical Mechanism of the Increase in Intestinal Glucose Absorption by Portal Insulin

There are numerous reports describing long-term alterations in the capacity of intestinal glucose absorption.[3–6] The common underlying mechanism was an increase in the number of the SGLT1s by *de novo* synthesis. However, this mechanism can be excluded as a possible explanation for the rapid short-term increase by portal insulin of intestinal glucose absorption observed here.

The rapid stimulatory effect of portal insulin on intestinal glucose absorption can be regarded as a novel regulatory pathway to acutely regulate the absorptive function of the small intestine. Following the nutrient-stimulated release by the pancreas, portal insulin stimulates intestinal glucose absorption in a retrograde manner via cholinergic hepatoenteral nerves (FIG. 3). The intraenterocytic messenger is cAMP. Since muscarinic receptors are not known to increase intracellular cAMP concentrations, the nerves must reach cells other than enterocytes and there stimulate the release of prostaglandin E_2, which in turn stimulates glucose absorption via EP_2 and/ or EP_4 by the enterocytes (FIG. 3).

ACKNOWLEDGMENTS

We wish to thank B. Döring, A. Hunger, and F. Rhode for their excellent technical assistance. This study was supported by a grant from the Deutsche Forschungsgemeinschaft through the SFB 402 "Molekulare und Zelluläre Hepatogastroenterologie," Göttingen, Teilprojekt B3.

REFERENCES

1. WRIGHT, E.M. 1993. The intestinal Na+/glucose cotransporter. Annu. Rev. Physiol. **55:** 575–589.
2. THORENS, B. 1993. Facilitated glucose transporters in epithelial cells. Annu. Rev. Physiol. **55:** 591–608.
3. SHIRAZI-BEECHEY, S.P., B.A. HIRAYAMA, Y. WANG et al. 1991. Ontogenetic development of the lamp intestinal Na+/glucose cotransporter is regulated by diet. J. Physiol. **437:** 699–708.
4. PHILPOTT, D.J., J.D. BUTZNER & J.B. MEDDINGS. 1992. Regulation of intestinal glucose transport. Can. J. Physiol. Pharmacol. **70:** 1201–1207.
5. KERSHAW, T.G., K.D. NEAME & G. WISEMAN. 1960. The effect of semistarvation on absorption by the rat small intestine in vivo and in vitro. J. Physiol. **152:** 182–190.
6. FEDORAK, R.N. 1990. Adaptation of small intestinal membrane transport processes during diabetes mellitus in rats. J. Physiol. Pharmacol. **68:** 630–635.
7. STÜMPEL, F., K. KUCERA, A. GARDEMANN & K. JUNGERMANN. 1996. Acute increase by portal insulin in intestinal glucose absorption via hepatoenteral nerves in the rat. Gastroenterology **110:** 1863–1869.
8. GARDEMANN, A., Y. WATANABE, V. GROSSE et al. 1992. Increase in intestinal glucose absorption and hepatic glucose uptake elicited by luminal but not vascular glutamine in the jointly perfused small intestine and liver of the rat. Biochem. J. **283:** 759–765.
9. STÜMPEL, F., T. KUCERA, A. GARDEMANN & K. JUNGERMANN. 1996. Involvement of enterohepatic and hepatoenteral nerves in the regulation of intestinal glucose absorption and hepatic glucose uptake in rats: studies with the isolated, jointly perfused intestine and the liver. In Liver Innervation. T. Shimazu, Ed.: 157–165. John Libbey. London.
10. RITCHIE, J.M. 1980. Tetrodotoxin and saxitoxin and the sodium channels of excitable tissues. Trends Pharmacol. Sci. **1:** 275–279.
11. CAULFIELD, M.P. 1993. Muscarinic receptors—characterisation, coupling and function. Pharmacol. Ther. **58:** 319–379.
12. SCHOLTKA, B., F. STÜMPEL & K. JUNGERMANN. 1999. Acute increase in glucose absorption via the sodium-dependent glucose transporter-1 by prostaglandin E_2 in rat intestine. Gut **44:** 490–496.

Effects of HIV Protease Inhibitors on Barrier Function in the Human Intestinal Cell Line HT-29/B6

H. BODE,[a,b] W. SCHMIDT,[a] J.D. SCHULZKE,[a] M. FROMM,[c]
E.O. RIECKEN,[a] AND R. ULLRICH[a]

*Departments of [a]Gastroenterology/Infectious Diseases and [c]Clinical Physiology,
Universitätsklinikum Benjamin Franklin, Freie Universität Berlin,
12200 Berlin, Germany*

INTRODUCTION

HIV protease inhibitors are standard components of highly active antiretroviral therapy (HAART). HAART effectively reduces HIV load,[1] increases CD4[+] lymphocyte count,[1] and reduces the incidence of opportunistic diseases and mortality in HIV-infected patients.[2,3] Because these effects require high compliance of the patients, they may be seriously hampered by adverse effects. Diarrhea is a common adverse effect of protease inhibitors, occurring in 16–72% of patients.[4–7] The pathogenesis of diarrhea elicited by protease inhibitors is unknown. Two important pathomechanisms of diarrhea are an increased active ion secretion (secretory type of diarrhea) or a passive back-leak of ions caused by a diminished epithelial barrier function (leak-flux type of diarrhea). Therefore, we investigated the effects of HIV protease inhibitors on ion secretion and barrier function in the colonic epithelial cell line HT-29/B6.

METHODS

Cell Culture of HT-29/B6 Cells

This subclone of the human colorectal carcinoma cell line HT-29 grows as a highly differentiated, polarized epithelial monolayer, has properties of mucus and Cl^--secreting cells, and shows functional responses to mediators acting via cAMP, protein kinase C, and calcium.[8] Cell culture was performed in RPMI 1640 medium containing 2% stabilized L-glutamine and supplemented with 10% fetal calf serum at 37°C in a 95% air, 5% CO_2 atmosphere. Cells were seeded on Millicell-PCF filters (Millipore; effective membrane area 0.6 cm^2), and two inserts were placed together into one conventional culture dish. Confluence of the cells was reached after 7 days, and protease inhibitors were added to the monolayer on day 8.

[b]Address for correspondence: Dr. Hagen Bode, Medizinische Klinik I, Gastroenterologie und Infektiologie, Universitätsklinikum Benjamin Franklin, 12200 Berlin, Germany. Voice: +49 30 8445-2392; fax: +49 30 8445-4239.
bode@medizin.fu-berlin.de

Solutions and Drugs

After sterilization in ethanol, protease inhibitor capsules were dissolved in RPMI 1460 medium. The drugs were added to the basolateral compartment of the incubation chamber in multiples of 1 and 10, to the apical compartment in multiples of 10 and 100 of their *in vivo* serum concentrations, which is 2.5 μg/mL for saquinavir (Fortovase™, Roche),[5] 11 μg/mL for ritonavir (Norvir™, Abbott),[9,10] and 7 μg/mL for nelfinavir (Viracept™, Agouron);[4,11] for indinavir (Crixivan™, MSD) a concentration of 10 μg/mL was chosen from the range of 7 to 17 μg/mL.[7]

Monitoring of Transepithelial Conductivity

Transepithelial conductivity (G) of the cell sheets was measured by a modification of the method as previously described.[8] Briefly, electrical measurements were performed in the culture dishes by two fixed pairs of electrodes (STX-2, World Precision Instruments, USA) connected with an impedance meter (D. Sorgenfrei, Inst. Klinische Physiologie, Berlin). Transepithelial conductivity was calculated from the voltage deflections caused by an external ±10 μA, 21 Hz rectangular current. Conductivity values were corrected for the conductivity of the empty filter and the bathing solution. The experiments were performed under sterile conditions at 37°C.

Ussing Experiments

Filter inserts were mounted into modified Ussing chambers,[8] which were connected with a computer controlled voltage clamp device (Fiebig, Berlin, Germany). The bathing solution consisted of RPMI 1460 medium containing 135 mM sodium, and 9 or 10 mM mannitol. The bathing solution was gassed with 95% air, 5% CO_2, and the temperature was maintained at 37°C using water jacketed reservoirs. Short-circuit current (I_{sc}) and G were recorded to the hard disk and corrected for bath resistance. Exposed tissue area was 0.6 cm^2. Tracer flux studies from basolateral to apical compartment were performed under short-circuit conditions with $^{22}Na^+$ and [^{3}H]mannitol (Du Pont de Nemours, Wilmington, USA). Four flux periods of 15 min were taken.

Statistical Analysis

Results are given as means ±SEM. Significance was tested by the unpaired Student's *t* test. $p < 0.05$ was considered significant.

RESULTS

Saquinavir

A concentration of 250 μg/mL saquinavir in the apical compartment (a) increased the transepithelial conductivity (G) of the cell layer (FIG. 1), and the basolateral 25 μg/mL concentration (b) enhanced the transepithelial permeability to 373 ± 111% of the initial conductivity after 48 hours. The apical 25 μg/mL and basolateral

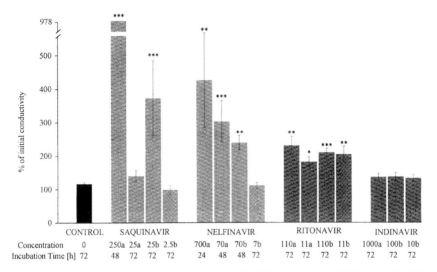

FIGURE 1. The effects of four HIV-protease inhibitors or transepithelial conductivity of HT-29/B6 cells. Saquinavir, nelfinavir, ritonavir, and indinavir were added to the apical (a) or basolateral (b) compartment of the cell layer in indicated concentrations in μg/mL. All values are means ± SEM. $n = 8$ for 700a nelfinavir, 4 for 2.5b added saquinavir, and 6 for all others. * $p < 0.05$; ** $p < 0.01$; *** $p < 0.001$.

2.5 μg/mL addition had no effect on G. The highest basolateral and apical addition of saquinavir did not alter the I_{sc} but increased the sodium and mannitol fluxes (TABLE 1).

Nelfinavir

Seventy μg/mL nelfinavir in the apical (10a) or basolateral (10b) compartment increased the transepithelial conductivity to about 250–300% of the initial value after 48 h (FIG. 1). With the apical 700 μg/mL dosage a pronounced effect occurred even one day earlier. The Ussing experiments revealed no effect of nelfinavir on short-circuit current (TABLE 1). 70b and 700a μg/mL nelfinavir increased as well the permeability of the monolayers for sodium and mannitol.

Ritonavir

Eleven μg/mL ritonavir in the basolateral or apical compartment as well as ten-fold higher concentrations doubled the transepithelial permeability of the monolayer after 72 h compared to the initial value (FIG. 1). Ritonavir had no effect on I_{sc}, and basolateral 11 and apical 110 μg/mL increased the permeability of the cellular layers for sodium and mannitol (TABLE 1).

Indinavir

Indinavir had no effect on the conductivity of the cell layers at any concentrations tested (FIG. 1).

TABLE 1. Effects of saquinavir, nelfinavir, and ritonavir on I_{sc} and unidirectional basolateral-to-apical $^{22}Na^+$ and [^{3}H]mannitol fluxes in HT-29/B6 cell layers[a]

	Concentration [μg/mL]	I_{sc}		J_{Na}		$J_{mannitol}$	
		[μA/cm^2]					
Saquinavir	0	−6.7 ± 0.6		50 ± 1		2 ± 0	
	25 b	−6.1 ± 0.6	n.s.	86 ± 4	***	3 ± 0	**
	250 a	−5.6 ± 1.9	n.s.	1309 ± 183	**	52 ± 8	**
Nelfinavir	0	−7.2 ± 0.6		43 ± 2		2 ± 0	
	70 b	−4.4 ± 1.7	n.s.	117 ± 11	**	4 ± 0	**
	700 a	−9.5 ± 7.0	n.s.	1107 ± 126	***	47 ± 7	**
Ritonavir	0	−3.9 ± 0.8		48 ± 3		2 ± 0	
	11 b	−2.8 ± 0.8	n.s.	93 ± 4	***	4 ± 0	***
	110 a	−2.2 ± 0.6	n.s.	108 ± 7	***	4 ± 0	***

[a]The values are means ± SEM of six independent measurements after an incubation period of 24 h for apical 700 μg/mL nelfinavir and 48 h for all others. a, apical; b, basolateral; I_{sc}, short-circuit current; J, basolateral to apical fluxes; *, $p < 0.05$; **, $p < 0.01$; ***, $p < 0.001$ vs. control.

DISCUSSION

HAART, on the one hand, reduced the incidence of the wasting syndrome and has been found effective in patients with cryptosporidiosis or microsporidiosis,[12,13] two AIDS-associated intestinal infections for which specific therapy is not available. On the other hand, diarrhea is a common adverse affect of HIV protease inhibitors. To investigate the mechanisms by which protease inhibitors could induce diarrhea, we used HT-29/B6 cells as a model epithelium to study two important types of diarrhea, i.e., secretory diarrhea, which is driven by active ion secretion, and leak-flux diarrhea, which is caused by diminished barrier function of the epithelium. In this model no evidence was found for an increased epithelial ion secretion, because the I_{sc} was not affected by the protease inhibitors tested. Thus, the diarrhea in patients receiving protease inhibitors seems not to be a secretory type of diarrhea.

By contrast, the transepithelial conductivity and basolateral-to-apical fluxes of sodium and mannitol were increased by saquinavir, nelfinavir, and ritonavir, demonstrating an impaired barrier function of the HT-29/B6 monolayers. These effects were observed when the substances were added to the basolateral side at onefold *in vivo* serum concentration for ritonavir, and at tenfold *in vivo* serum concentration for saquinavir and nelfinavir. Tissue concentrations of protease inhibitors *in vivo* are not known but may be considerably higher than those in serum,[14] especially in the intestinal mucosa where a concentration gradient within the epithelium is likely to develop after oral uptake. When added to the apical side of the HT-29/B6 monolayers, a diminished barrier function was observed at concentrations of 250 μg/mL saquinavir, 11 μg/mL ritonavir, and 70 μg/mL nelfinavir. A single oral uptake of the recommended dosage calculated for an intestinal content of one liter would result to luminal concentrations of 1200 μg/mL saquinavir, 600 μg/mL ritonavir, and 750 μg/mL nelfinavir. Therefore, the added concentrations are well below the *in vivo* concentrations within the intestinal lumen.

There is no transcellular basolateral-to-apical pathway for sodium and mannitol. Therefore, these solutes can only move passively through paracellular pathways or epithelial lesions. The barrier defects observed could be caused by necrotic cell death leading to permeable microerosions. Another mechanism leading to enhanced epithelial permeability could be the induction of apoptosis, inasmuch as we previously demonstrated that the basic epithelial conductivity directly above a single apoptotic enterocyte is increased by a factor of 12.[15] Finally, protease inhibitors could, like, for example, zonula occludens toxin,[16] TNFα,[17] and IFNγ[18] act directly on the tight junctional network that regulates paracellular permeability.

Although our experiments were limited by an observation time of approximately 3 days, the absence of barrier defects in indinavir-treated monolayers strongly indicates that epithelial damage is no necessary side effect of HIV protease inhibitors. Additional studies could reveal the exact mechanisms of such epithelial damage in protease inhibitor therapy.

In conclusion, HIV protease inhibitors show no effect on intestinal ion secretion. However, saquinavir, nelfinavir, and ritonavir, but not indinavir, diminish the epithelial barrier function in HT-29/B6 monolayers. These results indicate that protease inhibitors may induce diarrhea by an epithelial leak-flux mechanism.

ACKNOWLEDGMENTS

The excellent assistance of Ursula Lempart, Sieglinde Lüderitz, Ursula Schreiber, and Anja Fromm and the great support of Ing. grad. Detlef Sorgenfrei are gratefully acknowledged.

REFERENCES

1. KAKUDA, T.M., K.A. STRUBLE & S.C. PISCITELLI. 1998. Protease inhibitors for the treatment of human immunodefiency virus infection. Am. J. Health-Syst. Pharm. **55:** 233–254.
2. PALELLA, F.J., K.M. DELANEY, A.C. MOORMAN *et al.* 1998. Declining morbidity and mortality among patients with advanced human immunodeficiency virus infection. N. Engl. J. Med. **338:** 853–860.
3. HOGG, R.S., M.V. O'SHAUGHNESSY, N. GATARIC *et al.* 1997. Decline in deaths from AIDS due to new antiretrovirals. Lancet **349:** 1294.
4. Viracept prescribing information, Agouron Pharmaceuticals, Inc., La Jolla, CA 92037, USA, Item #634200MV, issued Mar. 14, 1997.
5. Fortovase prescribing information, Roche Pharmaceuticals, Roche Laboratories Inc., 340 Kingsland Street, Nutley, NJ 07110-1199, USA, issued Nov. 1997.
6. GILL, M.J. 1998. Safety profile of soft gel gelatin formulation of saquinavir of combination with nucleosides in a broad patient population. AIDS **12:** 1400–1402.
7. Crixivan prescribing information, Merck Sharp & Dohme, Hoddesdon, Hertfordshire EN11 9BU, UK, INR 6744, issued 12-97-CRX-96-D-730-B.

8. KREUSEL, K.M., M. FROMM, J.D. SCHULZKE *et al.* 1991. Cl⁻ secretion in epithelial monolayers of mucus-forming human colon cells (HT-29/B6). Am. J. Physiol. **261:** C574–C582.
9. Norvir prescribing information, Abbott GmbH, Max-Planck-Ring 2, 65205 Wiesbanden Delkenheim, Stand: Sept. 1997.
10. HSU, A., G.R. GRANNEMAN, G. WITT *et al.* 1997. Multiple-dose pharmacokinetics of ritonavir in human immunodeficiency virus-infected patients. Antimicrob. Agents Chemother. **41:** 898–905.
11. WU, E.Y., J.M. WILKINSON, D.G. NARET *et al.* 1997. High-performance liquid chromatographic method for the determination of nelfinavir, a novel HIV-1 protease inhibitor, in human plasma. J. Chromatogr. B **695:** 373–380.
12. FOUDRAINE, A.N., G.A. WEVERLING, T. VAN GOOL *et al.* 1998. Improvement of chronic diarrhoea in patients with advanced HIV-1 infection during potent antiretroviral therapy. AIDS **12:** 35–41.
13. CARR, A., D. MARRIOTT, A. FIELD *et al.* 1998. Treatment of HIV-associated microsporidiosis and cryptosporidiosis with combination antiretroviral therapy. Lancet **351:** 256–261.
14. SHETTY, B.V., M.B. KOSA, D.A. KHALIL *et al.* 1996. Preclinical pharmacokinetics and distribution to tissue of AG1343, an inhibitor of human immunodeficiency virus type 1 protease. Antimicrob. Agents Chemother. **40:** 110–114.
15. BENDFELDT, K., A.H. GITTER, J.D. SCHULZKE *et al.* 1999. Leaks in the epithelial barrier caused by apoptosis. Pflügers Arch. **437:** R95.
16. FASANO, A., B. BAUDRY, D.W. PUMPLIN *et al.* 1991. *Vibrio cholerae* produces a second enterotoxin, which affects intestinal tight junctions. Proc. Natl. Acad. Sci. USA **88:** 5242–5246.
17. SCHMITZ, H., M. FROMM, C.J. BENTZEL *et al.* 1998. Tumor necrosis factor-alpha (TNFα) regulates the epithelial barrier in the human intestinal cell line HT-29/B6. J. Cell Sci. **112:** 137–146.
18. MADARA, J.L. & J. STAFFORD. 1989. Interferon-γ directly affects barrier function of cultured intestinal epithelial monolayers. J. Clin. Invest. **83:** 724–727.

Induction of Endothelial Barrier Function *in Vitro*

G. KOCH, S. PRÄTZEL, M. RODE, AND B. M. KRÄLING[a]

German Cancer Research Center, Division of Cell Biology/A0100, 69120 Heidelberg, Germany

The main function of the endothelium is to provide a selective barrier between the blood stream and solid tissues that allows for a regulated transvascular flux of liquids and solutes.[1] This barrier function is achieved by the establishment of specific, but still functionally poorly understood, endothelial adherens and tight junctional complexes.[2] These endothelial junctional complexes play an important role in the integration of the tissue and vessel type–specific functional heterogeneity of the endothelial barrier.

We established an *in vitro* model for human dermal microvascular endothelial cell (HDMEC) maturation and differentiation[3] in which we could also induce the tightening of endothelial cell-to-cell contacts. In this model, endothelial maturation was induced by the addition of 0.5 µM dibutyryl cyclic AMP (Bt2) (Sigma) and 1 µg/ml hydrocortisone (HC) (Sigma) to confluent HDMEC cultures, while control cultures were maintained in basal conditions (without Bt2 and HC).[3] After addition of Bt2 and HC, HDMEC profoundly increased their interdigitation and cell membrane contacts, which we could demonstrate by electron microscopy. This resulted in prominent, densely delineated cell-to-cell contacts in Bt2- and HC-treated HDMEC as observed by phase contrast microscopy (FIG. 1b). By contrast, confluent HDMEC maintained in basal conditions formed small gaps between neighboring endothelial cells (FIG. 1a). In Bt2- and HC-treated cultures, the actin bundle-anchoring adherens junctions, as identified by the endothelium-specific VE-cadherin,[4] became organized into broader arrays of ordered network-like structures between contacting endothelial cells. In contrast, cultures maintained in basal conditions displayed small junctional discontinuities appearing as gaps (FIG. 1; compare c to d). Immunohistologically, these structures generally showed colocalization of the major components of the cytoplasmic plaque of this type of junction:[5] α-catenin, β-catenin, plakoglobin, and p120[ctn] both in basal medium and after Bt2 and HC treatment.

Functionally, Bt2- and HC-treated HDMEC became more resistant to EDTA treatment than did cultures grown in basal medium. Within 5 minutes after addition of 5 mM EDTA to the culture medium, HDMEC in basal conditions rapidly retracted, while Bt2- and HC-treated cells barely started to separate at their cell-to-cell contacts (FIG. 1; compare e to f). The tightening of the endothelial cell-to-cell contacts after Bt2 and HC treatment was further confirmed by electrical resistance measurements and permeability assays. The electrical resistance of confluent HDMEC

[a]Address for correspondence: Dr. B.M. Kräling, Division of Cell Biology/A0100, German Cancer Research Center, Im Neuenheimer Feld 280, 69120 Heidelberg, Germany. Voice: +49-6221-423443; fax: +49-622-423404.
b.kraling@dkfz-heidelberg.de

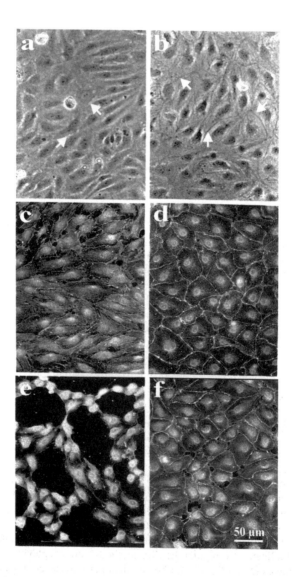

FIGURE 1. (**a**), (**c**), and (**e**) HDMEC in basal conditions and (**b**), (**d**), and (**f**) after Bt2 and HC treatment. Phase contrast microscopy revealed prominent endothelial cell-to-cell contacts after induction of *in vitro* differentiation (*arrows* in **b**), which displayed a more regular and more intense junctional localization of VE-cadherin (**d**) compared to cells in basal conditions (**c**). *Arrows* in (**a**) point to interendothelial gaps. Addition of 5 mM EDTA (for 5 min) caused rapid retraction of HDMEC in basal conditions (**e**), while Bt2- and HC-treated cells barely started to separate at their sites of contact (**f**). (**c–f**) Immunofluorescence labeling with mouse monoclonal antibody (mAb) BV9 [5] to VE-cadherin.

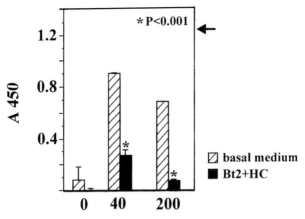

FIGURE 2. Bt2- and HC-treated cell cultures were significantly less permeable for biotin-dextran than HDMEC cultures in basal medium (*). Time point 0 was taken right after the addition of biotin-dextran to the top compartment. The *arrow* indicates the permeability of transwell chambers for biotin-dextran in the absence of cells.

monolayers was measured with a Millicell[®]-Electrical Resistance System (Millipore) in porous transwell chambers (0.4-μm pores; Falcon) and was found to be significantly increased after treatment with Bt2 and HC (26 ± 2 ohm $\times$ cm^2) compared to cells in basal medium (8 ± 4 ohm $\times$ cm^2) ($p < 0.002$). Treatment with Bt2 alone or HC alone showed an intermediate resistance (19 ± 1 ohm $\times$ cm^2 and 16 ± 2 ohm $\times$ cm^2, respectively).

For the permeability assays, HDMECs were plated on top of gelatinized transwell chambers (0.4-μm pores). After reaching confluence, HDMECs were treated with Bt2 and HC, and control cultures were maintained in basal medium for at least 4 days. Biotin-albumin (Sigma) or biotin-dextran (Sigma) (100–500 μg/mL, ~70 kDa each) were added to the top compartment of the transwells. The permeability of the endothelial monolayers for these tracer molecules was measured by collecting aliquots from the bottom compartment. For this purpose, the solution from the bottom compartment was aspirated and replaced by 0.5 mL Hepes-buffered saline (10 mM Hepes, pH 7.4, 135 mM NaCl, 5 mM KCl, 1 mM MgSO$_4$ and 1.8 mM CaCl$_2$) for 10 min, after which aliquots for sample detection were removed. Sample detection occurred in duplicates by regular ELISA methods[3] on streptavidin-coated 96-well microtiter plates (MicroCoat) with horseradish peroxidase–conjugated mouse anti-biotin antibody (Sigma). Bt2- and HC-treated HDMEC monolayers were significantly less permeable for the tracer molecules than cells in basal conditions ($p < 0.001$), which is representatively shown for biotin-dextran in FIGURE 2. Thus, this permeability assay can serve as a direct functional measure for the formation of the endothelial barrier during the process of endothelial differentiation *in vitro*.

Immunohistological studies revealed that the tight junction marker protein ZO-1 changed from discontinuous to apparently continuous lines along the tightening en-

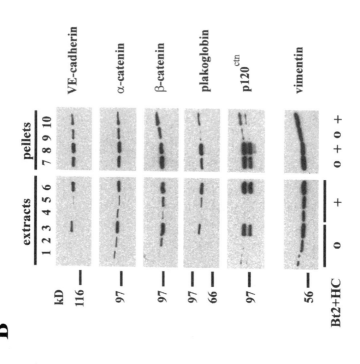

FIGURE 3. *See following page for caption.*

dothelial cell-to-cell contacts (FIG. 3A; compare a to b). Vinculin became spatially closely associated with protein ZO-1 during this process (Fig. 3A, a′, b′), which was confirmed by confocal laser scan analysis. It was recently reported that the functional interaction between vinculin and α-catenin is required for the induction of functional tight junctions and apical localization of ZO-1 in epithelial cells.[6] Our finding of a similar distribution of vinculin and ZO-1 during endothelial *in vitro* differentiation after Bt2 and HC treatment indicates that a similar, though perhaps indirect functional relationship may exist in HDMEC between vinculin and tight junction organization.

Confluent HDMEC cultures were treated for three days with Bt2 and HC, while controls were left in basal conditions. After three days, these monolayers were lysed with lysis buffer (10 mM Tris–HCl, pH 7.4, containing 50 µg/ml digitonin and 1 tablet Complete Mini protease inhibitors [Boehringer Mannheim]). The remaining pellets were extracted either with buffer A (25 mM Tris–HCl, pH 7.4, 2% NP-40) or buffer B (PBS with 1% Triton X-100, 0.5% DOC, 0.005 % SDS, and 1 mM EDTA). The detergent-insoluble pellets were directly boiled in $2 \times$ SDS sample buffer containing 4%-β-mercaptoethanol.[3] All these fractions were subjected to 8% SDS-PAGE and subsequent Western blot analysis, and the specific gel loading was normalized to relative cell densities as determined by methylene blue assays.[3] This Western blot analysis indicated that the process of tightening of the endothelial cell-to-cell contacts was accompanied by increased protein levels and cytoskeletal association of major components of the endothelial junctional complexes, namely, plakoglobin and p120[ctn] (FIG. 3B). The proportion of these two plaque proteins clearly increased in the detergent-insoluble pellets after Bt2 and HC treatment, which was most prominent after extraction with buffer B. This finding confirmed previous reports that plakoglobin plays an important role in tightening endothelial cell-to-cell junctions.[5,7] Especially the resistance of the endothelium to fluid shear stress seemed to be mediated by plakoglobin.[7]

This cell culture assay will serve as a baseline model to investigate aspects of endothelial cell differentiation at the molecular level, aiming at the identification of

---

FIGURE 3. (**A**) Confluent HDMEC in basal conditions (**a** and **a′**) and after treatment with Bt2 and HC (**b** and **b′**) were stained by immunofluorescence for ZO-1 (rabbit polyclonal Ab; Zymed) (**a,b**) and for vinculin (mouse mAb; Sigma) (**a′,b′**). Note the discontinuous staining for ZO-1 at the endothelial cell-to-cell contacts in (**a**) in comparison to the continuous pattern in (**b**). Vinculin and ZO-1 became partially colocalized at the cell-to-cell contacts of Bt2- and HC-treated HDMEC (compare *arrows* in **b** to **b′**). (**B**) SDS-PAGE (8%) and Western blot analysis of detergent soluble and insoluble fractions of constituents of HDMEC junctional complexes (for details see text): *Lanes* 1, 2, 3, 7 and 9 represent basal conditions; *lanes* 4, 5, 6, 8, and 10 represent Bt2 and HC treatment. *Lanes* 1 and 4 show cell lysates; *lanes* 2 and 5 are extracts with buffer A (25 mM Tris–HCl, pH 7.4, 2% NP-40) and *lanes* 3 and 6 with buffer B (PBS with 1% Triton X-100, 0.5% DOC, 0.005 % SDS, and 1 mM EDTA). The insoluble pellets after extraction with buffer A are shown in *lanes* 7 and 8; the insoluble pellets from buffer B in *lanes* 9 and 10. Vimentin levels are shown to allow a comparison for protein content in corresponding lanes for basal conditions (o) and after Bt2- and HC-treatment (+) (e.g., 1 with 4; 7 with 8, etc.). Source of antibodies: VE-cadherin, mouse mAb clone BV9; α-catenin, mouse mAb clone 5 (Transduction Laboratories); β-catenin, rabbit antiserum (Sigma); plakoglobin, mouse mAb clone 11E4; p120, mouse mAb clone 98 (Transduction Laboratories); vimentin, mouse mAb clone 3B4.

junctional proteins that are important in establishing and maintaining tight endothelial cell-to-cell contacts and mediators of the underlying cellular signaling pathways.

ACKNOWLEDGMENTS

We would like to thank Prof. Dr. Werner W. Franke of the German Cancer Research Center for his support and helpful discussions.

REFERENCES

1. LUM, H. & A.B.MALIK. 1994. Am. J. Physiol. **267:** L223–L241.
2. LAMPUGNANI, M.G. & E. DEJANA. 1997. Curr. Opin. Cell Biol. **9:** 674–682.
3. KRÄLING, B.M. *et al.* 1999. J. Cell Sci. **112:** 1599–1609.
4. LAMPUGNANI, M.G. *et al.* 1992. J. Cell Biol. **118:** 1511–1522.
5. LAMPUGNANI, M.G. *et al.* 1995. J. Cell Biol. **129:** 203–217.
6. WATABE-UCHIDA, M. *et al.* 1998. J. Cell Biol. **142:** 847–857.
7. SCHNITTLER, H.-J. *et al.* 1997. Am. J. Physiol. **273:** H2396–H2405.

The Structure and Function of Claudins, Cell Adhesion Molecules at Tight Junctions

SHOICHIRO TSUKITA[a] AND MIKIO FURUSE

Department of Cell Biology, Faculty of Medicine, Kyoto University, Yoshida-Konoe, Sakyo-ku, Kyoto 606-8501, Japan

ABSTRACT: Tight junctions (TJs) play a pivotal role in compartmentalization in multicellular organisms by sealing the paracellular pathway in epithelial and endothelial cell sheets. Recently, novel integral membrane proteins, claudins, have been identified as major cell adhesion molecules working at TJs. Claudins comprise a multigene family, and each member of ~23 kDa bears four trans-membrane domains. To date, 15 members of this gene family have been identified. When expression vectors of each species of claudins were transfected into fibroblasts lacking endogenous claudins or TJs, well-developed TJs were observed between adjacent transfectants. Furthermore, claudins were shown to be directly involved in the barrier function of TJs by experiments using *Clostridium perfringens* enterotoxin. Now that claudins have been identified, the structure and functions of TJs should be determined in detail in molecular terms.

INTRODUCTION

Tight junctions (TJs) are located at the most apical region of the lateral membranes of epithelial cells and are thought to function as a primary barrier to the diffusion of solutes through the paracellular pathway.[1,2] On ultrathin-section electron microscopy, TJs appear as a series of discrete sites of apparent fusion (kissing points), involving the outer leaflet of the plasma membrane of adjacent cells.[3] On freeze-fracture electron microscopy, TJs appear as a set of continuous, anastomosing intramembranous strands or fibrils (TJ strands) in the P-face, with complementary grooves in the E-face.[4] Recent technical progress has enabled the identification of several TJ-associated peripheral membrane proteins, such as ZO-1, ZO-2, ZO-3, cingulin, 7H6 antigen, and symplekin.[5] Although detailed analyses of these proteins have led to better understanding of the structure and function of TJs, lack of information concerning the TJ-specific integral membrane proteins has hampered more direct assessment of the function of TJs at the molecular level.

Occludin with four transmembrane domains was identified as the first TJ-specific integral membrane protein.[6,7] Occludin is thought to be not only a structural component but also a functional component of TJs;[8] occludin was shown to be directly involved in barrier functions, in fence functions of TJs, and in cell adhesion. Recently,

[a]Address for correspondence: Shoichiro Tsukita, M.D., Ph.D., Department of Cell Biology, Faculty of Medicine, Kyoto University, Yoshida-Konoe, Sakyo-ku, Kyoto 606-8501, Japan. Voice: +81-75-753-4372; fax: +81-75-753-4660.

htsukita@mfour.med.kyoto-u.ac.jp

the occludin gene was disrupted in embryonic stem cells.[9] Unexpectedly, occludin-deficient visceral endoderm cells that were differentiated from occludin-knockout embryonic stem cells still bore a well-developed network of TJ strands, pointing to the existence of as-yet-unidentified TJ-specific integral membrane proteins.

IDENTIFICATION AND CHARACTERIZATION OF CLAUDIN-1 AND -2

Two related (~23-kDa) integral membrane proteins, claudin-1 and -2 (38% identical at the amino acid sequence level), were then identified as components of isolated junctional fractions from the liver that were copartitioned with occludin.[10] Both claudin-1 and -2 also possess four transmembrane domains, but do not show any sequence similarity to occludin (FIG. 1a). The cytoplasmic domain and the second extracellular loop of claudins are significantly shorter than those of occludin. When FLAG-tagged, claudin-1 or -2 are expressed in cultured epithelial cells, such as MDCK cells, they are correctly targeted to, and incorporated into, preexisting TJ strands.

Next, claudin-1 or -2 cDNAs were introduced into mouse L fibroblasts lacking the expression of endogenous claudins or TJs. Cell dissociation and aggregation assays revealed that these transfectants showed fairly strong cell adhesion activity when compared to parental L cells.[11] When these cells were cultured on coverslips, expressed claudins were highly concentrated at cell-cell adhesion sites as planes,

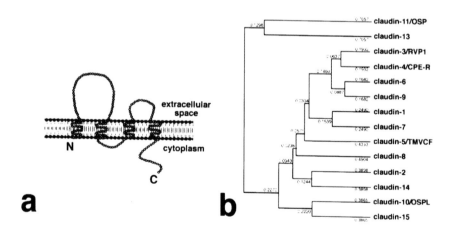

FIGURE 1. Claudins. (**a**) Membrane-folding model of claudin-1. Both NH_2 (N) and COOH (C) termini are located in the cytoplasm. This molecule has four transmembrane domains and two extracellular loops. (**b**) A potential phylogenetic tree of the claudin gene family (claudin-1~-11, -13~-15; see Ref. 14). The tree was constructed by the unweighed pair group method, using the calculated genetic distances (presented numerically) between pairs of members.

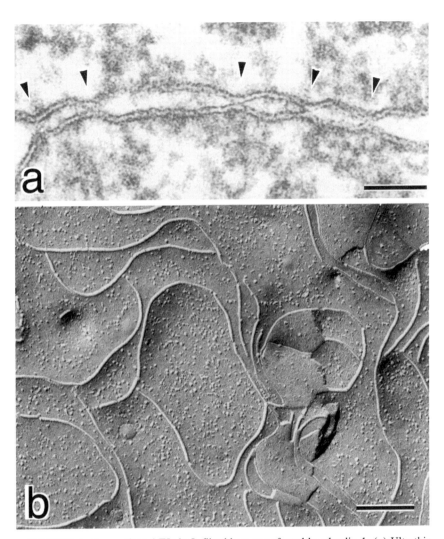

FIGURE 2. Reconstituted TJs in L fibroblasts transfected by claudin-1. (**a**) Ultrathin section electron microscopy. Note kissing points of TJs between adjacent transfectants (*arrowheads*) where the extracellular space is completely obliterated. (**b**) Freeze-fracture replica electron microscopy. Well-developed networks of TJ strands were reconstituted at the cell–cell contact planes of L transfectants. *Bars*, (**a**) 100 nm; (**b**) 200 nm.

where they were not distributed diffusely but in an elaborate network pattern.[12] Ultrathin-section electron microscopy identified many "kissing points of TJs" at these cell-cell contact planes, where the extracellular space was completely obliterated[11] (FIG. 2a). In freeze-fracture replicas of these contact sites, well-developed networks of strands were identified that were similar to TJ strand networks *in situ*[12] (FIG. 2b). In glutaraldehyde-fixed samples, claudin-1–induced strands were largely associated with the P-face as mostly continuous structures, whereas claudin-2–induced strands were discontinuous at the P-face, with complementary grooves at the E-face that were occupied by chains of particles. When occludin was cotransfected with claudin-1, it was concentrated at cell contact sites as planes to be incorporated into well-developed claudin-1–based strands.[12] We concluded that claudins are novel intercellular adhesion molecules that are responsible for TJ-specific obliteration of the intercellular space, and that they are responsible for TJ strand formation.

THE CLAUDIN FAMILY

Several sequences similar to claudin-1 and -2 were found in the databases, indicating the existence of a new gene family that can be called the *claudin family*. Based on sequence similarity, 15 members of this gene family have been identified in mice, as well as humans, and these have been designated as claudin-1 to -15[13,14] (see FIG. 1b). Among these, several members (claudin-3, -4, -5, -10, and -11) have been reported previously (RVP1, CPE-R, TMVCF/mBEC-1, OSPL, and OSP, respectively), although their physiological functions have not been elucidated. All of these claudin family members showed similar patterns on hydrophilicity plots, which predicted four transmembrane domains in each molecule. Similarly to claudin-1 and -2, hemagglutinin-tagged claudin-3 to -8 introduced into cultured MDCK cells are concentrated at pre-existing TJs.[13] Claudin-9 to -15 (except for claudin-11) have not been well characterized, but preliminary data favor the notion that these uncharacterized claudins are also directly involved in the formation of TJ strands *in situ*. Although the antigenecity of claudins is generally fairly low, specific antibodies for some claudin members have been obtained. These antibodies exclusively label the TJ strand itself *in situ* on freeze-fracture replicas.

The expression pattern of each claudin species varies considerably among tissues.[13] For example, claudin-3 mRNA is detected in large amounts in the lung and liver, and in small amounts in the kidney and testis. Claudin-4, -7, and -8 are primarily expressed in the lung and kidney. Claudin-6 expression is detected in large amounts in embryos, but not in adult tissues. Interestingly, claudin-5 was expressed in all tissues examined to date and was recently shown to be specifically expressed in endothelial cells of blood vessels, suggesting that this claudin species plays a key role in the regulation of blood vessel permeability.[15] Furthermore, claudin-11, which is expressed specifically in the brain and testis, was shown to constitute TJ strands between lamellae of myelin sheets of oligodendrocytes in the brain, and those between adjacent Sertoli cells in the testis.[16] These findings suggested that claudin-11 is responsible for the salutatory conduction along myelinated axons as well as the blood–testis barrier.

Claudins comprise a multigene family, and more than two distinct claudins are coexpressed in single cells, raising the questions of whether heterogeneous claudins form heteromeric TJ strands and whether claudins interact between each of the paired strands in a heterophilic manner. Recent detailed studies using L transfectants led to the conclusions that distinct species of claudins can be copolymerized into individual TJ strands and that distinct species of claudins can adhere to each other in a heterophilic manner.[17] This mode of assembly of claudins could increase the diversity of the structure and functions of TJ strands.

BARRIER FUNCTION AND CLAUDINS

Claudins have been shown to constitute TJ strands. However, evidence was, until recently, still lacking for the direct involvement of claudins in the barrier function of TJs. Reconstituted TJs in L transfectants did not surround individual cells continuously, making it difficult to measure their tightness as barriers. Coexpression of multiple species of claudins in epithelial cells complicated the evaluation of functions of claudins in this type of cell. Then, to clarify the function of claudins, we used the bacterial toxin *Clostridium perfringens* enterotoxin (CPE).[18] This enterotoxin, which consists of a single polypeptide chain with a molecular mass of ~35 kDa, is the causative agent of symptoms associated with *Clostridium perfringens* food poisoning in humans. Katahira and colleagues cloned a cDNA encoding the receptor for CPE (CPE-R) from an expression library of enterotoxin-sensitive monkey Vero cells.[19] They further found that previously reported RVP1 (rat ventral prostate-1) showed marked similarity to CPE-R and that RVP1 also functioned as a receptor for CPE, although the physiological functions of RVP1 and CPE-R remained unclear.[20] Interestingly, we noted significant sequence similarity between claudin-1/-2 and RVP1/CPE-R, and therefore we designated RVP-1 and CPE-R as claudin-3 and -4, respectively. We speculated that CPE could be used to modulate the function of claudins from the exterior of cells. There is accumulating evidence that the COOH-terminal half of this toxin (C-CPE) binds to CPE-R and that its NH_2 terminal half increases membrane permeability by forming small pores in the plasma membrane. Therefore, we used C-CPE as a peptide that specifically binds to claudin-3 and –4, and examined the effects of C-CPE on TJs in MDCK I cells expressing claudin-1 and -4. C-CPE bound to claudin-3 and -4 with high affinity as previously reported but not to claudin-1 or -2. In MDCK I cells incubated with C-CPE, claudin-4 was selectively removed from TJs with its concomitant degradation. After a 4-h incubation with C-CPE, TJ strands were disintegrated, and the number of TJ strands and the complexity of their network were markedly decreased. In good agreement with the time course of these morphological changes, the TJ barrier (transepithelial electric resistance and paracellular flux) of MDCK I cells was downregulated by C-CPE in a dose-dependent manner. These findings first provided evidence for the direct involvement of claudins in the barrier functions of TJs.[18]

PERSPECTIVE

Now that claudins have been identified and that TJ strands can be reconstituted from these molecules, the structure and functions of TJs should be able to be determined in detail in molecular terms within the next decade. The molecular mechanisms of the vectorial transport across the epithelial/endothelial cell sheets through the "transcellular" pathway have been examined by analyzing various channels and pumps. The "paracellular" pathway also plays an important role in vectorial transport, and this pathway shows some selectivity for transported materials such as ions. Morphological and physiological studies predicted that TJs are not absolute seals but that they contain "aqueous pores" that fluctuate between open and closed states.[21] Interestingly, recent identification of the gene responsible for hereditary hypomagnesemia suggested that claudins themselves are directly involved in the formation of these putative paracellular channels.[22] Thus, the further elucidation of how claudin-based TJ strands function as barriers that allow selective leakage will be very important for future studies.

The mechanism of regulation of TJ functions is another important issue for future studies. At present, no information is available regarding the transcriptional regulation of each species of claudin or of posttranslational regulation by various signaling pathways. Of course, detailed analyses of the interaction and mechanism of regulation between claudins and underlying cytoskeletal proteins, such as ZO-1/ZO-2/ZO-3 and other signaling molecules, are also required.[23,24] Elucidation of the mechanisms of regulation of TJ functions will lead not only to better understanding of various physiological and pathological events in molecular terms but also to development of methods to modulate TJ functions, which will be important for clinical medicine.

REFERENCES

1. SCHNEEBERGER, E.E. & R.D. LYNCH. 1992. Structure, function, and regulation of cellular tight junctions. Am. J. Physiol. **262:** L647–L661.
2. GUMBINER, B. 1993. Breaking through the tight junction barrier. J. Cell Biol. **123:** 1631–1633.
3. FARQUHAR, M.G. & G.E. PALADE. 1963. Junctional complexes in various epithelia. J. Cell Biol. **17:** 375–412.
4. STAEHELIN, L.A. 1974. Structure and function of intercellular junctions. Int. Rev. Cytol. **39:** 191–283.
5. MITIC, L.L. & J.M. ANDERSON. 1998. Molecular architecture of tight junctions. Annu. Rev. Physiol. **60:** 121–142.
6. FURUSE, M., T. HIRASE, M. ITOH et al. 1993. Occludin: a novel integral membrane protein localizing at tight junctions. J. Cell Biol. **123:** 1777–1788.
7. ANDO-AKATSUKA, Y., M. SAITOU, T. HIRASE et al. 1996. Interspecies diversity of the occludin sequence: cDNA cloning of human, mouse, dog, and rat-kangaroo homologues. J. Cell Biol. **133:** 43–47.
8. MATTER, K. & M. BALDA. 1999. Occludin and the function of tight junctions. Int. Rev. Cytol. **186:** 117–146.
9. SAITOU, M., K. FUJIMOTO, Y. DOI et al. 1998. Occludin-deficient embryonic stem cells can differentiate into polarized epithelial cells bearing tight junctions. J. Cell Biol. **141:** 397–408.
10. FURUSE, M., K. FUJITA, T. HIIRAGI et al. 1998. Claudin-1 and -2: novel integral membrane proteins localizing at tight junctions with no sequence similarity to occludin. J. Cell Biol. **141:** 1539–1550.
11. KUBOTA, K., M. FURUSE, H. SASAKI et al. 1999. Ca^{2+}-independent cell adhesion activity of claudins, integral membrane proteins of tight junctions. Curr. Biol. **9:** 1035–1038.
12. FURUSE, M., H. SASAKI, K. FUJIMOTO & SH. TSUKITA. 1998. A single gene product, claudin-1 or -2, reconstitutes tight junction strands and recruits occludin in fibroblasts. J. Cell Biol. **143:** 391–401.
13. MORITA, K., M. FURUSE, K. FUJIMOTO & SH. TSUKITA. 1999. Claudin multigene family encoding four-transmembrane domain protein components of tight junction strands. Proc. Natl. Acad. Sci. USA **96:** 511–516.
14. TSUKITA, SH. & M. FURUSE. 1999. Occludin and claudins in tight junction strands: leading or supporting players? Trends Cell Biol. **9:** 268–273.
15. MORITA, K., H. SASAKI, M. FURUSE & SH. TSUKITA. 1999. Endothelial claudin: claudin-5/TMVCF constitutes tight junction strands in endothelial cells. J. Cell Biol. **147:** 185–194.

16. MORITA, K., H. SASAKI, K. FUJIMOTO *et al.* 1999. Claudin-11/OSP-based tight junctions in myelinated sheaths of oligodendrocytes and Sertoli cells in testis. J. Cell Biol. **145:** 579–588.

17. FURUSE, M., H. SASAKI & SH. TSUKITA. 1999. Manner of interaction of heterogeneous claudin species within and between tight junction strands. J. Cell Biol. **147:** 891–903.

18. SONODA, N., M. FURUSE, H. SASAKI *et al.* 1999. *Clostridium perfringens* enterotoxin fragment removes specific claudins from tight junction strands: evidence for direct involvement of claudins in tight junction barrier. **147:** 195–204.

19. KATAHIRA, J., N. INOUE, Y. HORIGUCHI *et al.* 1997. Molecular cloning and functional characterization of the receptor for *Clostridium perfringens* enterotoxin. J. Cell Biol. **136:** 1239–1247.

20. KATAHIRA, J., H. SUGIYAMA, N. INOUE *et al.* 1997. *Clostridium perfringens* enterotoxin utilizes two structurally related membrane proteins as functional receptors *in vivo.* J. Biol. Chem. **272:** 26652–26658.

21. CLAUDE, P. 1978. Morphological factors influencing transepithelial permeability: a model for the resistance of the zonula occludens. J. Membr. Biol. **10:** 219–232.

22. SIMON, D.B., Y. LU, K.A. CHOATE *et al.* 1999. Paracellin-1, a renal tight junction protein required for paracellular Mg^{2+} resorption. Science **285:** 103–106.

23. TSUKITA, SH., M. FURUSE & M. ITOH. 1999. Structural and signaling molecules come together at tight junctions. Curr. Opin. Cell Biol. **11:** 628–633.

24. ITOH, M., M. FURUSE, K. MORITA *et al.* 1999. Direct binding of three tight junction-associated MAGUKs, ZO-1, ZO-2 and ZO-3, with the COOH-termini of claudins. J. Cell Biol. **147:** 1351–1363.

Intestinal Cell Adhesion Molecules

Liver-Intestine Cadherin

REINHARD GESSNER[a] AND RUDOLF TAUBER[b,c]

[a]*Institut für Laboratoriumsmedizin und Pathobiochemie, Charité, Campus Virchow-Klinikum, Augustenburger Platz 1, 13353 Berlin, Germany*

[b]*Institut für Klinische Chemie und Pathobiochemie, Universitätsklinikum Benjamin Franklin, Freie Universität Berlin, Hindenburgdamm 30, 12200 Berlin, Germany*

ABSTRACT: The cadherin superfamily comprises a large number of cell adhesion molecules, several of which are expressed in the gastrointestinal tract. LI-cadherin represents a novel type of cadherin within the cadherin superfamily distinguished from other cadherins by structural and functional features described in this review. In the mouse and human, LI-cadherin is selectively expressed on the basolateral surface of enterocytes and goblet cells in the small and large intestine, whereas in the rat this cadherin is additionally detectable in hepatocytes. LI-cadherin is capable of mediating Ca^{2+}-dependent homophilic cell–cell adhesion independent of interactions with the cytoskeleton, indicating that the adhesive function of this novel cadherin is complementary to that of E-cadherin and desmosomal cadherins co expressed in the intestinal mucosa.

THE CADHERIN SUPERFAMILY OF CELL ADHESION RECEPTORS

Cadherins represent a multigene family of cell adhesion molecules mediating Ca^{2+}-dependent adhesion of adjacent cells in a homophilic manner (for review, see Refs. 1–3). Cadherins play a key role as morphogenetic regulators during embryogenesis (for review, see Refs. 1 and 4), in the formation of junctional complexes and in the induction of the polarized cell type,[5] and, among other functions, in developing physical cell–cell associations.[6] Distinct members of the cadherin superfamily have been shown to function as tumor invasion suppressors,[7] whereas other cadherins—although this is a matter of controversy—have been reported to be involved in transmembrane transport.[8, 9]

Members of the cadherin superfamily share common structural features, having an N-terminal extracellular portion composed of tandemly repeated domains of approximately 110 amino acid residues, a single transmembrane region, and a cytoplasmic portion at the C-terminal side. The increasing number of cadherins identified during recent years has been classified into subfamilies according to their structural features—in particular, according to the number of the cadherin-type re-

[c]Address for correspondence: Dr. Rudolf Tauber, Institut für Klinische Chemie und Pathobiochemie, Universitätsklinikum Benjamin Franklin, Freie Universität Berlin, Hindenburgdamm 30, 12200 Berlin, Germany. Voice: +49-30-8445-2555; fax +49-30-8445-4152.

Tauber@ukbf.fu-berlin.de

peated domains, to homologies in the cell adhesion recognition (CAR) region within the N-terminal EC1 repeat, and to homologies in the cytoplasmic portion mediating the interaction with cytoskeletal and cytoplasmic proteins (for review, see Ref. 10). Members of the subfamily of classical cadherins, including B-, E-, EP-, N-, P-, and R-cadherin, contain a strongly conserved N-terminal EC1 domain harboring a HAV sequence motif in the CAR site, and in addition four other repeated cadherin-type domains EC2-5 in their extracellular portion and a highly conserved C-terminal cytoplasmic portion (for review, see Refs. 2 and 11). Other cadherins closely related to classical cadherins, such as cadherin-5,[12] OB-cadherin,[13] and M-cadherin,[14] exhibit a highly similar structure, but contain an EC1-domain in which the HAV motif is replaced by modified sequences and a less-well-conserved cytoplasmic portion.

Desmosomal cadherins, desmocollins and desmogleins, represent another cadherin family (for review, see Ref. 15). Desmocollins and desmogleins resemble the structure of classical cadherins in their extracellular region, whereas their cytoplasmic domains are only poorly conserved. Desmogleins have a C-terminal cytoplasmic domain containing, in addition to the cadherin-type segment, a number of repeats of a 29 ± 1 residue sequence not present in the other cadherins.

An additional type of cadherin is presumedly represented by T cadherin from chicken and its human analogue, cadherin-13, which exhibit a structure of their extracellular portion similar to that of classical cadherins, but entirely lacking the transmembrane domain and with the cytoplasmic portion being attached to the outer membrane surface via a glycosylphosphatidylinositol anchor.[16,17] Protocadherins identified in the central nervous system are distinguished from classical cadherins both by one or two additional extracellular cadherin-type repeats and by a complete lack of homology of their cytoplasmic domain.[18]

To date, the structural basis of the adhesive function of cadherins has been studied in detail for two members of the family of classical cadherins. The high-resolution 3-D structure of the N-terminal domains of E- and N-cadherin revealed by X-ray crystallography and nuclear magnetic resonance (NMR) spectroscopy suggested that the cadherin binding activity resides in the outermost (EC1) domain.[19,20] These studies, furthermore, proposed that the single cadherin molecules on the surface of one cell form "strand dimers" that are arranged as a linear "zipper," forming interdigitations with the strand dimers on the surface of the adjacent cell within the intercellular contact zones. The particular role of the EC1 domain in the formation of the intercellular linkage has also been elegantly shown by recent electron microscopic studies.[21] The conserved cytoplasmic domain of classical cadherins plays a central role for cadherin function both in adhesion and signaling (for review, see Refs. 22 and 23). As has been shown for E-cadherin, the cytoplasmic domain forms dynamic complexes with either β-catenin or plakoglobin (γ-catenin), both exhibiting homology to the product of the *Drosophila* segment polarity gene *armadillo*. β-Catenin and plakoglobin directly associate with α-catenin, which presumably links the E-cadherin/β-catenin complex or the E-cadherin/γ-catenin complex to actin filaments and thus provides an intracellular anchorage of the cadherin molecules. Recent studies demonstrate that β-catenin also plays a central role in the Wnt/Wingless signaling pathway.[24] Increase in the level of cytoplasmic β-catenin results in a translocation of β-catenin to the nucleus, where it forms a heterodimeric complex with members of the family of HMG-box LEF/TCF transcription factors[25,26] and regulates the transcription of different target genes, including c-Myc.[27,28] This dual role of β-catenin explains why changes in the level of β-catenin binding to E-cadherin can affect signaling via the Wnt/Wingless pathway.

Expression of each of the different cadherins is strictly regulated. Since most cell types are able to express more than one cadherin, characteristic patterns of tissue and cell distribution emerge. Reflecting their role as morphogenetic regulators, expression, at least of distinct cadherins, is under developmental control and correlates with steps of cell differentiation and tissue morphogenesis.[4]

In the intestinal mucosal epithelium, E-cadherin is the principal classical cadherin. E-cadherin is expressed in the enterocytes along the entire length of the crypt–villus axis and is subcellularly concentrated in adherens junctions.[29,30] Studies employing a dominant-negative N-cadherin mutant showed that the loss of E-cadherin function in enterocytes causes a loss of the differentiated polarized phenotype and results in the development of the histomorphology of inflammatory bowel disease,[30] highlighting the biological importance of this cadherin in the intestinal mucosa. Whether other cadherin cell adhesion receptors are coexpressed and are functionally active in enterocytes had been a matter of debate.

LIVER-INTESTINE CADHERIN

In an attempt to identify novel cell adhesion molecules expressed in the hepato-gastrointestinal system, a novel member of the cadherin superfamily representing a new subtype within this multigene family has recently been identified in the rat.[31] In this species, the novel cadherin is solely expressed in the liver and intestine, and was thus assigned the name liver-intestine (LI) cadherin. Molecular cloning from a rat liver cDNA expression library and analysis of the nucleotide and deduced amino acid sequence revealed that LI-cadherin is distinguished from the other cadherins by structural differences both in the extracellular and the cytoplasmic portion. In contrast to classical cadherins and to desmosomal cadherins, the extracellular portion of LI-cadherin consists of seven instead of five structurally defined cadherin repeats. Moreover, in the CAR region of the first ectodomain (EC1), LI-cadherin contains the sequence motif AAL in place of the HAV motif conserved among classical cadherins (FIG. 1). The AAL motif of LI-cadherin thus resembles more closely the CAR region of M-cadherin[14] and desmogleins[32] containing a FAL sequence or a RAL sequence, respectively. The second major structural difference concerns the cytoplasmic portion of LI-cadherin. Whereas the cytoplasmic tail is highly conserved among classical cadherins and consists of 150 to 160 amino acid residues, that of LI-cadherin has only 20 amino acid residues and displays no significant homology to classical cadherins. LI-cadherin also differs structurally from the group of protocadherins that contain a cytoplasmic portion that is considerably longer than that of LI-cadherin.[18] In summary, with respect to structure LI-cadherin is distinguished from classical cadherins, cadherins closely related to classical cadherins such as M-cadherin[14] or OB-cadherin,[13] desmosomal cadherins,[15] as well as protocadherins[18] (FIG. 2). Sequence comparison of LI-cadherin from rat,[31] human (Ref. 8 and Zitt et al., submitted for publication), and mouse (unpublished material) revealed that the structural features that distinguish this novel cadherin from classical cadherins are conserved between different species. Recently, another novel cadherin has been

```
LI(r)     LYHTRVLDRETRAVHHLQLAALDSQ-GAIVDGPVPIIIEVK--DINDNRPTFLQ

E(m)      LKVTQPLDREAIAKYILYSHAVSSN-GEAVEDPMEIVITVT--DQNDNRPEFTQ

N(m)      LSVTKPLDRELIARFHLRAHAVDIN-GNQVENPIDIVINVI--DMNDNRPEFLH

M(m)      YLNATTLDREKTDRFRLRAFALDLDGGSTLEDPTDLEIVVV--DQNDNRPAFLQ

DSG-3(h)  INITAIVDREETPSF-LITCRALNAQGLDVEKPLILTVKIL--DINDNPPVSQQ

5(h)      FAIERRLDRENISEYHLTAVIVDKDTGENLETPSSFTIKVH--DVNDNWPVETH

T(c)      VSVTRPLDREAIANYELEVEVTDLS-GKIIDGPVRLDISVI--DQNDNRPMFKE
```

FIGURE 1. Alignment of the amino acid sequence of the EC1 domain of rat LI-cadherin with that of other cadherins. The alignment shows the part of the extracellular domain EC1 including the putative cell adhesion recognition (CAR) site, which is *encircled*. Conserved motifs and amino acids are indicated by *black boxes*. LI(r), LI-cadherin (rat);[31] E(m), E-cadherin (mouse);[33] N(m), N-cadherin (mouse);[34] M(m), M-cadherin (mouse);[14] DSG-3(h), desmoglein-3 (human);[38] 5(h), cadherin 5 (human);[12] T(c), T-cadherin (chicken).[16]

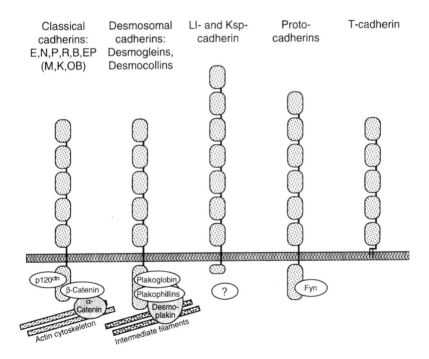

FIGURE 2. Schematic representation of the major structural features of members of the cadherin superfamily.

identified in rabbit kidney that has the same molecular structure as LI-cadherin,[9] indicating that LI-cadherin and Ksp (kidney-specific) cadherin represent a novel cadherin subfamily.

ADHESIVE FUNCTION OF LI-CADHERIN

Transfection experiments employing mouse L cells showed that LI-cadherin is not associated with catenins and is not tightly connected to the actin cytoskeleton.[35] This can be explained by the lack of homology of its cytoplasmic domain to the recently identified β-catenin binding region of E-cadherin.[36] Nevertheless, cell adhesion assays employing rat MH 7777 cells expressing LI-cadherin physiologically,[31] as well as transfected mouse L cells[35] and *Drosophila* S2 cells,[31] showed that LI-cadherin acts as a functional Ca^{2+}-dependent homophilic cell adhesion molecule. Independence of its adhesive function from cytoskeletal anchorage clearly distinguishes LI-cadherin from E-cadherin and other classical cadherins. By contrast to LI-cadherin, the adhesive function of E-cadherin crucially depends on the formation of the cadherin/catenin compex and on the anchorage to the actin cytoskeleton. For example, E-cadherin molecules lacking segments of the cytoplasmic domain, including the β-catenin binding site, are unable to mediate cell–cell adhesion.[37]

Furthermore, according to present data, the adhesive function of LI-cadherin is independent of any interaction with cytoplasmic components. This was shown by constructing a GPI-anchored form of LI-cadherin that, despite the obvious lack of cytoplasmic interactions, mediated Ca^{2+}-dependent adhesion to the same extent as the transmembrane (wild-type) form of LI-cadherin.[35] It is, therefore, very likely that by contrast to classical cadherins, the extracellular portion of LI-cadherin alone is capable of mediating the adhesive properties of this cell adhesion molecule. Why the adhesive mechanism of LI-cadherin is able to compensate for the missing intracellular linkage to the cytoskeleton is still unknown.

EXPRESSION IN CELLS AND TISSUES AND
SUBCELLULAR LOCALIZATION

As has been shown by Western and Northern blotting, LI-cadherin expression in the rat is restricted to the small and large intestine and the liver.[31] Interestingly, LI-cadherin was detectable in neither rat esophagus nor rat stomach. In rat liver and intestine LI-cadherin is expressed solely in polarized epithelial cells—that is, in hepatocytes and in enterocytes and goblet cells, respectively. By contrast, E-cadherin that is co-expressed in these cell types is found in a wide range of other epithelial cells, including nonpolarized epithelia (for review, see Ref. 2).

As was shown by immune electron microscopy, LI-cadherin is located solely on the basolateral surface of hepatocytes, enterocytes, and goblet cells but is absent on the apical plasma membrane. Within the basolateral surface of enterocytes and goblet cells, LI-cadherin is restricted almost entirely to the lateral area forming the intercellular adhesive site; whereas only very few LI-cadherin molecules are detectable on the basal cell surface facing the basement membrane.[31] Distinguished from E-cadherin and desmosomal cadherins that are strongly enriched in the adherens junctions and in desmosomes, respectively,[29,15] LI-cadherin is almost evenly distributed within the lateral plasma membrane and is strictly excluded from the junctional complex and from desmosomes.[31] The extrajunctional localization of LI-cadherin most likely reflects the lack of association with the actin cytoskeleton and possibly a high lateral mobility within the plasma membrane.

The biological significance of the fact that LI-cadherin expression is restricted to polarized epithelia of the gut and liver of the rat is still unknown. Preliminary data indicate that in the rat small intestine LI-cadherin may be involved in the control of the width of the lateral intercellular space during fluid absorption (unpublished material) and that the adhesive function of LI-cadherin is complementary to that of the coexpressed E-cadherin. The latter assumption is supported by data obtained in transgenic mice expressing a dominant-negative N-cadherin mutant lacking the extracellular portion under the control of an enterocyte-specific promoter.[30] Despite the loss of E-cadherin function in enterocytes, and although cell–cell adhesion in the intestinal mucosa is severely compromised, enterocytes partly retain lateral intercellular adherence, indicating that apart from E-cadherin other cell–cell adhesion molecules must be functional in these mice.

Recent cloning and expression studies of human (Zitt *et al.*, submitted for publication) and mouse LI-cadherin (unpublished material) revealed that the hepatogastrointestinal expression appears to be species specific. Whereas in the rat LI-cadherin is strongly expressed in the liver as well as in the small and large intestine,

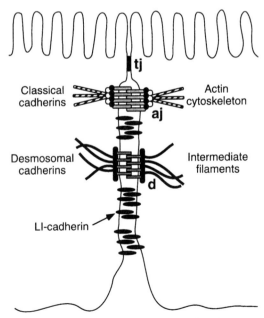

FIGURE 3. Schematic representation of adhesive cell–cell contacts by classical cadherins, desmosomal cadherins, and LI-cadherin in the intestinal mucosa (tj, tight junctions; aj, adherens junctions; d, desmosomes).

LI-cadherin is expressed exclusively in the intestinal mucosa in the human and mouse, but not in the liver. The biological relevance of the species-specific expression of LI-cadherin in hepatocytes is under current investigation.

According to present data, LI-cadherin appears to constitute a third Ca^{2+}-dependent cell adhesive system in the intestinal mucosa, in addition to E-cadherin concentrated in adherens junctions and to desmosomal cadherins present in desmosomes. Whereas E-cadherin and desmosomal cadherins mediate mechanically stable contacts in the upper region of the lateral cell surface domain due to their association with the actin cytoskeleton or with intermediate filaments, respectively, and due to their preferred location in circumscript junctional structures, LI-cadherin most likely is responsible for flexible intercellular adhesive contacts outside the junctional complexes (FIG. 3).

ACKNOWLEDGMENTS

Studies on LI-cadherin have been supported by the Deutsche Forschungsgemeinschaft (SFB 366) and by the Sonnenfeld-Stiftung.

REFERENCES

1. TAKEICHI, M. 1991. Cadherin cell adhesion receptors as a morphogenetic regulator. Science **251:** 1451–1455.
2. GEIGER, B. & O. AYALON. 1992. Cadherins. Annu. Rev. Cell Biol. **8:** 307–332.
3. KEMLER, R. 1993. From cadherins to catenins: cytoplasmic protein interactions and regulation of cell adhesion. Trends Genet. **9:** 317–321.
4. HUBER, O., C. BIERKAMP & R. KEMLER. 1996. Cadherins and catenins in development. Curr. Opin. Cell Biol. **8:** 685–691.
5. MCNEILL, H., M. OZAWA, R. KEMLER & W.J. NELSON. 1990. Novel function of the cell adhesion molecule uvomorulin as an inducer of cell surface polarity. Cell **62:** 309–316.
6. ANGRES, B., A. BARTH & W.J. NELSON. 1996. Mechanism for transition from initial to stable cell-cell adhesion: kinetic analysis of E-cadherin-mediated adhesion using a quantitative adhesion assay. J. Cell Biol. **134:** 549–557.
7. BIRCHMEIER, W. & J. BEHRENS. 1994. Cadherin expression in carcinomas: role in the formation of cell junctions and the prevention of invasiveness. Biochem. Biophys. Acta **1198:** 11–26.
8. DANTZIG, A.H., J. HOSKINS, L.B. TABAS et al. 1994. Association of intestinal transport with a protein related to the cadherin superfamily. Science **264:** 430–433.
9. THOMSON, R.B., P. IGARASHI, D. BIEMESDERFER et al. 1995. Isolation and cDNA cloning of Ksp-cadherin, a novel kidney-specific member of the cadherin mutigene family. J. Biol. Chem. **270:** 17594–17601.
10. POULIOT, Y. 1992. Phylogenetic analysis of the cadherin superfamily. Bioessays **14:** 743–748.
11. HERRENKNECHT, K. 1996. Cadherins. In Molecular Biology of Cell Adhesion Molecules. M.A. Horton, Ed.: 45-70. John Wiley. Chichester.
12. SUZUKI, S., K. SANO & H. TANIHARA. 1991. Diversity of the cadherin family: evidence for eight new cadherins in nervous tissue. Cell Regul. **2:** 261–270.
13. OKAZAKI, M., S. TAKESHITA, S. KAWAI et al. 1994. Molecular cloning and characterization of OB-cadherin, a new member of cadherin family expressed in osteoblasts. J. Biol. Chem. **269:** 12092–12098.
14. DONALIES, M., M. CRAMER, M. RINGWALD & A. STARZINSKI-POWITZ. 1991. Expression of M-cadherin, a member of the cadherin multigene family, correlates with differentiation of skeletal muscle cells. Proc. Natl. Acad. Sci. USA **88:** 8024–8028.

15. KOCH, P.J. & W.W. FRANKE. 1994. Desmosomal cadherins: another growing multigene family of adhesion molecules. Curr. Opin. Cell Biol. **6:** 682–687.
16. RANSCHT, B. & M.T. DOURS-ZIMMERMANN. 1991. T-cadherin, a novel cadherin cell adhesion molecule in the nervous system, lacks the conserved cytoplasmic region. Neuron **7:** 391–402.
17. TANIHARA, H., K. SANO, R.L. HEIMARK *et al.* 1994. Cloning of five human cadherins clarifies characteristic features of cadherin extracellular domain and provides further evidence for two structurally different types of cadherin. Cell Adh. Commun. **2:** 15–26.
18. SANO, K., H. TANIHARA, R.L. HEIMARK *et al.* 1993. Protocadherins: a large family of cadherin-related molecules in central nervous system. EMBO J. **12:** 2249–2256.
19. SHAPIRO, L., A.M. FANNON, P.D. KWONG *et al.* 1995. Structural basis of cell-cell adhesion by cadherins. Nature **374:** 327–337.
20. OVERDUIN, M., T.S. HARVEY, S. BAGBY *et al.* 1995. Solution structure of the epithelial cadherin domain responsible for selective cell adhesion. Science **267:** 386–389.
21. TOMSCHY, A., C. FAUSER, R. LANDWEHR & J. ENGEL. 1996. Homophilic adhesion of E-cadherin occurs by a cooperative two-step interaction of N-terminal domains. EMBO J. **15:** 3507–3514.
22. ABERLE, H., H. SCHWARTZ & R. KEMLER. 1996. Cadherin-catenin complex: protein interactions and their implications for cadherin function. J. Cell. Biochem. **61:** 514–523.
23. BEN ZE'EV, A. & B. GEIGER. 1998. Differential molecular interactions of β-catenin and plakoglobin in adhesion, signaling and cancer. Curr. Opin. Cell Biol. **10:** 629–639.
24. BROWN, J.D. & R.T. MOON. 1998. Wnt signaling: why is everything so negative? Curr. Opin. Cell Biol. **10:** 182–187.
25. HUBER, O., R. KORN, J. MCLAUGHLIN *et al.* 1996. Nuclear localization of β-catenin by interaction with transcription factor LEF-1. Mech. Dev. **59:** 3–11.
26. BEHRENS, J., J.P. VON KRIES, M. KUHL *et al.* 1996. Functional interaction of β-catenin with the transcription factor LEF-1. Nature **382:** 638–642.
27. HE, T.C., A.B. SPARKS, C. RAGO, H. HERMEKING, L. ZAWEL, L.T. DA COSTA, P.J. MORIN, B. VOGELSTEIN & K.W. KINZLER. 1998. Identification of c-MYC as a target of the APC pathway. Science **281:** 1509–1512.
28. MANN, B., M. GELOS, A. SIEDOW *et al.* 1999. Target genes of beta-catenin-T-cell-factor/lymphoid-enhancer-factor signaling in human colorectal carcinomas. Proc. Natl. Acad. Sci. USA. **96:** 1603–1608.
29. BOLLER, K., D. VESTWEBER & R. KEMLER. 1985. Cell-adhesion molecule uvomorulin is localized in the intermediate junctions of adult intestinal epithelial cells. J. Cell Biol. **100:** 327–332.
30. HERMISTON, M.L. & J.I. GORDON. 1995. *In vivo* analysis of cadherin function in mouse small intestinal enterocytes: essential role in adhesion, maintenance of differentiation, and regulation of programmed cell death. J. Cell Biol. **129:** 489–506.
31. BERNDORFF, D., R. GESSNER, B. UREFT *et al.* 1994. Liver-intestine cadherin: molecular cloning and characterization of a novel Ca^{2+}-dependent cell adhesion molecule expressed in liver and intestine. J. Cell Biol. **125:** 1353–1369.
32. KOCH, P.J., M.J. WALSH, M. SCHMELZ *et al.* 1990. Identification of desmoglein, a constitutive desmosomal glycoprotein, as a member of the cadherin family of cell adhesion molecules. Eur. J. Cell Biol. **53:** 1–12.
33. MANSOURI, A., N. SPURR, P.N. GOODFELLOW & R. KEMLER. 1988. Characterization and chromosomal localization of the gene encoding the human cell adhesion molecule uvomorulin. Differentiation **38:** 67–71.
34. MIYATANI, S., K. SHIMAMURA, M. HATTA *et al.* 1989. Neural cadherin: role in selective cell-cell adhesion. Science **245:** 631–635.
35. KREFT, B., D. BERNDORFF, A. BÖTTINGER *et al.* 1996. LI-cadherin mediated cell-cell adhesion does not require cytoplasmic interactions. J. Cell Biol. **136:** 1109–1121.
36. STAPPERT, J. & R. KEMLER. 1994. A short core region of E-cadherin is essential for catenin binding and is highly phosphorylated. Cell Adh. Commun. **2:** 319–327.
37. NAGAFUCHI, A. & M. TAKEICHI. 1988. Cell binding function of E-cadherin is regulated by the cytoplasmic domain. EMBO J. **7:** 3679–3684.
38. AMAGAI, M., V. KLAUS-KOVTUN & J.R. STANLEY. 1991. Autoantibodies against a novel epithelial cadherin in pemphigus vulgaris, a disease of cell adhesion. Cell **67:** 869–877.

Molecular Diversity of Plaques of Epithelial-Adhering Junctions

CAROLA M. BORRMANN, CLAUDIA MERTENS, ANSGAR SCHMIDT,
LUTZ LANGBEIN, CAECILIA KUHN, AND WERNER W. FRANKE[a]

*Division of Cell Biology, German Cancer Research Center,
69120 Heidelberg, Germany*

ABSTRACT: In biochemical and immunocytochemical comparisons of adhering junctions of different epithelia, we have observed differences in molecular composition not only between the intermediate filament–attached desmosomes and the actin filaments–anchoring adherens junctions but also between desmosomes of different tissues and of different strata in the same stratified epithelium. In addition we now report cell type–specific differences of molecular composition and immunoreactivity in both desmosomes and adherens junctions of certain simple epithelia. Whereas the *zonula adhaerens* of human intestinal and colonic epithelial cells, and of carcinomas derived therefrom, contains the additional *armadillo*-type plaque protein ARVCF, this protein has not been detected in the *zonula adhaerens* of hepatocytes. Similarly, plakophilin 3 is present in the desmosomal plaques of intestinal and colonic cells but appears to be absent from the hepatocytic desmosomes. We suggest that these profound compositional differences in the junctions of related simple epithelia are correlated to functional differences of the specific type of epithelium.

INTRODUCTION

The architecture and the functions of cells and tissues are established and maintained by intercellular adhesion structures connected to cytoskeletal meshworks of cytoplasmic filament bundles. These filament bundles anchor at specific dense plaques locally coating the cytoplasmic surface of cell-cell–adhesion junctions. Four classes of filament-associated intercellular junctions have been identified between epithelial cells: (1) gap junctions (*nexus*); (2) tight junctions (*zonulae occludentes* or *puncta occludentia*); "adhering junctions" comprising two major categories, i.e., (3) adherens junctions of different morphological forms (*zonula adhaerens, fascia adh.* or *punctum adh.*) usually anchoring bundles of actin-containing microfilaments (MFs), and (4) desmosomes (*maculae adhaerentes*) at which typically intermediate-sized filaments (IFs) are attached. This classification, originally proposed on sheerly morphological grounds,[1,2] has recently been justified and supported by discoveries of junction class-specific transmembrane and plaque-forming proteins.[3–9]

Studies of the molecular composition of adhering junctions from different tissues have confirmed the existence of two major categories, adherens junctions and des-

[a]Address for correspondence: Dr. Werner W. Franke, Division of Cell Biology, German Cancer Research Center/DKFZ, Im Neuenheimer Feld 280, 69120 Heidelberg, Germany. Voice: +49-6221-423212; fax: +49-6221-423404.
w.franke@dkfz-heidelberg.de

mosomes, and also some common principles. Both kinds of adhering junctions are characterized by clusters of transmembrane glycoproteins of the cadherin type, and their plaques can contain the widespread *armadillo*-repeat family[10] protein, plakoglobin[11] (for an apparent exception, see Ref. 12).

Otherwise, typical desmosomes and adherens junctions can be distinguished by their specific subsets of cadherins and plaque constituents. Desmosomes contain one or several desmogleins (Dsg1-3) and desmocollins (Dsc1-3),[4,13-16] and their plaques are characterized by desmoplakin[17] that may occur in one or two splice variant forms,[18] together with one or two members of the plakophilin subfamily of *armadillo*-repeat proteins (PKP1–3), which are also known for their dual location both in the nucleus and in desmosomal plaques.[9,19–25] By contrast, adherens junctions contain "classical" cadherins such as E-, VE-, M-, N-, P- or R-cadherin, the cytoplasmic carboxyterminal portions of which assemble plaques made from α- and β-catenins, protein p120^{ctn}, and several actin-associating proteins.[3,5,6,10,26–30]

In addition to these two major categories of adhering junctions, diverse other cell type–specific subforms of junctions have been distinguished by their distinct composition, such as the *complexus adhaerentes* of certain lymphatic endothelia,[31,32] the *contactus adhaerens* of cerebellar granule cells,[33] and the heterotypic junctions of the "outer limiting zone" of the retina, which contain the *armadillo*-repeat family protein, neurojungin (protein NPRAP[12]). Recently, we have identified another *armadillo*-repeat protein, termed ARVCF protein,[34–36] which also occurs in nuclei of several cell types, to be a constituent of the complex plaques of the adhering junctions connecting the myocardiac cells of the heart in the intercalated disks, together with α- and β-catenin as well as with the desmosomal marker proteins, desmoplakin and plakophilin 2 ("*areae compositae*").[37]

JUNCTION HETEROGENEITY IN STRATIFIED EPITHELIA

Immunocytochemically, compositionally different subtypes of desmosomes have been identified in different strata of stratified epithelia. Whereas cadherins Dsg2 and Dsc2 and the plaque protein, plakophilin 2, have been detected only in the basal layer of several stratified epithelia, Dsg1 and Dsc1 as well as plakophilin 1 are usually restricted to suprabasal layers, and Dsg3 and Dsc3 as well as plakophilin 3 display a broader and intermediate pattern of tissue distribution, often showing colocalization with one of the two other members of the same subfamily.[9,13–16,23,38–43]

One has to conclude from such observations of strata- or cell type–specific differences of immunoreaction that desmosomes are similar but not identical in composition and organization and probably also in their functions. This recognition of systematic desmosomal differences in different kinds of cells also provides a good explanation for the well-known cell- and tissue-type specificities of autoimmune diseases of the pemphigus type and of certain lesions observed on ablation of—or defects in—genes of constituent proteins of desmosomal membranes or plaques.[44–49]

Similar systematic differences have not yet been reported for the adherens junctions, mostly *puncta adhaerentia*, present in stratified or complex epithelia.

TABLE 1. **Compositional differences of human adherens junctions and desmosomes, respectively, between two types of polar epithelia, that is, intestinal and colonic epithelium on the one hand and hepatocytes on the other**

	Adherens Junctions (*zonula adhaerens, fascia adh., punctum adh.*)		Desmosomes *maculae adhaerentes*	
	Intestinal Epithelium	Hepatocyte	Intestinal Epithelium	Hepatocyte
Transmembrane glycoproteins	E-cadherin	E-cadherin	desmoglein 2/ desmocollin 2	desmoglein 2/ desmocollin 2
Plaque		plakoglobin α-catenin β-catenin		plakoglobin desmoplakin I plakophilin 2
	ARVCF-protein	—	plakophilin 3	—

NOTE: Both kinds of adhering junctions have only one constituent protein in common, plakoglobin, but differ in their transmembrane glycoproteins (E-cadherin vs. Dsg2 and Dsc2) and major plaque proteins, such as α- and β-catenin, often together with protein p120, vis-à-vis desmoplakin and plakophilin 2. Surprisingly, both epithelial junctions differ in a major plaque protein: although the *zonulae adhaerentes* of intestinal and colonic cells contain considerable amounts of ARVCF protein and their desmosomes plakophilin 3, neither of these proteins has been detected in the corresponding junctions of hepatocytes.

JUNCTION HETEROGENEITY IN SIMPLE EPITHELIA

In the single-layered ("simple") epithelial cells of the gastrointestinal tract and the liver, a particularly simple molecular composition has been found in both categories of adhering junctions, and it has been assumed that this composition would be identical in these closely related epithelia. However, recent careful comparisons of human and animal tissues by immunohistochemistry and biochemical analyses have revealed several striking differences between hepatocytes on the one hand and intestinal or colonic cells on the other, including colon carcinomas.

Thus, the plaque of the *zonula adhaerens* of the highly polarized colonic or intestinal epithelia contains, besides α- and β-catenin, protein p120[ctn] and plakoglobin, an additional—fourth—*armadillo*-repeat protein, the ARVCF protein (FIG. 1). Remarkably, this protein has only been detected in a minor fraction of the plaques of the lateral *puncta adhaerentia* (FIG. 1). This surprising observation also appears to indicate differences in the presence of ARVCF protein between adherens junction plaques of the same cell.

By contrast, however, using the same antibodies and techniques, ARVCF protein has not been detected in the *zonulae adhaerentes* lining the canaliculi of hepatocytes (TABLE 1).

Similarly, we have identified plakophilin 3, in addition to plakophilin 2, in the desmosomes of intestinal colonic and colon carcinoma cells but not in hepatocytic desmosomes (TABLE 1; for details, see Ref. 9). This finding corresponds to the absence—or very low concentrations—of plakophilin 3 and its mRNA in hepatocytes and hepatomas, as determined by immunoblot and RT-PCR techniques.

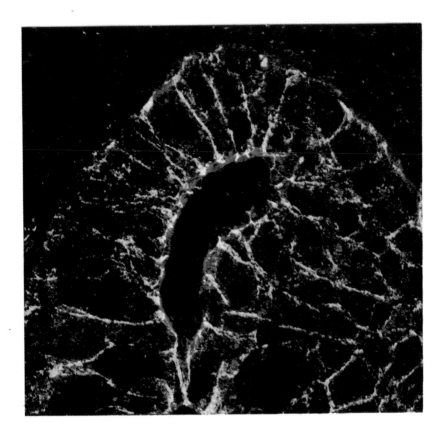

FIGURE 1. Double-label immunofluorescence microscopy of a cryostat section through a human colon carcinoma, using guinea pig antibodies (*red*) raised against a synthetic unde-capeptide (AVGSQRRRDD) of human ARVCF protein, in comparison with monoclonal murine antibodies against β-catenin (*green*). Note the occurrence of β-catenin at multiple membrane-membrane attachment sites ("junctions") along the lateral membranes of the highly polarized and differentiated tumor cells, including numerous *puncta adhaerentia* and the apical *zonula adhaerens*, whereas most of the ARVCF protein is enriched at the *zonula adhaerens*, displaying frequent double localization of both proteins (*yellow*). Similar double-labeling results have been obtained in comparisons of ARVCF protein with α-catenin and E-cadherin but not with desmosomal proteins. Magnification: 1000 ×.

We have to conclude from these observations that even between related polar simple epithelial cell types, such as hepatocytes on the one hand and intestinal-colonic cells on the other, systematic differences of molecular composition can occur in both kinds of adhering junctions. The functional implications of such profound differences between polar epithelia will have to be elucidated in future experiments, including targeted deletions or inactivations of the specific genes.

REFERENCES

1. FARQUHAR, M.G. & G.E. PALADE. 1963. Junctional complexes in various epithelia. J. Cell Biol. **17:** 375–412.
2. STAEHELIN, L.A. 1974. Structure and function of intercellular junctions. Int. Rev. Cytol. **39:** 191–283.
3. TAKEICHI, M. 1991. Cadherin cell adhesion receptors as a morphogenetic regulator. Science **251:** 1451–1455.
4. KOCH, P.J. & W.W. FRANKE. 1994. Desmosomal cadherins: another growing multigene family of adhesion molecules. Curr. Opin. Cell Biol. **6:** 682–687.
5. HUBER, O., C. BIERKAMP & R. KEMLER. 1996. Cadherins and catenins in development. Curr. Opin. Cell Biol. **8:** 685–691.
6. YAP, A.S., W.M. BRIEHER & B.M. GUMBINER. 1997. Molecular and functional analysis of cadherin-based adherens junctions. Annu. Rev. Cell Dev. Biol. **13:** 119–146.
7. MORITA, K., M. ITOH, M. SAITOU et al. 1998. Subcellular distribution of tight junction-associated proteins (occludin, ZO-1, ZO-2) in rodent skin. J. Invest. Dermatol. **110:** 862–866.
8. MORITA, K., M. FURUSE, K. FUJIMOTO & SH. TSUKITA. 1999. Claudin multigene family encoding four-transmembrane domain protein components of tight junction strands. Proc. Natl. Acad. Sci. USA **96:** 511–516.
9. SCHMIDT, A., L. LANGBEIN, S. PRÄTZEL et al. 1999. Plakophilin 3: a novel cell-type-specific desmosomal plaque protein. Differentiation **64:** 291–306.
10. PEIFER, M., S. BERG & A.B. REYNOLDS. 1994. A repeating amino acid motif shared by proteins with diverse cellular roles. Cell **76:** 789–791.
11. COWIN, P., H.-P. KAPPRELL, W.W. FRANKE et al. 1986. Plakoglobin: a protein common to different kinds of intercellular adhering junctions. Cell **46:** 1063–1073.
12. PAFFENHOLZ, R., S. STEHR, C. KUHN et al. 1999. The arm-repeat protein NPRAP (neurojungin) is a constituent of the plaques of the outer limiting zone in the retina, defining a novel type of adhering junction. Exp. Cell Res. **250:** 452–464.
13. NUBER, U.A., S. SCHÄFER, A. SCHMIDT et al. 1995. The widespread human desmocollin Dsc2 and tissue-specific patterns of synthesis of various desmocollin subtypes. Eur. J. Cell Biol. **66:** 69–74.
14. NUBER, U.A., S. SCHÄFER, S. STEHR et al. 1996. Patterns of desmocollin synthesis in human epithelia: immunolocalization of desmocollins 1 and 3 in special epithelia and in cultured cells. Eur. J. Cell Biol. **71:** 1–13.
15. NORTH, A.J., M.A.J. CHIDGEY, J.P. CLARKE et al. 1996. Distinct desmocollin isoforms occur in the same desmosomes and show reciprocally graded distributions in bovine nasal epidermis. Proc. Natl. Acad. Sci. USA **93:** 7701–7705.
16. SCHÄFER, S., S. STUMPP & W.W. FRANKE. 1996. Immunological identification and characterization of the desmosomal cadherin Dsg2 in coupled and uncoupled epithelial cells and in human tissues. Differentiation **60:** 99–108.
17. FRANKE, W.W., R. MOLL, D.L. SCHILLER et al. 1982. Desmoplakins of epithelial and myocardial desmosomes are immunologically and biochemically related. Differentiation **23:** 115–127.
18. GREEN, K.J., D.A.D. PARRY, P.M. STEINERT et al. 1990. Structure of the human desmoplakins: implications for function in the desmosomal plaque. J. Biol. Chem. **265:** 2603–2612.
19. KAPPRELL, H.-P., K. OWARIBE & W.W. FRANKE. 1988. Identification of a basic protein of M_r 75,000 as an accessory desmosomal plaque protein in stratified and complex epithelia. J. Cell Biol. **106:** 1679–1691.
20. HATZFELD, M., G.I. KRISTJANSSON, U. PLESSMANN & K. WEBER. 1994. Band 6 protein, a major constituent of desmosomes from stratified epithelia, is a novel member of the armadillo multigene family. J. Cell Sci. **107:** 2259–2270.
21. SCHMIDT, A., H.W. HEID, S. SCHÄFER et al. 1994. Desmosomes and cytoskeletal architecture in epithelial differentiation: cell type-specific plaque components and intermediate filament anchorage. Eur. J. Cell Biol. **65:** 229–245.
22. SCHMIDT, A., L. LANGBEIN, M. RODE et al. 1997. Plakophilins 1a and 1b: widespread nuclear proteins recruited in specific epithelial cells as desmosomal plaque components. Cell Tissue Res. **290:** 481–499.

23. MERTENS, C., C. KUHN & W.W. FRANKE. 1996. Plakophilins 2a and 2b: constitutive proteins of dual location in the karyoplasm and the desmosomal plaque. J. Cell Biol. **135:** 1009–1025.
24. BONNE, S., J. VAN HENGEL, F. NOLLET *et al.* 1999. Plakophilin-3, a novel *armadillo*-like protein present in nuclei and desmosomes of epithelial cells. J. Cell Sci. **112:** 2265–2276.
25. HATZFELD, M. 1999. The *armadillo* family of structural proteins. Int. Rev. Cytol. **186:** 179–224.
26. NÄTHKE, I.S., L. HICK & W.J. NELSON. 1995. The cadherin/catenin complex: connections to multiple cellular processes involved in cell adhesion, proliferation and morphogenesis. Semin. Dev. Biol. **6:** 89–95.
27. SHIBAMOTO, S., M. HAYAKAWA, K. TAKEUCHI *et al.* 1995. Association of p120, a tyrosine kinase substrate, with E-cadherin/catenin complexes. J. Cell Biol. **128:** 949–957.
28. STADDON, J.M., C. SMALES, C. SCHULZE *et al.* 1995. p120, a p120-related protein (p100), and the cadherin/catenin complex. J. Cell Biol. **130:** 369–381.
29. REYNOLDS, A.B. & J.M. DANIELS. 1997. p120: a src-substrate turned catenin. *In* Cytoskeletal-Membrane Interactions and Signal Transduction. P. Cowin & M.W. Klymkowsky, Eds.: 31–48. Chapman & Hall, New York, NY.
30. BEHRENS, J. 1999. Cadherins and catenins: role in signal transduction and tumor progression. Cancer Metastasis Rev. **18:** 15–30.
31. SCHMELZ, M. & W.W. FRANKE. 1993. *Complexus adhaerentes*, a new group of desmoplakin-containing junctions in endothelial cells: the syndesmos connecting retothelial cells of lymph nodes. Eur. J. Cell Biol. **61:** 274–289.
32. SCHMELZ, M., R. MOLL, C. KUHN & W.W. FRANKE. 1994. *Complexus adhaerentes*, a new group of desmoplakin-containing junctions in endothelial cells. II. Different types of lymphatic vessels. Differentiation **57:** 97–117.
33. ROSE, O., C. GRUND, S. REINHARDT *et al.* 1995. *Contactus adherens*, a special type of plaque-bearing adhering junction containing M-cadherin, in the granule cell layer of the cerebellar glomerulus. Proc. Natl. Acad. Sci. USA **92:** 6022–6026.
34. SIROTKIN, H., H. O'DONNELL, R. DAS GYPTA *et al.* 1997. Identification of a new human catenin gene family member (ARVCF) from the region deleted in velo-cardio-vacial syndrome. Genomics **41:** 75–83.
35. MARINER, D.J., H. SIROTKIN, J.M. DANIEL *et al.* 1999. Production and characterization of monoclonal antibodies to ARVCF. Hybridoma **18:** 343–349.
36. YAMAGISHI, H., V. GARG, R. MATSUOKA *et al.* 1999. A molecular pathway revealing a genetic basis for human cardiac and craniofacial defects. Science **283:** 1158–1161.
37. BORRMANN, C.M., C. KUHN & W.W. FRANKE. 1999. Human *armadillo* repeat protein ARVCF localized to myocardiac intercalated discs. Collected abstracts of the EMBO Workshop on "Molecular Genetics of Muscle Development and Neuromuscular Diseases": no. 9. Kloster Irsee, Germany, September 26–October 1, 1999.
38. ARNEMANN, J., K.H. SULLIVAN, A.I. MAGEE *et al.* 1993. Stratification-related expression of isoforms of the desmosomal cadherins in human epidermis. J. Cell Sci. **104:** 741–750.
39. THEIS, D.G., P.J. KOCH & W.W. FRANKE. 1993. Differential synthesis of type 1 and type 2 desmocollin mRNAs in human stratified epithelia. Int. J. Dev. Biol. **37:** 101–110.
40. LORIMER, J.E., L.S. HALL, J.P. CLARKE *et al.* 1994. Cloning, sequence analysis and expression pattern of mouse desmocollin 2 (DSC2), a cadherin-like adhesion molecule. Mol. Membr. Biol. **11:** 229–236.
41. KING, I.A., K.H. SULLIVAN, R. BENNETT, JR. & R.S. BUXTON. 1995. The desmocollins of human foreskin epidermis: identification and chromosomal assignment of a third gene and expression patterns of the three isoforms. J. Invest. Dermatol. **105:** 314–321.
42. YUE, K.K., J.L. HOLTON, J.P. CLARKE *et al.* 1995. Characterisation of a desmocollin isoform (bovine DSC3) exclusively expressed in lower layers of stratified epithelia. J. Cell Sci. **108:** 2163–2173.
43. MERTENS, C., C. KUHN, R. MOLL *et al.* 1999. Desmosomal plakophilin 2 as a differentiation marker in normal and malignant tissues. Differentiation **64:** 277–290.
44. ROH, J.-Y. & J.R. STANLEY. 1995. Intracellular domain of desmoglein 3 (*Pemphigus vulgaris* antigen) confers adhesive function on the extracellular domain of E-cadherin without binding catenins. J. Cell Biol. **128:** 939–947.

45. STANLEY, J.R. 1995. Autoantibodies against adhesion molecules and structures in blistering skin diseases. J. Exp. Med. **181:** 1–4.
46. ALLEN, E., Q.C. YU & E. FUCHS. 1996. Mice expressing a mutant desmosomal cadherin exhibit abnormalities in desmosomes, proliferation, and epidermal differentiation. J. Cell Biol. **133:** 1367–1382.
47. AMAGAI, M., P.J. KOCH, T. NISHIKAWA & J.R. STANLEY. 1996. *Pemphigus vulgaris* antigen (desmoglein 3) is localized in the lower epidermis, the site of blister formation in patients. J. Invest. Dermatol. **102:** 351–355.
48. KOCH, P.J., M.G. MAHONEY, H. ISHIKAWA *et al.* 1997. Targeted disruption of the *Pemphigus vulgaris* antigen (desmoglein 3) gene in mice causes loss of keratinocyte cell adhesion with a phenotype similar to *Pemphigus vulgaris*. J. Cell Biol. **137:** 1091–1102.
49. MCGRATH, J.A., P.H. HOEGER, A.M. CHRISTIANO *et al.* 1999. Skin fragility and hypohidrotic ectodermal dysplasia resulting from ablation of plakophilin 1. Br. J. Dermatol. **140:** 297–307.

Neutrophil Migration across Intestinal Epithelium

DAVID L. JAYE AND CHARLES A. PARKOS[a]

Department of Pathology and Laboratory Medicine, Emory University Medical School, Atlanta, Georgia 30322, USA

ABSTRACT: Transmigration of neutrophils across epithelial surfaces is the hallmark of inflammatory mucosal diseases of diverse organs. In disorders such as Crohn's disease, ulcerative colitis, pyelonephritis, and bronchitis, for example, neutrophil transmigration correlates with clinical disease activity, is associated morphologically with injury to the epithelium, and is central to disease pathophysiology. The mechanisms by which neutrophils transmigrate across epithelia are, therefore, of considerable significance for numerous pathologic states. In this paper, we discuss current evidence that defines these mechanisms in intestinal epithelium, emphasizing the structural constituents determining adhesive interactions and a subset of the complex regulatory signals between neutrophils and epithelium.

INTRODUCTION

Transepithelial migration of neutrophils (PMN) is a prominent component of many inflammatory mucosal diseases. Examples of such diseases are quite common in both the gastrointestinal and hepatobiliary systems and include ulcerative colitis, Crohn's disease, bacterial enterocolitis, gastritis, cholangitis, and acute cholecystitis. The increased mucosal permeability and fluid loss into the lumen subsequent to ischemic bowel injury or necrotizing enterocolitis are likewise directly related to PMN transepithelial migration. Notable examples in other organ systems include bronchial pneumonia, bronchitis, and bronchiectasis in the respiratory tract; and pyelonephritis and cystitis in the urinary tract. In fact, PMN transepithelial migration in these pathological processes is associated with epithelial injury, correlates with disease activity and symptoms, and is critical to pathophysiology.[1–4] Thus, elucidation of the mechanisms governing PMN transmigration across epithelia is of great significance for a diversity of diseases. In this article, we discuss the current understanding of these mechanisms, focusing on the structural components governing adhesive interactions and a subset of the complex signaling interplay between PMN and intestinal epithelium.

[a]Address for correspondence: Charles A. Parkos, Department of Pathology and Laboratory Medicine, Emory University, Woodruff Memorial Research Building, Room 2309, Atlanta, GA 30322. Voice: 404-727-8536; fax: 404-727-8538.

cparkos@emory.edu

MODEL EXPERIMENTAL SYSTEMS

Despite a wealth of evidence supporting a central role of PMN in epithelial dysfunction in inflammatory diseases, the molecular details of PMN adhesive interactions with mucosal surfaces are only beginning to be defined. Phenomenologically, PMN move across an epithelium from the basolateral epithelial membrane, then through the paracellular space before disruption of tight junctions occurs. Multiple adhesive interactions likely take place given the considerable length of the paracellular space (up to 20 μm). To begin to dissect out crucial components of this process, important reductionistic *in vitro* model systems have been developed. These models employ various transformed and nontransformed epithelial cell lines (e.g., T84 and CaCo2 intestinal carcinoma cells) cultured on permeable filters with pores of sufficient diameter to afford passage of PMN.[5] For the setup most commonly used (FIG. 1, panel A), epithelial cells are cultured to confluence on the upper surface of a collagen-coated, permeable support. PMN transepithelial migration may be assayed in an apical-to-basolateral direction by introduction of PMN to the upper chamber and a solution of chemoattractant such as N-formylated peptides to the lower chamber of the device. Although useful for characterization of direct interactions between PMN and epithelia, PMN transmigration in panel A occurs in the reverse direction from that which occurs under natural conditions, and thus necessitated modification of the setup. Shown in panel B of FIGURE 1, the modification consists of using epithelial mono-

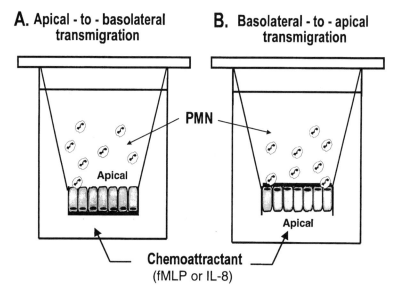

A. Apical - to - basolateral transmigration

B. Basolateral - to - apical transmigration

PMN

Apical

Apical

Chemoattractant
(fMLP or IL-8)

FIGURE 1. Model *in vitro* system used to study neutrophil transepithelial migration. Confluent epithelial monolayers are cultured on the surface (**A**) or underside (**B**) of a porous filter. Neutrophils (PMN) migrate from the upper chamber across the cell layer in response to a chemoattractant, such as f-met-leu-phe (fMLP) or the chemokine IL-8, placed in the lower reservoir. In the inverted configuration (**B**), transmigration occurs in the physiologically relevant basolateral-to-apical direction.

TABLE 1. Contrasting adhesive interactions during neutrophil (PMN) transepithelial and transendothelial migration

Adhesion Molecule	PMN-Epithelium	PMN-Endothelium
CD11a/CD18 (LFA-1)	−	+
CD11b/CD18 (Mac-1)	+	+
CD11c/CD18 (p150)	±?	±?
CD11d/CD18	?	?
CD47	+	+
CD31 (PECAM)	−	+
CD54 (ICAM-1)	−[a]	+
CD62L (L-selectin)	−	+
CD62E (E-selectin)	−	+
CD62P (P-selectin)	−	+
Other (carbohydrate-mediated)	+	+

[a]Following cytokine stimulation or bacterial exposure, epithelia express CD54 (ICAM-1) on the apical epithelial surface, where it may serve as a PMN adhesive receptor.[7,30] Apical compartmentalization of this receptor, however, precludes it from being used as a PMN receptor during the process of transepithelial migration.[7]

layers cultured in an inverted configuration on the underside of the permeable filter.[5] In this configuration, PMN transmigration occurs in a pathophysiologically relevant, basolateral-to-apical direction.

Using these and similar model systems, surface molecules expressed on PMN and/or intestinal epithelium have been identified that are of importance in PMN transepithelial movement (summarized in TABLE 1). For contrast, adhesion molecules significant for PMN interactions with endothelium are included. As displayed in TABLE 1, there are key differences between the repertoire of adhesive molecules employed in PMN interactions with epithelial cells and those employed with endothelial cells. Due to marked differences in the microenvironments of these two cellular barriers, substantial distinctions in the adhesive interactions governing transendothelial and transepithelial migration are not unexpected. Therefore, generalizations cannot necessarily be drawn from studies on PMN transendothelial migration, as will become clear below.

ADHESIVE MOLECULES THAT PROMOTE NEUTROPHIL TRANSMIGRATION

As displayed in TABLE 1, β2 integrins are significant in PMN migration across both epithelia and endothelia. β2 integrins are heterodimeric integral membrane glycoproteins expressed solely on leukocytes and consist of a common β chain (CD18) that associates with one of four α chains, either CD11a (LFA-1), CD11b (Mac-1), CD11c (p150), or CD11d.[6] Experiments with PMN from leukocyte adhesion deficiency (LAD) patients, who lack β2 integrins, highlight the importance of β2 integrins in PMN transepithelial migration. PMN from LAD patients fail to migrate across intestinal epithelial monolayers in the presence of a chemotactic gradient.[5] Moreover, in studies using blocking antibodies specific for β2 integrin α chains, PMN adhesion to epithelial cells was shown to be mediated exclusively by CD11b/CD18;[5] antibodies against the CD11a/CD18 and CD11c/CD18 failed to recapitulate the LAD defect. By contrast, PMN-endothelial interactions employ additional β2 integrin α chains. These studies confirmed the centrality of CD11b/CD18 in PMN transepithelial migration in transformed epithelial cell lines. Importantly, the findings have been verified in adhesion experiments with isolated normal human colonic epithelial cells and purified CD11b/CD18.[7] In addition to providing the first indication of the differences between PMN adhesive interactions with endothelia versus epithelia, these studies served to affirm the use of transformed intestinal epithelia as *in vitro* models of PMN-epithelial interactions.

Recent functional mapping studies of PMN CD11b have employed domain-specific monoclonal antibodies to delimit the regions important for interactions with epithelium.[8] These studies demonstrate that the I domain, a stretch of 200 amino acids toward the amino-terminal end of the extracellular portion of CD11b, and a discontinuous region, including the amino-terminal and cation-binding domains, play central roles in PMN adhesion to and transmigration across epithelia. Although the I domain is known to bind directly to a number of ligands, including iC3b, factor X,[9] fibrinogen,[10] ICAM-1, ICAM-2, heparin,[11] elastase,[12] and a hookworm-derived PMN inhibitory factor,[13,14] none appear to mediate PMN transepithelial migration. Thus, the epithelial counterreceptor for CD11b/CD18 currently remains undefined.

Additional moieties participating in PMN adhesive interactions with intestinal epithelium have been identified by screening panels of monoclonal antibodies (mAbs) for ability to block PMN transmigration[15] (TABLE 1). mAbs were discovered that abrogate PMN migration across monolayers of T84 intestinal epithelial cells,[15] vascular endothelium,[16] and collagen-coated filters.[15] Characterization showed that the antigen is CD47, a member of the immunoglobulin gene superfamily.[17] Analysis of the primary structure encoded by the gene for CD47 suggests several N-glycosylation sites, five transmembrane α helices, and an extracellular loop with homology to immunoglobulin V (IgV). CD47 is expressed by most cells in the body, and four isoforms differing in the composition of the putative intracellular C-terminal domain have been described.[18]

Because CD47 is expressed on both epithelia and PMN, experiments have been performed to discern the relative contributions of CD47 from each cell type in the transmigration process. These experiments disclosed the importance of PMN CD47 in transmigration, and a likely role for epithelial-expressed CD47 as well.[15] *In vivo* verification of the importance of CD47 in PMN migration has been demonstrated in a homozygous knockout mouse model.[19] Although mice devoid of CD47 are viable, they rapidly succumb to infection after intraabdominal challenge with *E. coli* due in part to delayed recruitment of PMN to the site of infection.

The mechanism by which CD47 influences PMN migration remains undefined. *In vitro* studies, however, indicate that contributions of CD47 to PMN migration may be as important as those of the β2 integrins. Additional studies have shown quite convincingly that CD47 regulates the avidity of αvβ3-integrin for its ligand vitronectin.[17,20] Furthermore, biochemical analyses have shown that CD47 associates directly with $\alpha v\beta 3$ and that the extracellular IgV-like loop is essential for this function. Although these studies suggest a refined mechanism for CD47 function, they do not provide a basis for CD47 regulation of PMN transepithelial migration. That is, despite the fact that PMN express $\beta 3$ integrins, functionally inhibitory anti-$\beta 3$ monoclonal antibodies alone or in combination fail to perturb PMN transepithelial migration (C.A. Parkos, unpublished observations).

Another possible mechanism by which CD47 may function is directly as an adhesion receptor in homotypic or heterotypic interactions. The structure of the extracellular loop is consistent with that of an adhesion protein. Yet, epithelial cells do not adhere to CD47 purified by different immunoaffinity methods (C.A. Parkos, unpublished observations). CD47 reportedly binds to thrombospondin-1, an adhesive glycoprotein that can mediate cell–cell and cell-matrix interactions;[21] and in platelets, CD47 binding by thrombospondin-1 has been reported to activate the integrin $\alpha IIb\beta 3$.[22] However, intestinal epithelium does not stain for thrombospondin, and antithrombospondin antibodies have no effect on PMN transepithelial migration (C.A. Parkos, unpublished observations). An expression cloning strategy recently identified CD47 as a binding partner for a heterophilic adhesive membrane protein termed P84.[23] P84 is involved in receptor tyrosine kinase signaling and is found at synapses in the mammalian central nervous system. A role for P84-CD47 interactions in PMN transepithelial migration, though, has yet to be identified.

Rather than promoting adhesion directly, CD47 may serve to regulate de-adhesion during CD11b/CD18-mediated transepithelial migration. Studies have shown that inhibition of CD47-mediated transmigration results in accumulation of PMN within the epithelium,[15] which would be consistent with this hypothesis. Moreover, examples of regulated de-adhesion of integrins exist in other systems. Direct interaction of CD47 within the plane of the membrane may also serve to modulate CD11b/CD18 function. The urokinase receptor (CD87), another PMN membrane protein, has been shown to modulate CD11b/CD18 function and seems to act through direct interactions within the plane of the membrane. Although very little is known about intracellular signaling associated with CD47 function, recent data suggest that CD47 may be functionally coupled to a heterotrimeric G protein.[24] The relevance of these observations to PMN-epithelial interactions, however, has yet to be demonstrated.

Based on observations of PMN transendothelial migration, other candidate adhesive receptors have been examined for their influence on transepithelial migration. Selectins (CD62 E, P, and L) are clearly important in PMN interactions with vascular endothelium (TABLE 1). Although selectins mediate initial adhesive interactions between PMN and endothelium and account for the phenomenon of PMN "rolling,"[6] antiselectin antibodies fail to inhibit transepithelial migration by PMN.[25] These findings are not unexpected given the dependence of selectin-mediated events on fluid shear forces and the lack of such forces in the transepithelial migration assays, as described above. A role for carbohydrate-mediated interactions in PMN transepithelial migration, though, is suggested by experiments in which certain simple or complex sugars, such as mannose 6-phosphate, glucose 6-phosphate, fucoidin, and the yeast-derived phosphomannan oligosaccharide PPME, potently inhibit PMN transepithelial migration.[25] Therefore, there may be a role for other, as-yet-undefined carbohydrate-based interactions, though classic selectin-mediated adhesion does not appear to be relevant to PMN-epithelial adhesive interactions.

Other important endothelial ligands for PMN do not appear to participate in PMN-epithelial interactions. For example, CD31 (platelet endothelial adhesion molecule, PECAM),[26] colocalizes to endothelial cell–cell lateral junctions and has been shown to mediate transendothelial migration of both PMN and monocytes. Both homotypic and heterotypic interactions have been described. CD31, however, does not appear to be expressed by colonic epithelial cells. Additionally, inhibitory anti-CD31 antibodies have no effect on PMN transepithelial migration (C.A. Parkos, unpublished observations).

A crucial PMN adhesive counterreceptor on endothelial cells is CD54 (intercellular adhesion molecule–1, ICAM-1),[27] which is differentially expressed under inflammatory conditions.[28] ICAM-1 localizes to the apical (lumenal) membrane of vascular endothelium, where it serves as an important binding partner for both CD11a/CD18 and CD11b/CD18 present on adherent PMN (Table 1). Similarly, ICAM-1 recently has been shown to play a role in eosinophil adhesion and transmigration across endothelium in the pulmonary microvasculature.[29] Inflammatory mediators likewise regulate the expression of ICAM-1 on intestinal epithelial cells. These cells do not normally express ICAM-1 in a basal state, but active inflammation[7] or invasion by bacterial pathogens[30] induces apically compartmentalized expression. With these stimuli, PMN can adhere to apically expressed ICAM-1 in a CD11b/CD18-dependent manner. However, the physiological relevance of this interaction is unclear, as transmigrating PMN would presumably not access apically expressed ICAM-1 until after traversing the paracellular space.[7] At that point, the process of transmigration per se would be over.

Although the potential role of lumenally expressed ICAM-1 is purely speculative, some interesting possibilities exist. One possible role might be in holding inflammatory cells in or near sites of inflammation. Given the increased lumenal fluid flow under many of these conditions, partly in response to electrogenic chloride secretion from PMN-derived 5'-AMP,[31] ICAM-1 might tether PMN to specific locations. In fact, this adhesive foothold might also aid in PMN-mediated clearance of pathogens from mucosal surfaces. Recent studies of transepithelial T cell movement have shown that ICAM-1 expression is upregulated by airway epithelial cells following inflammatory stimuli.[32] In contrast to the purely apical expression in intestinal epithelia, ICAM-1 appears to be expressed basolaterally and apically in airway epithelial cells. Here, T cell transmigration appears to be driven by a gradient of an apically secreted chemokine, RANTES; the role of ICAM-1 may similarly be to tether transmigrated T cells to the apical epithelium by β integrin–dependent adhesion.[32]

Given the paucity of described intestinal epithelial counterreceptors for PMN transmigration, additional epithelial adhesive structures that participate in the transmigration process likely await discovery. As noted previously, data suggest that carbohydrate structures may be involved in PMN-epithelial adhesive interactions. Because both heparin and heparan sulfate bind to CD11b/CD18,[11] candidate ligands may be found among proteoglycan family members containing heparin or heparan sulfate moieties such as syndecan, glycipan, and perlecan. Another candidate counterreceptor, junction adhesion molecule (JAM), is a recently described protein in the mouse that represents a component of both endothelial and epithelial intercellular junctions.[33] Like CD47, JAM contains IgV-like domains, though it is selectively expressed at junctions and appears to play a role in regulating monocyte

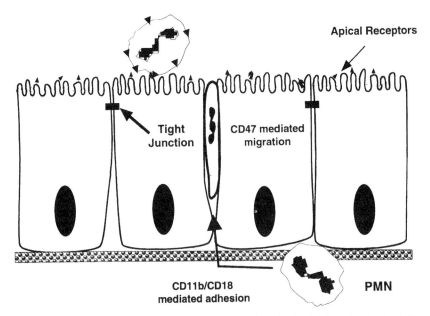

FIGURE 2. Model of neutrophil migration across intestinal epithelium. Under the influence of chemotactic gradients set up by epithelial cell secretion or from lumenal sources, neutrophils (PMN) are driven to migrate to the basolateral aspect of the epithelium and transmigrate across to the apical/lumenal side. A series of sequential molecular interactions at the level of the basal, lateral, and apical epithelial membrane, respectively, transpires. Although these interactions are incompletely understood, initial adhesion is dependent on the b2-integrin, CD11b/CD18. PMN then migrate through the paracellular space in a CD47-dependent manner. Reversible abrogation of tight junction integrity occurs via an unknown mechanism; however, PMN-epithelial signaling may be important. This affords PMN access to the apical surface where binding to apical receptors, possibly ICAM-1 and others, may serve to tether PMN to mucosal sites to facilitate clearance of, for example, pathogenic organisms.

transmigration.[33] The human homologue of JAM awaits description; thus, any potential role that this molecule may play as an epithelial ligand in regulating PMN transmigration is speculative. Last, dissecting out the contributions that these epithelial receptors may have will likely be a formidable task. If more than one epithelial counterreceptor exists, which is probable, ligand identification by standard techniques such as antibody inhibition assays will be difficult. Consistent with this idea, adhesion of ICAM-1–expressing epithelial cells to CD11b/CD18-coated surfaces is not inhibited by anti-ICAM-1 antibodies. Thus, it is likely that functional antibodies to important epithelial ligands will yield only partial inhibitory effects on adhesion and transmigration.

Based on the results discussed above, a model of PMN transepithelial migration is displayed in FIGURE 2. In this model, PMN cross the epithelium in a series of sequential molecular interactions at the level of the basal, lateral, and apical epithelial membranes, respectively.

ADDITIONAL REGULATORY FACTORS

Additional factors that are important in regulating PMN transmigration across intestinal epithelia have been described, as have the functional consequences of PMN transmigration on epithelial function. These are reviewed in depth elsewhere[34,35] but will be briefly mentioned here, as together they highlight the complexity of PMN-epithelial interactions.

The role of PMN-epithelial cross-signaling in facilitating transmigration is only beginning to be understood. At the intercellular junctions, for example, evidence suggests that such signaling events facilitate the passage of PMN. In particular, modeling studies have shown that the efficiency of PMN transmigration is enhanced in the direction in which it occurs *in vivo* across polarized epithelia (the basolateral-to-apical direction).[5] Furthermore, such differences can be ablated by disruption of epithelial F actin function.[36] Tight junction function is intimately linked to the apical actin ring in epithelia,[37,38] and events that alter these microfilaments might facilitate PMN transjunctional movement; the tight junction serves as a rate-limiting factor in transmigration. Such results indicate that complex signaling events likely take place between PMN and epithelial cells.

Models of vascular inflammation support these observations where PMN adhesion to endothelial monolayers has been shown to result in modification of endothelial adherens junctions and increased monolayer permeability.[39,40] Specifically, disorganization of cell–cell junctional components β-catenin, E-cadherin, and plakoglobin occurred adjacent to sites of firm adhesion between PMN and endothelium. Although similar findings have not been observed between PMN and epithelial cells, transepithelial electrical resistance, which correlates with epithelial barrier integrity, reversibly diminishes during PMN transepithelial migration. In the basolateral-to-apical direction, the decrease in transepithelial resistance actually precedes PMN migration but does not correlate with stimulated epithelial ion secretion (Ref. 5 and C.A. Parkos, unpublished observations). It will be of great interest to see whether the human homologue of the recently described JAM protein[33] plays a role in regulation of tight junction function during PMN transmigration.

Increased PMN adhesion to epithelia, activation, and tissue necrosis occurs under the influence of inflammatory mediators or in states of hypoxia. Potent modifiers of PMN-epithelial interactions include interferon-γ (IFN-γ)[41] and interleukin-4 (IL-4); by contrast, IL-1, lipopolysaccharide, tumor necrosis factor–α, and others strongly modulate PMN-endothelial interactions.[28] Based on the direction in which PMN transmigration is assayed (FIG. 1), epithelial proteins expressed after IFN-γ or IL-4 stimulation may result in increased or diminished PMN transepithelial migration. Upregulation of epithelial adhesive ligands also occurs after transient exposure of epithelial monolayers to low oxygen tension. Enhanced PMN transepithelial migration occurs, which is CD11b/CD18 and protein synthesis dependent.[42] Although the identity of these upregulated epithelial receptors remains unclear, they appear to be distinct from receptors upregulated on endothelial cells.

Signals derived from epithelial cells can also direct transepithelial migration. After invasion by bacterial pathogens, exposure to hypoxia, or stimulation with cytokines, the intestinal epithelium secretes potent chemoattractants, such as IL-8[42–44] from the basolateral surface,[45] and other, as yet incompletely characterized, agents from the apical surface.[45,46]

SUMMARY

This article highlights the current understanding of adhesive interactions and notes certain regulatory aspects of PMN migration across intestinal epithelial cells. CD11b/CD18-mediated adhesion has been shown to be an initial event in PMN interactions with epithelial cells that is followed by CD47-mediated transepithelial migration. Epithelial counterreceptors for migrating PMN have yet to be identified, though further studies will likely yield new receptors. Specific inflammatory mediators and still incompletely defined PMN-epithelial signaling mechanisms and interactions at the tight junctions appear to be important regulators of PMN transmigration as well. Comparison of PMN interactions with epithelia from other organ systems and with endothelia indicates that PMN interactions with intestinal epithelia engender many distinctive features. These distinctive features are likely to be of great importance in the design of targeted therapies for the amelioration of active inflammatory disease of the intestinal epithelium and of epithelia in a variety of other organ systems. They likewise emphasize the importance of continued research in this area.

ACKNOWLEDGMENTS

This work was supported by grants from the National Institutes of Health (HL54229 and HL60540) and a Biomedical Science Grant from the Arthritis Foundation (C.P.).

REFERENCES

1. HAWKER, P.C., J.S. MCKAY & L. A. TURNBERG. 1980. Electrolyte transport across colonic mucosa from patients with inflammatory bowel disease. Gastroenterology 79: 508–511.
2. KOYAMA, S., S.I. RENNARD, G.D. LEIKAUF et al. 1991. Endotoxin stimulates bronchial epithelial cells to release chemotactic factors for neutrophils: a potential mechanism for neutrophil recruitment, cytotoxicity, and inhibition of proliferation in bronchial inflammation. J. Immunol. 147: 4293–4301.
3. NUSRAT, A., C. DELP & J.L. MADARA. 1992. Intestinal epithelial restitution. J. Clin. Invest. 89: 1501–1511.
4. WEILAND, J.E., W.B. DAVIS, J.F. HOLTER et al. 1986. Lung neutrophils in the adult respiratory distress syndrome. Clinical and pathophysiologic significance. Am. Rev. Respir. Dis. 133: 218–225.
5. PARKOS, C.A., C. DELP, M.A. ARNAOUT & J.L. MADARA. 1991. Neutrophil migration across a cultured intestinal epithelium: dependence on a CD11b/CD18-mediated event and enhanced efficiency in the physiologic direction. J. Clin. Invest. 88: 1605–1612.
6. SPRINGER, T.A. 1995. Traffic signals on endothelium for lymphocyte recirculation and leukocyte emigration. Annu. Rev. Physiol. 57: 827–872.
7. PARKOS, C.A., S.P. COLGAN, M.S. DIAMOND et al. 1996. Expression and polarization of intercellular adhesion molecule–1 on human intestinal epithelia: consequences for CD11b/CD18-mediated interactions with neutrophils. Mol. Med. 2: 489–505.
8. BALSAM, L.B., T.W. LIANG & C.A. PARKOS. 1998. Functional mapping of CD11b/CD18 epitopes important in neutrophil-epithelial interactions: a central role of the I domain. J. Immunol. 160: 5058–5065.

9. ALTIERI, D.C. & T.S. EDGINGTON. 1988. The saturable high affinity association of factor X to ADP-stimulated monocytes defines a novel function of the Mac-1 receptor. J. Biol. Chem. **263:** 7007–7015.
10. ALTIERI, D.C., F.R. AGBANYO, J. PLESCIA et al. 1990. A unique recognition site mediates the interaction of fibrinogen with the leukocyte integrin Mac-1 (Cd11b/Cd18a). J. Biol. Chem. **265:** 12119–12122.
11. DIAMOND, M.S., R. ALON, C.A. PARKOS et al. 1995. Heparin is an adhesive ligand for the leukocyte integrin Mac-1 (CD11b/CD18). J. Cell Biol. **130:** 1473–1482.
12. CAI, T.Q. & S.D. WRIGHT. 1996. Human leukocyte elastase is an endogenous ligand for the integrin CR3 (CD11b/CD18, Mac-1, alpha M beta 2) and modulates polymorphonuclear leukocyte adhesion. J. Exp. Med. **184:** 1213–1223.
13. MOYLE, M., D.L. FOSTER, E.E. MCGRATH et al. 1994. A hookworm glycoprotein that inhibits neutrophil function is a ligand of the integrin CD11b/CD18. J. Biol. Chem. **269:** 10008–10015.
14. RIEU, P., T. UEDA, I. HARUTA et al. 1994. The A-domain of beta 2 integrin CR3 (CD11b/CD18) is a receptor for the hookworm-derived neutrophil adhesion inhibitor NIF. J. Cell Biol. **127:** 2081–2091.
15. PARKOS, C.A., S.P. COLGAN, T.W. LIANG et al. 1996. CD47 mediates post-adhesive events required for neutrophil migration across polarized intestinal epithelia. J. Cell Biol. **132:** 437–450.
16. COOPER, D., F.P. LINDBERG, J.R. GAMBLE et al. 1995. Transendothelial migration of neutrophils involves integrin-associated protein (CD47). Proc. Natl. Acad. Sci. USA **92:** 3978–3982.
17. LINDBERG, F.P., H.D. GRESHAM, E. SCHWARZ & E.J. BROWN. 1993. Molecular cloning of integrin-associated protein: an immunoglobulin family member with multiple membrane-spanning domains implicated in alpha v beta 3–dependent ligand binding. J. Cell Biol. **123:** 485–496.
18. REINHOLD, M.I., F.P. LINDBERG, D. PLAS et al. 1995. In vivo expression of alternatively spliced forms of integrin-associated protein (CD47). J. Cell Sci. **108:** 3419–3425.
19. LINDBERG, F.P., D.C. BULLARD, T.E. CAVER et al. 1996. Decreased resistance to bacterial infection and granulocyte defects in Iap-deficient mice. Science **274:** 795–798.
20. LINDBERG, F.P., H.D. GRESHAM, M.I. REINHOLD & E.J. BROWN. 1996. Integrin-associated protein immunoglobulin domain is necessary for efficient vitronectin bead binding. J. Cell Biol. **134:** 1313–1322.
21. GAO, A.G., F.P. LINDBERG, M.B. FINN et al. 1996. Integrin-associated protein is a receptor for the C-terminal domain of thrombospondin. J. Biol. Chem. **271:** 21–24.
22. CHUNG, J., A.G. GAO & W.A. FRAZIER. 1997. Thrombspondin acts via integrin-associated protein to activate the platelet integrin alpha IIb beta 3. J. Biol. Chem. **272:** 14740–14746.
23. JIANG, P., C.F. LAGENAUR & V. NARAYANAN. 1999. Integrin-associated protein is a ligand for the P84 neural adhesion molecule. J. Biol. Chem. **274:** 559–562.
24. FRAZIER, W.A., A.G. GAO, J. DIMITRY et al. 1999. The thrombospondin receptor integrin-associated protein (CD47) functionally couples to heterotrimeric Gi. J. Biol. Chem. **274:** 8554–8560.
25. COLGAN, S.P., C.A. PARKOS, D. MCGUIRK et al. 1995. Receptors involved in carbohydrate binding modulate intestinal epithelial-neutrophil interactions. J. Biol. Chem. **270:** 10531–10539.
26. MULLER, W.A., S.A. WEIGL, X. DENG & D.M. PHILLIPS. 1993. PECAM-1 is required for transendothelial migration of leukocytes. J. Exp. Med. **178:** 449–460.
27. DIAMOND, M.S., D.E. STAUNTON, A.R. DE FOUGEROLLES et al. 1990. ICAM-1 (CD54): a counter-receptor for Mac-1 (CD11b/CD18). J. Cell Biol. **111:** 3129–3139.
28. POBER, J.S. & R.S. COTRAN. 1990. The role of endothelial cells in inflammation. Transplantation **50:** 537–544.
29. YAMAMOTO, H., J.B. SEDGWICK & W.W. BUSSE. 1998. Differential regulation of eosinophil adhesion and transmigration by pulmonary microvascular endothelial cells. J. Immunol. **161:** 971–977.
30. HUANG, G.T., L. ECKMANN, T.C. SAVIDGE & M.F. KAGNOFF. 1996. Infection of human intestinal epithelial cells with invasive bacteria upregulates apical intercellular adhe-

sion molecule–1 (ICAM)-1) expression and neutrophil adhesion. J. Clin. Invest. **98:** 572–583.
31. MADARA, J.L., T.W. PATAPOFF, B. GILLECE-CASTRO *et al.* 1993. 5'-AMP is the neutrophil-derived paracrine factor that ilicits chloride secretion from T84 epithelial monolayers. J. Clin. Invest. **91:** 2320–2325.
32. TAGUCHI, M., D. SAMPATH, T. KOGA *et al.* 1998. Patterns for RANTES secretion and intercellular adhesion molecule 1 expression mediate transepithelial T cell traffic based on analyses *in vitro* and *in vivo.* J. Exp. Med. **187:** 1927–1940.
33. MARTIN-PADURA, I., S. LOSTAGLIO, M. SCHNEEMANN *et al.* 1998. Junctional adhesion molecule, a novel member of the immunoglobulin superfamily that distributes at intercellular junctions and modulates monocyte transmigration. J. Cell Biol. **142:** 117–127.
34. PARKOS, C.A. 1997. Cell adhesion and migration. I. Neutrophil adhesive interactions with intestinal epithelium. Am. J. Physiol. **273:** G763–768.
35. PARKOS, C.A. 1997. Molecular events in neutrophil transepithelial migration. Bioessays **19:** 865–873.
36. HOFMAN, P., L. D'ANDREA, D. CARNES *et al.* 1996. Intestinal epithelial cytoskeleton selectively constrains lumen-to-tissue migration of neutrophils. Am. J. Physiol. **271:** C312–C320.
37. MADARA, J.L., D. BARENBERG & S. CARLSON. 1986. Effects of cytochalasin D on occluding junctions of intestinal absorptive cells: further evidence that the cytoskeleton may influence paracellular permeability and junctional charge selectivity. J. Cell Biol. **102:** 2125–2136.
38. NUSRAT, A., M. GIRY, J.R. TURNER *et al.* 1995. Rho protein regulates tight junctions and perijunctional actin organization in polarized epithelia. Proc. Natl. Acad. Sci. USA **92:** 10629–10633.
39. DEL MASCHIO, A., A. ZANETTI, M. CORADA *et al.* 1996. Polymorphonuclear leukocyte adhesion triggers the disorganization of endothelial cell-to-cell adherens junctions. J. Cell Biol. **135:** 497–510.
40. LUSCINSKAS, F.W., M.I. CYBULSKY, J.M. KIELY *et al.* 1991. Cytokine-activated human endothelial monolayers support enhanced neutrophil transmigration via a mechanism involving both endothelial-leukocyte adhesion molecule–1 and intercellular adhesion molecule–1. J. Immunol. **146:** 1617–1625.
41. COLGAN, S.P., C.A. PARKOS, C. DELP *et al.* 1993. Neutrophil migration across cultured intestinal epithelial monolayers is modulated by epithelial exposure to IFN-gamma in a highly polarized fashion. J. Cell. Biol. **120:** 785–798.
42. COLGAN, S.P., A.L. DZUS & C.A. PARKOS. 1996. Epithelial exposure to hypoxia modulates neutrophil transepithelial migration. J. Exp. Med. **184:** 1003–1015.
43. ECKMANN, L., M.F. KAGNOFF & J. FIERER. 1993. Epithelial cells secrete the chemokine interleukin-8 in response to bacterial entry. Infect. Immun. **61:** 4569–4574.
44. MCCORMICK, B.A., S.P. COLGAN, C. DELP-ARCHER *et al.* 1993. *Salmonella typhimurium* attachment to human intestinal epithelial monolayers: transcellular signalling to subepithelial neutrophils. J. Cell. Biol. **123:** 895–907.
45. MCCORMICK, B.A., P.M. HOFMAN, J. KIM *et al.* 1995. Surface attachment of *Salmonella typhimurium* to intestinal epithelia imprints the subepithelial matrix with gradients chemotactic for neutrophils. J. Cell Biol. **131:** 1599–1608.
46. MCCORMICK, B.A., C.A. PARKOS, S.P. COLGAN *et al.* 1998. Apical secretion of a pathogen-elicited epithelial chemoattractant activity in response to surface colonization of intestinal epithelia by *Salmonella typhimurium.* J. Immunol. **160:** 455–466.

Expression and Function of Death Receptors and Their Natural Ligands in the Intestine

JÖRN STRÄTER AND PETER MÖLLER[a]

Department of Pathology, University of Ulm, 89081 Ulm, Germany

ABSTRACT: The tumor necrosis factor receptor (TNFR) family is a still-growing group of homologous transmembrane proteins, some of which bear an intracellular "death domain" and are able to directly mediate apoptosis. Apoptosis is induced upon trimerization of the receptors by their natural ligands' constituting the complementary TNF family. The best-characterized apoptosis-mediating TNFR family member is CD95 (APO-1/Fas). CD95 is functionally expressed on the basolateral surface of colonic epithelial cells regardless of their position along the crypt axis. The biological significance of this CD95 expression in the gut, however, is still under discussion. Although it is unlikely that the CD95/CD95L system is involved in the physiologic regeneration of the intestinal epithelium, this system may play an important role in the pathogenesis of inflammatory bowel diseases. In contrast to the normal epithelium, colon carcinoma cell lines are mostly resistant to CD95-induced apoptosis. The detection of CD95L expression in colon carcinoma cell lines has led to the concept of carcinomas as "immunoprivileged sites," where invading immune cells are killed by CD95L-expressing tumor cells. A more recently described member of the TNF family is TRAIL, which is also able to induce apoptosis. As yet, four TRAIL receptors have been cloned, two of which (TRAIL-R1 and 2) bear a death domain and mediate apoptosis, whereas two others (TRAIL-R3 and 4) lack (functional) death domains and are supposed to act as decoy receptors. Because many tumor cell lines *in vitro* are sensitive to TRAIL-induced apoptosis while their normal counterparts are not, TRAIL is currently under discussion as a possible anticancer therapeutic agent.

INTRODUCTION

The tumor necrosis factor receptor (TNFR) family is a still growing group of type I transmembrane proteins that are homologous to each other in two to four extracellular cysteine-rich domains. Being involved in the regulation of diverse biological processes, they act as growth factors or contribute to the activation/stimulation of cells, especially in the immune system.[1] Some family members are of special interest due to a particular property: they are directly able to induce apoptosis.[2] These apoptosis-mediating family members are characterized by a highly conserved intracellular signaling domain, the "death domain," which recruits certain adaptor molecules to the "death-inducing signaling complex" (DISC), finally activating the apoptotic caspase cascade.[3]

[a]Address for correspondence: Peter Möller, Department of Pathology, University of Ulm, Albert-Einstein-Allee 11, 89081 Ulm, Germany. Voice: +49-731-50-23320; fax: +49-731-50-3884. peter.moeller@medizin.uni-ulm.de

Their ligands, on the other hand, constitute their own family of homologous type II transmembrane proteins, the TNF family.[2] They typically exist in a trimeric form, and it is by trimerizing their receptors that they perform their deadly work. Ligand binding may be simulated by agonistic antireceptor antibodies *in vitro* and *in vivo*.

Why are these apoptosis-inducing TNF/TNFR systems of interest for gastroenterologists? In the normal situation, senescent epithelial cells in the gut mucosa are eliminated by apoptosis, but it is not yet clear how apoptosis is induced in these cells.[4,5] Thus, death receptors present on gut epithelial cells may be involved in this elimination process. Next, apoptosis-inducing receptor/ligand systems are important regulators of inflammatory processes,[6] also making them potential players in the pathogenesis of chronic inflammatory bowel diseases. Finally, some of them, particularly the CD95/CD95L system, have been implicated in colorectal cancer.[7]

In the following, the possible function of two apoptosis-inducing receptor/ligand systems, the CD95/CD95L and the TRAIL/TRAIL receptor system, in the gastrointestinal tract will be discussed.

THE CD95/CD95L SYSTEM

Probably the best-studied apoptosis-inducing TNFR family member is CD95 (Fas/APO-1).[8,9] CD95 is quite broadly expressed in a variety of normal organs and tissues.[10] It is also constitutively present on the basolateral surface of intestinal and colon epithelial cells all along the crypt(/villus) axis.[11] Moreover, isolated colonic crypt cells readily enter apoptosis when incubated with an agonistic CD95 antibody *in vitro*.[12] The biological significance of this functional CD95 expression in the normal gut, however, is not yet completely understood. Some clarification on this point was expected from addressing the question of where the CD95 ligand is expressed in the gut mucosa.

In contrast to CD95, the tissue distribution of its ligand, CD95L, is much more restricted. It was at first detected in activated T cells[13–15] and natural killer cells[16] but also in B lymphocytes[17] and finally plasma cells.[18] In the normal colon, it is also exclusively found in a few scattered lamina propria cells, whereas gut epithelial cells do not express CD95L, with the only, but remarkable, exception of Paneth cells.[19] Numbers and distribution of CD95L-expressing cells in the colon mucosa suggest that CD95-mediated apoptosis of epithelial cells is a rare event in the normal colon and not the way senescent epithelial cells are eliminated in the gut.[12] This view is further substantiated by studies on *lpr/lpr* and *gld/gld* mice having severe defects in CD95 function and CD95L expression, respectively. These mice do not show altered apoptotic rates in the colonic surface epithelium compared to wild-type animals (our unpublished data).

The predominant expression of CD95L in lymphatic cells, together with functional *in vitro* data, has made clear that CD95L plays an important role in the regulation of immune processes and the elimination of virus-infected cells (for review, see Ref. 6). Thus, CD95L is crucially involved in the elimination of chronically activated T lymphocytes and the main cytotoxic effector of T helper type 1 cells. These findings drew the attention to chronic inflammatory bowel diseases. Actually, CD95L is significantly upregulated in lamina propria mononuclear cells in ulcerative colitis (UC).[12] In parallel, the number of apoptotic lamina propria, but also epithelial, cells in UC is dramatically increased, and a focal association of CD95L-expressing cells and epithelial apoptosis is observed. Although we do not yet completely understand the early processes leading to undue activation of inflammatory cells, the probable consequences are focally enhanced epithelial cell apoptosis via CD95 and the formation of epithelial microlesions. Epithelial damage may finally result in a breakdown of mucosal barrier function, allowing the invasion of pathogenic microorganisms that, in turn, may aggravate the inflammatory process.[12]

A particular role in the regulation of inflammatory responses was ascribed to the CD95 system when CD95L was detected in the testis[20] and some occular tissues.[21] In these organs, known for a long time as "immunoprivileged sites," immune processes may be cut down by CD95L expression in organ-constituting cells and induction of apoptosis in invading activated (CD95-expressing and -sensitive) immune cells.[22] Thus, the CD95L expression may act as a protective mechanism in organs, in which inflammation would cause irreversible tissue damage and loss of function.

Recently, this concept was extended to malignant tumors when it was noticed that different tumor cells lines[23-25] express CD95L and are able to induce apoptosis in CD95-sensitive target cells in vitro ("CD95 counterattack" hypothesis; for review, see Ref. 7). It has been proposed that tumor cells escape the attack of tumor-infiltrating lymphocytes, in a manner that is analogous to the concept of immunoprivileged sites, by expressing CD95L and inducing CD95-mediated apoptosis in the immune cells.

This concept of tumors as immunoprivileged sites implies that tumor cells themselves are resistant to CD95-mediated apoptosis. Actually, in contrast to the normal colonic epithelium, many colon carcinoma cell lines are relatively resistant to CD95 cross-link.[26,27] An important mechanism leading to CD95 resistance may be the loss of CD95 from the surface, which is seen immunohistochemically in most colorectal carcinomas compared to normal mucosa or adenomas.[11] We found that in 39% of colon carcinomas, CD95 expression was diminished; and in 48% of carcinomas, predominantly of the nonmucinous type, CD95 was reduced to beyond detection levels. Complete loss of CD95 was more frequent in carcinomas that had already metastasized. Comparative analysis revealed that the aberrant mode of CD95 expression correlated with that of abnormal MHC class I, class II, and invariant chain expression, and also with the secretory component.[11] Comparing the CD95 protein levels by immunochemistry of colon epithelial cells in situ and cytospin preparations of colon carcinoma cells, the latter have, as a rule, essentially severely downregulated CD95. Collectively, these data suggest that in colon carcinogenesis a constant and severe evolutionary pressure exists, selecting for abrogation of CD95 and implicating that resistance to CD95L confers a survival advantage in this type of malignant tumor. It is, thus, highly conceivable that, apart from the CD95 counterattack theory, tumor cells profit from their CD95 downregulation, in that they become resistant to an important effector of cytotoxic immune cells. Further support of this concept comes from functional in vitro studies. Quite in contrast to normal colonocytes, colon carcinoma cells constitutively show relative or absolute resistance to CD95-mediated death.[26,27] This was initially attributed to the relatively low levels of CD95 surface expression on colon carcinoma cells. This view seemed further supported by the fact that interferon-gamma increases both CD95 surface levels and susceptibility to CD95-triggered death.[26] However, cycloheximide and actinomycin D were also shown to enhance the effectivness of the CD95 antibody.[26] Thus inhibition of protein synthesis and gene transcription, respectively, also sensitized the cells. More detailed analysis revealed that the sensitizing effect of interferon-gamma is not explained by the increase of surface CD95 levels. Blocking protein export, for example, by brefeldin A, sensitized the cells without inducing changes in surface CD95. Thus, under pathophysiologically altered conditions, even low levels of CD95 are sufficient for effective signaling.[27]

Expression of CD95L, on the other hand, was detected not only in colon carcinoma cell lines, but also in primary colorectal carcinomas[28] and their metastases.[29,30] The latter observation is of particular interest because the formation of metastases in the liver may be facilitated by driving hepatocytes, which express and are highly sensitive to CD95, into apoptosis. However, most studies on CD95L expression in colorectal carcinomas rely on immunohistochemistry carried out with polyclonal antisera. We were not able to confirm these data using monoclonal antibodies (unpublished data). In addition, it is still an open question whether colon carcinoma cells express CD95L on their surface or whether it is secreted in its soluble form. The mode of CD95L expression, however, may influence considerably its apoptosis-inducing activity.[31,32] Thus, although it seems very likely that development of CD95 resistance is an important step in the pathogenesis of colorectal cancer, with respect to their immune escape, the significance of CD95L expression in tumors has to be substantiated further.

THE TRAIL/TRAIL-RECEPTOR SYSTEM

A member of the TNF family, just recently described, is the TNF-related apoptosis-inducing ligand, TRAIL,[33] also called APO-2 ligand.[34] Similar to CD95L, TRAIL is able to induce apoptosis but can also activate the transcription factor NFκB. TRAIL exists in a membrane-bound and, after proteolytic cleavage by cystein proteases, soluble form.[35] By Northern blot analysis, TRAIL was detected in a wide range of tissues, including small intestine and colon.[33]

So far, four homologous TRAIL receptors have been described (FIG. 1): TRAIL-R1[36] and TRAIL-R2[37–39] bear a cytoplasmic death domain and mediate apoptosis. TRAIL-R3 lacks a cytoplasmic domain and is linked to the cell membrane by a phosphatidyl inositol anchor.[37,38,40] TRAIL-R4, finally, has a defective death domain and is not able to transmit an apoptotic signal either.[41,42] The latter two are therefore thought to act as decoy receptors. Similar to TRAIL, TRAIL-R1,[36] TRAIL-R2,[37–39] and TRAIL-R4[41] have been shown by Northern blot analyses to be widely expressed in normal tissues, the gut included. TRAIL-R3, instead, is expressed in a much more restricted way, detectable on the transcriptional level mainly in peripheral blood cells and the spleen.[37,40]

A possible fifth receptor for TRAIL is osteoprotegerin (OPG). Also a member of the TNFR family, its main function known so far is the inhibition of osteoclast differentiation and osteoclast-mediated bone resorption.[43] Although Emery and coworkers[44] found OPG to bind to TRAIL in vitro, this was not confirmed by others.[45] OPG lacks a cytoplamic domain and exists only in a dimeric soluble form.

Thus, if OPG binding to TRAIL actually occurs *in vivo*, it should act as a decoy receptor for TRAIL.

Although TRAIL, TRAIL-R1, TRAIL-R2, and TRAIL-R4 were shown to be expressed in the gut on the transcriptional level, their cellular distribution as well as their physiological role has not been determined. To address these open questions, we studied the expression pattern of these apoptosis-mediating molecules in the gut and determined the *in vitro* sensitivity of crypt epithelial cells to TRAIL.

Performing RT-PCR on isolated crypt cells, we revealed transcriptional expression of TRAIL as well as TRAIL-R1, TRAIL-R2, and TRAIL-R4, whereas TRAIL-R3 message was not detected (Pukrop *et al.*, submitted for publication). These findings were confirmed by immunohistochemistry. The apoptosis-inducing TRAIL-R1 showed a tissue distribution very similar to that of CD95: it was expressed in epithelial cells all along the crypt axis and in some interstitial mononuclear cells. Interestingly, TRAIL and TRAIL-R2 were coexpressed mainly in epithelial cells of the crypt mouths and the surface epithelium where senescent cells undergo apoptosis. Although this coexpression of TRAIL and TRAIL-R2 was very suggestive for a role in the physiological turnover of gut epithelial cells, isolated crypt cells turned out to be completely resistant to TRAIL *in vitro* (Pukrop *et al.*, submitted for publication). This finding, however, may not necessarily exclude its involvement in the elimination of senescent cells. If the elimination of cells at the mucosal surface is, in fact, dependent on a signal directly inducing apoptosis, sensitivity towards this signal must be tightly regulated. Senescent cells may be prepared to die at the tips of the crypts by upregulation of TRAIL and TRAIL-R2 but ultimately have to be sensitized towards TRAIL by additional factors not active *in vitro*. A knockout system may prove to be more appropriate to determine the role of TRAIL in enterocyte turnover.

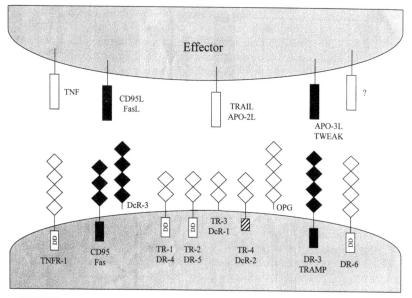

FIGURE 1. Apoptosis-inducing TNF family members (*above*) and their corresponding receptors of the TNFR family (*below*).

Restriction of TRAIL and TRAIL-R2 expression to the upper parts of the crypts and the surface epithelium may indicate another possible function of this apoptosis-inducing system in the gastrointestinal tract. The surface epithelium is the first line of defense against pathogenic microorganisms invading from the gut lumen. Interestingly, there is evidence that apoptosis-mediating TRAIL-R1 and TRAIL-R2 are upregulated in virus-infected cells, whereas interferon-gamma downregulates the expression of TRAIL receptors in uninfected cells, rendering them more resistant to TRAIL.[46] Perhaps epithelial cells at the mucosal surface, by coexpressing TRAIL and TRAIL-R2, dispose of a machinery that, upon infection, allows for their immediate elimination (together with their infectious load) by autocrine or T cell–derived TRAIL to avoid further spread of the infectious agent.

Another interesting observation on the TRAIL system is that TRAIL induces apoptosis in many cancer cell lines, whereas untransformed cells are largely TRAIL resistant.[33,34,47,48] This difference in TRAIL sensitivity has been ascribed to a selective expression of decoy receptors in untransformed cells.[38,49] Walczak and coworkers demonstrated that TRAIL induced regression of some mammary and colon cancer cell lines grown in SCID mice without causing detectable toxicity in the host.[50] This observation is in striking contrast to CD95 antibodies that lead to acute liver failure within a few hours following injection into mice.[51] Consequently, TRAIL has recently been attracting attention in cancer research as a possible anti-cancer therapeutic agent. We therefore wanted to test the TRAIL receptor expression pattern and *in vitro* sensitivity of current colon carcinoma cell lines and isolated cells from sporadic colorectal carcinomas.

Interestingly, most cell lines tested (CaCo2, HT-29, Colo205, SW480, and SW620) exhibit a very similar expression pattern on transcriptional and surface protein levels: with RT-PCR, they coexpress TRAIL and all four TRAIL receptors, whereas, on the cell surface, they bear only TRAIL-R1 and TRAIL-R2. The only exception to this rule is Colo205, which also faintly expresses TRAIL-R3 protein on its surface. Thus, although colon carcinoma cell lines express TRAIL-R3 and TRAIL-R4 on the transcriptional and protein levels (our unpublished data) in the cytoplasm, these "decoy receptors" do not appear on the cell surface.

Despite the very similar TRAIL receptor expression pattern, colon carcinoma cell lines exhibit considerable differences in TRAIL sensitivity *in vitro*, ranging from resistant to highly sensitive (Pukrop *et al.*, submitted for publication). Although Colo205 expresses TRAIL-R3 on its surface, Colo205 is effectively killed by TRAIL.

When colon carcinoma cell lines are, at least in part, sensitive to TRAIL, does this also apply to sporadic colorectal carcinomas? To test this, we isolated carcinoma cells from five patients who had had colorectal cancer resections, four of whom were immunophenotyped by FACS analysis. Interestingly, only two out of four tumors showed the same surface expression pattern as cell lines, being positive for TRAIL-R1 and TRAIL-R2 while lacking the nonapoptosis receptors. The other two tumors analyzed exhibited a complementary pattern, expressing TRAIL-R3 and TRAIL-R4 on their surface while lacking TRAIL-R1 and TRAIL-R2. All four tumors, as well as an additional fifth carcinoma that was not phenotyped, were completely resistant to TRAIL *in vitro*. These data led us to two conclusions:

(1) TRAIL sensitivity is not regulated by expression of "decoy receptors" alone. Griffith *et al.*[52] provided similar data on melanoma cell lines, showing that TRAIL

sensitivity in these cells did not depend on the TRAIL receptor status, and suggested FLIP as a critical regulator of TRAIL sensitivity.

(2) Sporadic colorectal carcinomas may be more resistant to TRAIL than cell lines. This observation may not be suprising, given the fact that TRAIL is expressed in activated T cells[53] and acts as one of the mediators of cytotoxic T-cell activity.[54] Thus, *in vivo*, carcinoma cells may be subjected to a strong selection due to TRAIL expression on tumor-infiltrating lymphocytes. Because only a limited number of carcinomas have been examined, further studies on TRAIL sensitivity of colorectal carcinomas are needed to assess the value of TRAIL as a potential anticancer therapeutic agent in colorectal cancers.

REFERENCES

1. BAKER, S.J. & E.P. REDDY. 1998. Modulation of life and death by the TNF receptor superfamily. Oncogene **17:** 3261–3270.
2. ASHKENAZI, A. & V.M. DIXIT. 1998. Death receptors: signaling and modulation. Science **281:** 1305–1308.
3. MEDEMA, J.P., C. SCAFFIDI, F.C. KISCHKEL *et al.* 1997. FLICE is activated by association with the CD95 death-inducing signaling complex (DISC). EMBO J. **16:** 2794–2804.
4. HALL, P.A., P.J. COATES, B. ANSARI & D. HOPWOOD. 1994. Regulation of cell number in the mammalian gastrointestinal tract: the importance of apoptosis. J. Cell Sci. **107:** 3569–3577.
5. STRÄTER, J., K. KORETZ, A.R. GÜNTHERT & P. MÖLLER. 1995. *In situ* detection of enterocytic apoptosis in normal colonic mucosa and in familial adenomatous polyposis. Gut **37:** 819–825.
6. LYNCH, D.H., F. RAMSDELL & M.R. ALDERSON. 1995. Fas and FasL in the homeostatic regulation of immune responses. Immunol. Today **16:** 569–574.
7. O'CONNELL, J., M.W. BENNETT, G.C. O'SULLIVAN *et al.* 1999. The F as counterattack: cancer as a site of immune privilege. Immunol. Today **20:** 46–52.
8. ITOH, N., S. YONEHARA, A. ISHII *et al.* 1991. The polypeptide encoded by the cDNA for human cell surface Fas can mediate apoptosis. Cell **66:** 233–243.
9. OEHM, A., I. BEHRMANN, W. FALK *et al.* 1992. Purification and molecular cloning of the APO-1 cell surface antigen, a member of the tumor necrosis/nerve growth factor receptor superfamily. J. Biol. Chem. **267:** 10709–10715.
10. LEITHÄUSER, F., J. DHEIN, G. MECHTERSHEIMER *et al.* 1993. Constitutive and induced expression of APO-1, a new member of the nerve growth factor/tumor necrosis factor receptor superfamily. Lab. Invest. **69:** 415–429.
11. MÖLLER, P., K. KORETZ, F. LEITHÄUSER *et al.* 1994. Expression of APO-1 (CD95), a member of the NGF/TNF receptor superfamily, in normal and neoplastic colon epithelium. Int. J. Cancer **57:** 371–377.
12. STRÄTER, J., I. WELLISCH, S. RIEDL *et al.* 1997. CD95 (APO-1/Fas)-mediated apoptosis in colon epithelial cells: a possible role in ulcerative colitis. Gastroenterology **113:** 160–167.
13. DHEIN, J., H. WALCZAK, C. BÄUMLE *et al.* 1995. Autocrine T-cell suicide mediated by APO-1/(Fas/CD95). Nature **373:** 438–440.
14. JU, S-T., D.J. PANKA, H. CUI *et al.* 1995. Fas (CD95)/FasL interactions required for programmed cell death after T-cell activation. Nature **373:** 444–448.
15. BRUNNER, T., R.J. MOGIL, D. LAFACE *et al.* 1995. Cell-autonomous Fas (CD95)/Fas-ligand interaction mediates activation-induced apoptosis in T-cell hybridomas. Nature **373:** 441–444.
16. MONTEL, A.H., M.R. BOCHAN, J.A. HOBBS *et al.* 1995. Fas involvement in cytotoxicity mediated by human NK cells. Cell. Immunol. **166:** 236–246.
17. HAHNE, M., T. RENNO & M. SCHRÖTER. 1996. Activated B cells express functional Fas ligand. Eur. J. Immunol. **26:** 721–724.

18. STRÄTER, J., S.M. MARIANI, H. WALCZAK et al. 1999. CD95 ligand (CD95L) in normal human lymphoid tissues: a subset of plasma cells are prominent producers of CD95L. Am. J. Pathol. **154:** 193–201.
19. MÖLLER, P., H. WALCZAK, S. RIEDL et al. 1996. Paneth cells express high levels of CD95 ligand transcripts. A unique property among gastrointestinal epithelia. Am. J. Pathol. **149:** 9–13.
20. BELLGRAU, D., D. GOLD, H. SELAWRY et al. 1995. A role for CD95 ligand in preventing graft rejection. Nature **377:** 630–632.
21. GRIFFITH, T.S., T. BRUNNER, S.M. FLETCHER et al. 1995. Fas ligand-induced apoptosis as a mechanism of immune privilege. Science **270:** 1189–1192.
22. GRIFFITH, T.S. & T.A. FERGUSON. 1997. The role of FasL-induced apoptosis in immune privilege. Immunol. Today **18:** 240–244.
23. O'CONNELL, J., G.C. O'SULLIVAN, J.K. COLLINS & F. SHANAHAN. 1996. The Fas counter-attack: Fas-mediated T cell killing by colon carcinoma cells expressing Fas ligand. J. Exp. Med. **184:** 1075–1082.
24. HAHNE, M., D. RIMOLDI, M. SCHRÖTER et al. 1996. Melanoma cell expression of Fas(Apo-1/CD95) ligand: implications for tumor immune escape. Science **274:** 1363–1366.
25. STRAND, S., W.J. HOFMANN, H. HUG et al. 1996. Lymphocyte apoptosis induced by CD95 (APO-1/Fas) ligand-expressing tumor cells—a mechanism of immune evasion? Nature Med. **2:** 1361–1366.
26. OWEN-SCHAUB, L.B., R. RADINSKY, E. KRUZEL et al. 1994. Anti-Fas on nonhematopoetic tumors: levels of Fas/APO-1 and bcl-2 are not predictive of biological responsiveness. Cancer Res. **54:** 1580–1586.
27. VON REYHER, U., J. STRÄTER, W. KITTSTEIN et al. 1998. Colon carcinoma cells use different mechanisms to escape CD95-mediated apoptosis. Cancer Res. **58:** 526–534.
28. O'CONNELL, J., M.W. BENNETT, G.C. O'SULLIVAN et al. 1998. Fas ligand expression in primary colon adenocarcinomas: evidence that the Fas counterattack is a prevalent mechanism of immune evasion in human colon cancer. J. Pathol. **186:** 240–246.
29. SHIRAKI, K., N. TSUJI, T. SHIODA et al. 1997. Expression of Fas ligand in liver metastases of human colonic adenocarcinomas. Proc. Natl. Acad. Sci. USA **94:** 6420–6425.
30. YOONG, K.F., S.C. AFFORD, S. RANDHAWA et al. 1999. Fas/Fas ligand interaction in human colorectal hepatic metastases. A mechanism of hepatocyte destruction to facilitate local tumor invasion. Am. J. Pathol. **154:** 693–703.
31. SUDA, T., H. HASHIMOTO, M. TANAKA et al. 1997. Membrane Fas ligand kills human peripheral blood T lymphocytes, and soluble Fas ligand blocks the killing. J. Exp. Med. **186:** 2045–2050.
32. TANAKA, M., T. ITAI, M. ADACHI & S. NAGATA. 1998. Downregulation of Fas ligand by shedding. Nature Med. **4:** 31–36.
33. WILEY, S.R., K. SCHOOLEY, P.J. SMOLAK et al. 1995. Identification and characterization of a new member of the TNF family that induces apoptosis. Immunity **3:** 673–682.
34. PITTI, R.M., S.A. MARSTERS, S. RUPPERT et al. 1996. Induction of apoptosis by APO-2 ligand, a new member of the tumor necrosis factor cytokine family. J. Biol. Chem. **271:** 12687–12690.
35. MARIANI, S.M. & P.H. KRAMMER. 1998. Differential regulation of TRAIL and CD95 ligand in transformed cells of the T and B lymphocyte lineage. Eur. J. Immunol. **28:** 973–982.
36. PAN, G., K. O'ROURKE, A.M. CHINNAIYAN et al. 1997. The receptor for the cytotoxic ligand TRAIL. Science **276:** 111–113.
37. PAN, G., J. NI, Y-F.WEI et al. 1997. An antagonist decoy receptor and a death domain-containing receptor for TRAIL. Science **277:** 815–818.
38. SHERIDAN, J.P., S.A. MARSTERS, P.M. PITTI et al. 1997. Control of TRAIL-induced apoptosis by a family of signaling and decoy receptors. Science **277:** 818–821.
39. WALCZAK, H., M.A. DEGLI-EPOSTI, R.S. JOHNSON et al. 1997. TRAIL-R2: a novel apoptosis-mediating receptor for TRAIL. EMBO J. **16:** 5386–5397.
40. DEGLI-EPOSTI, M.A., P.J. SMOLAK, H. WALCZAK et al. 1997. Cloning and characterization of TRAIL-R3, a novel member of the emerging TRAIL receptor family. J. Exp. Med. **186:** 1165–1170.

41. MARSTERS, S.A., J.P. SHERIDAN, R.M. PITTI et al. 1997. A novel receptor for APO-2L/TRAIL contains a truncated death domain. Curr. Biol. **7:** 1003–1006.
42. DEGLI-EPOSTI, M.A., W.C. DOUGALL, P.J. SMOLAK et al. 1997. The novel receptor TRAIL-R4 induces NF-kB and protects against TRAIL-mediated apoptosis, yet retains an incomplete death domain. Immunity **7:** 813–820.
43. SIMONET, W.S., D.L. LACEY, C.R. DUNSTAN et al. 1997. Osteoprotegerin: a novel secreted protein involved in the regulation of bone densitiy. Cell **89:** 309–319.
44. EMERY, J.G., P. MCDONNELL, M.B. BURKE et al. 1998. Osteoprotegerin is a receptor for the cytotoxic ligand TRAIL. J. Biol. Chem. **273:** 14363–14367.
45. LACEY, D.L., E. TIMMS, T.L. TAN et al. 1998. Osteoprotegerin ligand is a cytokine that regulates osteoclast differentiation and activation. Cell **93:** 165–176.
46. SEDGER, L.M., D.M. SHOWS, R.A. BLANTON et al. 1999. IFN-gamma mediates a novel antiviral activity through dynamic modulation of TRAIL and TRAIL receptor expression. J. Immunol. **163:** 920–926.
47. MARIANI, S.M., B. MATIBA, E.A. ARMANDOLA & P.H. KRAMMER. 1997. Interleukin 1β-converting enzyme related proteases/caspases are involved in TRAIL-induced apoptosis of myeloma and leukemia cells. J. Cell Biol. **137:** 221–229.
48. GRIFFITH, T.S., W.A. CHIN, G.C. JACKSON et al. 1998. Intracellular regulation of TRAIL-induced apoptosis in human melanoma cells. J. Immunol. **161:** 2833–2840.
49. MACFARLANE, M., M. AHMAD, S.M. SRINIVASULA et al. 1997. Identification and molecular cloning of two novel receptors for the cytotoxic ligand TRAIL. J. Biol. Chem. **272:** 25417–25420.
50. WALCZAK, H., R.E. MILLER, K. ARIAIL et al. 1999. Tumoricidal activity of tumor necrosis factor–related apoptosis-inducing ligand in vitro. Nature Med. **5:** 157–162.
51. OGASAWARA, J., R. WATANABE-FUKUNAGA, M. ADACHI et al. 1993. Lethal effect of the anti-Fas antibody in mice. Nature **364:** 806–809.
52. GRIFFITH, T.S., W.A. CHIN, G.C. JACKSON et al. 1998. Intracellular regulation of TRAIL-induced apoptosis in human melanoma cells. J. Immunol. **161:** 2833–2840.
53. JEREMIAS, I., I. HERR, T. BOEHLER & K.M. DEBATIN. 1998. TRAIL/Apo-2-ligand–induced apoptosis in human T cells. Eur. J. Immunol. **28:** 143–152.
54. KAYAGAKI, N., N. YAMAGUCHI, M. NAKAYAMA et al. 1999. Involvement of TNF-related apoptosis-inducing ligand in human CD4+ T cell–mediated cytotoxicity. J. Immunol. 162: 2639–2647.

Role of M Cells in Intestinal Barrier Function

T. KUCHARZIK,[a,b] N. LÜGERING,[a] K. RAUTENBERG,[a] A. LÜGERING,[a] M.A. SCHMIDT,[c] R. STOLL,[a] AND W. DOMSCHKE[a]

[a]Department of Medicine B and [c]Center for Molecular Biology of Inflammation, University of Münster, 48129 Münster, Germany

ABSTRACT: M cells are known as specialized epithelial cells of the follicle-associated epithelium of the gastrointestinal tract. As M cells have a high capacity for transcytosis of a wide range of microorganisms and macromolecules, they are believed to act as an antigen sampling system. The primary physiological role of M cells seems to be the rapid uptake and presentation of particular antigens and microorganisms to the immune cells of the lymphoid follicle to induce an effective immune response. In contrast to absorptive enterocytes, M cells do not exert direct defense mechanisms to antigens and pathogens in the gut lumen. Therefore, they provide functional openings of the epithelial barrier. Although M cells represent a weak point of the epithelial barrier, even under noninflamed conditions, there seems to be a balance between antigen uptake and immunological response. The low number of M cells in the gastrointestinal tract and the direct contact to immune cells in the lamina propria usually prevent the occurrence of mucosal inflammation. During chronic intestinal inflammation we observe an increase of M cell number and apoptosis selectively in M cells. M cell damage seems to be responsible for the increase of the uptake of microorganisms that is observed during intestinal inflammation. Under inflammatory conditions in the intestine, the maintenance of the epithelial barrier is broken and M cells seem to play a major role during this process.

INTRODUCTION

The gastrointestinal tract is continuously under the surveillance of the immune system to prevent invasion of antigens and macromolecules. The intestinal tract has an enormous surface, with an exchange area of more than 100 m^2 and a thin epithelium, and it is populated by various microorganisms. An intact protective barrier is therefore important. The intestinal barrier is permeable to digested nutrients as well as fluids but is in general impermeable to macromolecules, particular antigens, and most microorganisms. However, the immune system needs direct contact with antigens or pathogens to generate specific immune responses. For this purpose, the gut provides specialized epithelial cells, called M cells, that are exclusively present in the follicle-associated epithelium (FAE) and responsible for the uptake of antigens and microorganisms. M cells deliver samples of foreign material by transepithelial transport from the lumen to organized lymphoid tissues within the mucosa.[1–3] Several pathogens and macromolecules exploit the M cell transport process to cross the

[b]Address for correspondence: Torsten Kucharzik, M.D., Department of Medicine B, University of Münster, Albert-Schweitzer-Str. 33, 48129 Münster, Germany. Voice: +49-251-8347661; fax: +49-251-8347570.

tkucharz@unimuenster.de

epithelial barrier and invade the mucosa (for review see Refs. 4,5). According to these properties M cells are believed to act as an antigen sampling system for the immunological surveillance of the mucosa.[2] As M cells have the facility to transport a variety of particular antigens and microorganisms, they seem to provide functional gaps in the protective intestinal barrier. They interact closely with immune cells of Peyer's patches and therefore have a key function in the initiation of immunological response and tolerance.

During intestinal inflammation—for example, in inflammatory bowel disease— we observe a breakdown of the intestinal barrier function.[6] Disruption of the epithelial barrier function by apoptosis of epithelial cells has been discussed as a possible mechanism for the induction of mucosal inflammation.[7] Although several studies in various animal systems have focused on the transport facilities of M cells, only little is known about the role of M cells during chronic intestinal inflammation.

COMPONENTS OF THE INTESTINAL EPITHELIAL BARRIER

The mucosal gastrointestinal surface is lined by an epithelial barrier composed of enterocytes that are joined by tight junctions. The epithelium that covers the lymphoid follicle generally differs from both crypt and villus epithelium in its cellular composition and function. The cells of the gastrointestinal tract are derived from stem cells in the crypt. Epithelial cells of each villus are derived from several surrounding crypts. In contrast to the villus epithelia of the small and large intestine that are responsible for absorption of digested nutrients and resorption of fluids, the epithelium overlying mucosal lymphoid follicles, the so-called follicle-associated epithelium (FAE), is designed for the uptake of macromolecules, particular antigens, and microorganisms. The follicle-associated crypts differ from villus crypts as they contain two distinct axes of differentiation and migration. Cells that migrate onto the villi differentiate into enterocytes, goblet cells, and enteroendocrine cells. In contrast, cells that migrate onto the dome differentiate into distinct follicle-associated enterocytes and M cells.[8] M cells are a specialized epithelial cell type that occurs only over mucosal lymphoid follicles (FIG. 1). They provide functional openings in the epithelial barrier through vesicular transport activity.

MECHANISMS OF TRANSEPITHELIAL TRANSPORT

The uptake of microorganisms and macromolecules is usually prevented by epithelial cells that are sealed by tight junctions and form the intestinal barrier of the gastrointestinal tract. This structure is effective in excluding peptides and macromolecules with antigenic potential. The uptake of macromolecules, particular antigens, and microorganisms across the epithelial barrier occurs almost entirely through active transepithelial vesicular transport. The majority of the cells of the intestinal barrier are absorptive enterocytes that are responsible for the absorption of digested nutrients and fluids. Epithelial cells are coated with closely packed microvilli that are covered by the glycocalix with a thick layer of glycoproteins, which constitute a very effective diffusion barrier.[9] Most macromolecules and microorganisms are prevented from contact with epithelial cells by these structures as well as by local se-

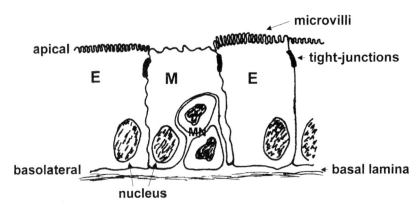

FIGURE 1. Diagram of an M cell with neighboring enterocytes. The M cell exhibits microfolds instead of microvilli, a thin cytoplasm and an intraepithelial pocket that contains T and B ymphocytes and macrophages. Enterocyte (E), M cell (M), mononuclear cells (MN).

cretion of mucin and secretory IgA antibodies.[10] The uptake and transport of macromolecules and microganisms through epithelial cells is possible only by using active vesicular transport.[11]

The epithelium that overlies the lymphoid follicles in the gut, the FAE, contains enterocyte-like cells and M cells that are responsible for the transport of microorganisms and antigens. The transport of particles and microorganisms into the mucosal lymphoid follicles seems to be an important first step in the induction of a secretory immune response.

Transport across epithelial cells requires formation and fusion of endosomes and recycling of membrane vesicles. For these mechanisms G proteins and the polarized cytoskeleton of intestinal epithelial cells are necessary.[12] In M cells, rapid vesicular transport and release of antigens in the M cell pocket is the major means of transport. Although the molecular mechanisms of the transport have not yet been characterized, membrane traffic by M cells depends on the polarized organization and signaling networks typical of polarized epithelial cells.[13]

MORPHOLOGICAL FEATURES OF M CELLS

Intestinal M cells differ from adjacent enterocytes by the absence of a typical brush border and the presence of variable microvilli or microfolds (FIG. 1). The basolateral surface of M cells includes a lateral subdomain that is involved in cell-cell adhesion, a basal subdomain that interacts with extracellular matrix and basal lamina, and, in contrast to epithelial cells, a deeply invaginated, intraepithelial pocket containing a variety of mononuclear cells.[13]

Several studies have been performed to study the cytoskeleton of M cells in different species. Vimentin, a marker commonly detected in mesenchymal cells, is coexpressed with cytokeratins in rabbit M cells. It could be demonstrated that intermediate filaments from rat M cells differ from absorptive enterocytes by strong staining for a specific monoclonal antibody clone against cytokeratin-8.[14] Cytokeratin 18, a cytoskeletal component of intestinal epithelial cells, is strongly expressed in porcine M cells.[15] In humans, we could demonstrate that the composition of the intermediate filaments in M cells is not different from that in enterocytes.[16] The M cell apical membranes show patterns that distinguish them from adjacent enterocyes. The molecular composition of the M cell membranes has not yet been determined, as it is very difficult to obtain a sufficient number of pure M cell membranes. However, we know that alkaline phosphatase, which is a positive marker of the brush border of enterocytes, is reduced or even absent in M cells.[17] Villin, a brush border protein of the actin cytoskeleton, shows an unusual distribution pattern in the cytosol of M cells from Balb/c mice.[18] This suggests that actin filaments are involved and reorganized when large, particulate antigens are phagocytosed by M cells. M cells lack the uniform thick glycocalyx that is seen on enterocytes, although their apical membranes display abundant glycoconjugates in a cell coat that varies widely in thickness and density.[8] In Peyer's patches of BALB/c-mice, lectins that recognize different carbohydrate structures containing $\alpha(1-2)$ fucose selectively stained M cells in the FAE.[19] In individual M cells within a single FAE, there are variations of the glycosylation patterns.[20] Glycoconjugates expressed on M cells in different intestinal regions have also been shown to be distinct, which could explain the regional specificity of certain pathogens that exploit M cells to invade the body.

M CELLS AS A WEAK POINT OF THE INTESTINAL EPITHELIAL BARRIER—TRANSPORT OF PATHOGENS

M cells have been shown to represent gaps in the intestinal barrier, since they are able to transport several antigens and microorganisms. Transport facilities of a variety of antigens always carry the risk that pathogens may use the M cell pathway to invade the body. In this context, the ability of invasive pathogens to undergo M cell transport must be considered as a virulence factor. Although M cell transport induces an immediate mucosal immune response that will subsequently limit mucosal disease, this may not always prevent intestinal inflammation. In different animal models, several pathogens have been found to use M cell transport to invade the mucosa (TABLE 1) (for review see Refs. 4,5).

Pathogenic microorganisms have developed diverse strategies for invading the mucosa, including selective adherence on M cell surfaces. It has been shown for several microorganims that M cell transport can induce a rapid infection of the mucosa and systemic spread. The interaction of M cells with different pathogens varies. Simple transit or destruction of the FAE can be observed. The interaction of bacteria with M cells includes recognition, processing of M cell surface molecules, activation of intracellular signaling pathways, and reorganization of the apical membrane cytoskeletal proteins. Viruses are unable to alter the M cell surface but can be processed by proteases in the gut lumen to an M cell–adherent form.[21] The selective adherence and transcytosis of several pathogens on M cells is not fully understood. Distinctive surface glycoconjugates that may be recognized by microorganisms, the apparent lack of a rigid brush border and a thick glycocalix on M cells, or the particular nature of the microorganisms have all been discussed as causes for M cell specificity. Re-

TABLE 1. Antigens and microorganisms that are transported by M cells

Macromolecules	Viruses	Bacterials	Other Micro-organisms
Ferritin	Reovirus	Yersinia	Cryptosporidia
HRP	Poliovirus	Mycobacteria	
RCA I + II	HIV	Vibrio cholerae	
WGA	Coxsackievirus	Campylobacter jejuni	
Latex particles	Astrovirus	Shigella flexneri	
Cholera toxin		*E. coli* (RDEC-1)	
		Streptococcus pyogenes	
		Salmonella	
		Chlamydia	

cently, studies with a range of cholera toxin B subunit conjugates have demonstrated that the glycocalyx of enterocytes acts as a size-selective barrier that limits access of particles, over a broad size range, to apical membrane glycolipids of rabbit enterocytes. The reduced glycocalix of M cells permits adherence of the smaller of these particles to the same membrane components.[22]

M CELLS DURING INTESTINAL INFLAMMATION

Although there are numerous studies on M cells in various animal systems, little is known about the role of M cells during intestinal inflammation. One group recently reported a strong increase of M cells in spondylarthropathy-associated ileal inflammation in humans.[23] The quantification of M cells in this study has been done by light microscopy, although the visualization of M cells is especially difficult as there are no markers known for human M cells. Another study reported a loss of M cells from the epithelium of colonic lymphoid follicles in Crohn's disease.[24] Thus, although M cells represent the "gateway" for mucosal immune responses, our knowledge regarding the fate of M cells during intestinal inflammation is rather rudimentary. Disruption of the epithelial barrier function by apoptosis of enterocytes has been discussed as a possible mechanism for the induction of mucosal inflammation.[7] During this process, the pathophysiological role of M cells has not yet been further evaluated. From recent studies, there is increasing evidence that M cells may be induced under various immunological conditions. This view is supported by the fact that SCID mice, which lack mucosal follicles with M cells, develop FAE and M cells after reconstitution with Peyer's patch lymphoid tissue from normal mice.[25] Lymphoid follicles and M cells of specific pathogen–free (SFP) mice increase in number after transfer to a normal animal house environment.[26] Kernéis *et al.* could demonstrate that coculture of Peyer's patch lymphocytes with Caco-2 cells induced M cell–like cells with reorganization of the brush border and transport facilities for microorganisms, suggesting that soluble mediators are responsible for conversion of enterocytes into M cells.[27]

TABLE 2A. Increase of M cells during indomethacin-induced ileitis in rats

	n	Surface of FAE (μm)	Total Number of M Cells	Percent of M Cells in Relation to FAE
Noninflamed	6	2.2×10^6	1.3×10^4	4 %
Inflamed	8	2.8×10^6	4.7×10^4	11 %*

TABLE 2B. Dependence of number of M cells on macroscopic degree of inflammation

Degree of Inflammation	0 Noninflamed Tissue	1 Hyperemia or Petechial Bleeding	2 Single Erosion/ Ulceration	3 Single Ulceration/ Adhesion	4 Multiple Ulcerations/ Necrosis
Percent of CK-8$^+$ M cells	4%	7%**	10%*	12%*	1%

NOTE TO TABLES 2A AND 2B: Six rats without intestinal inflammation and eight rats with chronic ileitis were examined. Ileitis in rats was induced by two injections of indomethacin given 24 hours apart. M cells could be visualized with a fluorescence microscope by staining with a mAb (4.1.18) against cytokeratin-8 (CK-8), which is known as a specific M cell marker in rats.[14] The degree of inflammation was determined by macroscopic staging. Tissue was selectively analyzed with regard to the degree of inflammation in different segments of the small bowel. There was a significant increase of M cell number during intestinal inflammation (**A**). Moreover, we observed an increase of M cells dependent on the degree of inflammation (**B**). In tissue with multiple ulcerations and necrosis, there was no CK-8 staining of M cells, probably due to cytokeratin degradation in apoptotic M cells.
*$p < 0.001$; **$p < 0.05$.

To study the role of M cells in the intestinal barrier during intestinal inflammation, we generated chronic ileitis in rats by subcutaneous injections of indomethacin. The ileitis induced by indomethacin is characterized by linear ulcerations, thickening of the small intestinal wall, adhesions, partial obstructions, transmural granulomatous inflammation, and crypt abscesses.[28] These lesions share clinical, histological, and pathophysiological characteristics with Crohn's enteritis.[29,30] The mechanisms underlying the intestinal inflammation induced by indomethacin are not totally understood and remain controversial. One mechanism may be the inhibition of cyclooxygenase, which leads to depletion of endogenous, protective prostaglandins.[31] It has also been suggested that biliary secretion, food intake, luminal bacterial and host genetic susceptibility play important roles in indomethacin-induced small bowel inflammation.[32,33] The M cell density of 4% of FAE in non-inflamed tissue as encountered in our study is comparable to results of other studies reporting about 5% M cells in rodents.[34] In inflamed tissue sections we observed a threefold increase of M cell number, which could be demonstrated by immunohistochemistry and SEM studies. Increase of M cell number strongly depended on the degree of inflammation (TABLE 2; FIGS. 2,3). The strong induction of M cells in inflamed FAE due to indomethacin-induced ileitis in rats demonstrates that M cells are involved in gut inflammation. Electron microscopic studies and the results obtained using the nick end labeling technique revealed selective apoptosis of M cells during intestinal inflammation. In contrast, absorptive enterocytes were usually not affected (FIG. 4).

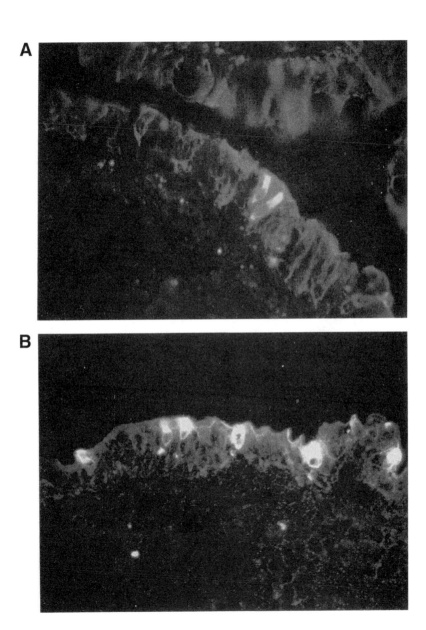

FIGURE 2. Induction of M cells during chronic intestinal inflammation. M cells in the FAE of rats with a normal appearing intestine (**A**) and rats with ileitis (**B**). M cells could be visualized by staining with a mAb against CK-8 (clone 4.1.1.8) (light fluorescence). There is a strong increase of M cell number during intestinal inflammation (**B**) (× 400).

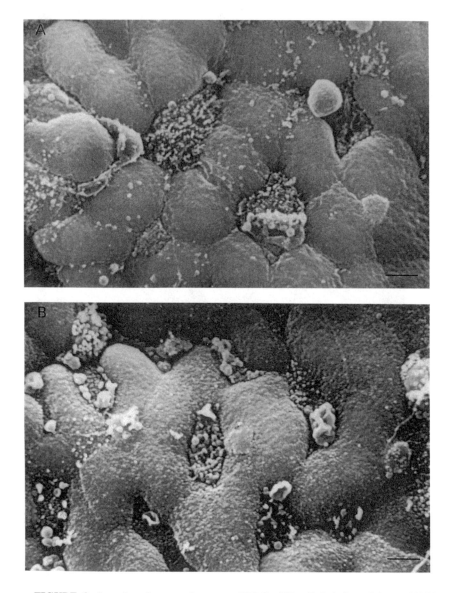

FIGURE 3. Scanning electron microscopy (SEM) of M cells in inflamed tissue. **(A)** M cells in noninflamed ileal mucosa of rats. **(B)** M cells in inflamed mucosa. M cells are characterized by fewer and shorter microvilli. There is an increase in the number of M cells during chronic intestinal ileitis **(B)**. (*Bars* = 2 μm).

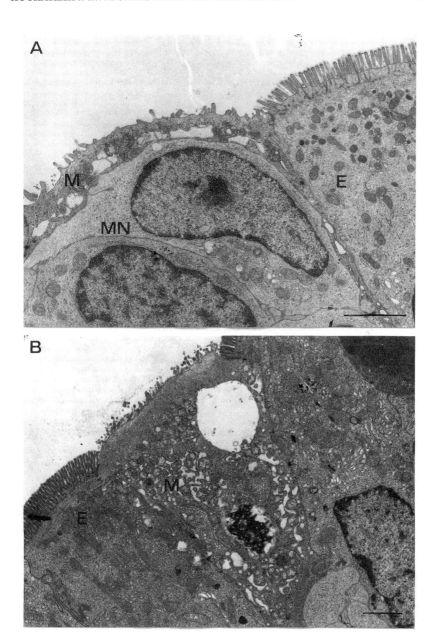

FIGURE 4. Apoptosis of M cells during intestinal inflammation. **(A)** Typical M cell (M) in Figure 4A with shorter and fewer microvilli in the brush border compared to the adjacent enterocyte (E) and an intraepithelial pocket containing mononuclear cells (MN). The apical cytoplasm is extremely thin and forms large vacuoles, whereas mitochondrias are swollen. These signs are typical of apoptosis. **(B)** shows an apoptotic M cell (M) with large vacuoles and chromatin condensation in the nucleus. Neighboring enterocytes (E) were usually not affected and appear normal. (*Bars* = 2 μm).

It is therefore conceivable that bacterial invasion of follicle-associated epithelium results in epithelial cell signal transduction events that stimulate underlying mono-nuclear cells. Under these conditions enterocytes of the FAE may convert into M cells. However, whether the induction of M cells is a direct response to luminal bacteria or secondarily induced by stimulation of lamina propria mononuclear cells remains unclear. We found a strong increase of CD4[+] mononuclear cells in the M cell pocket during ileitis, whereas other markers for T and B lymphocytes as well as monocytes and neutrophils did not show any increased expression compared to controls (data not shown). Conversion of enterocytes into M cells is a matter of controversy. With respect to the differentiation pathway of M cells, there are essentially two hypotheses. One group suggests that M cells have the same origin as other differentiated intestinal epithelial cells and may derive directly from potentially specific stem cells of surrounding crypts.[35,36] Others postulate that M cells derive from absorptive cells overlying the domes of the FAE, probably induced by the immunological environment or by bacteria.[37–39] The last hypothesis has recently been supported by Kérneis et al., who demonstrated that Caco-2 cells were able to convert into M cell–like cells following coculture with Peyer's patch cells.[27]

By electron microscopic studies and by employing the nick end labeling (TUNEL) technique, we demonstrated that in chronic ileitis stimulated M cells undergo apoptosis. Interestingly, the neighboring enterocytes were not affected by this process and usually appeared well preserved and normal.[40] During advanced inflammation, in addition to apoptotic cells, in some M cells we could also demonstrate morphological changes, although admittedly rare, that are typical of necrotic cells. Even in tissue segments with necrotic M cells, enterocytes usually appeared intact. In further SEM studies we observed that M cells during intestinal inflammation revealed several signs of cell damage, including large holes with various mononuclear cells migrating from the underlying mucosa into the gut lumen. In addition, several microorganisms used the M cell pathway to invade the mucosa (data not shown). The pathophysiological meaning underlying M cell induction followed by apoptosis remains unclear. It can only be speculated that increase of M cell number during inflammation should allow antigens and microorganisms to get in closer contact to the mucosal immune system to induce a protective immune response. Apoptosis of M cells may be an attempt to close the gaps between epithelial cells when the inflammation gets out of control. This "repair mechanism" is apparently not efficient enough to prevent further intensification of inflammation, as antigens and macromolecules are now able to invade the mucosa even more readily resulting in a consecutive complete breakdown of the epithelial barrier function (FIG. 5). Inappropriate regulation of M cell apoptosis may be the reason for uncontrolled inflammation. Tissue necrosis seems to be the last step in the cascade of uncontrolled intestinal inflammation. As aphthoid ulcers have been shown to occur as early changes in Crohn's disease[42] and probably originate from FAE,[41] it can be speculated that these mechanisms observed in our study may play some role in the pathogenesis of inflammatory bowl disease (IBD). Apoptosis of epithelial cells during IBD has been investigated in recent studies.[7,43] It has been shown that apoptosis in epithelial cells is increased during chronic intestinal inflammation and located predominantly in the crypts.[43] In experimental ileitis in rats M cells seem to be the first cells in the inflamed FAE that show signs of apoptosis.

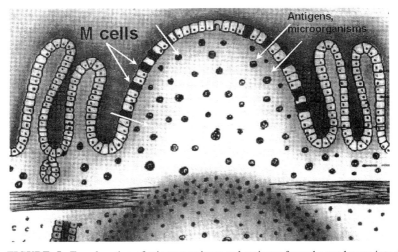

FIGURE 5. Translocation of microorganisms and antigens from the gut lumen into the mucosa during chronic ileitis in rats. During chronic intestinal inflammation in rats there is (1) an increase of M cell number, and (2) M cells undergo apoptosis. Mononuclear cells migrate from the mucosa into the gut lumen by passing damaged M cells. The diagram illustrates that there is a breakdown of the intestinal barrier during ileitis, with translocation of various microorganisms from the lumen into the mucosa by using the M cell pathway (*arrows*).

M cells, therefore, seem to be—at least in this animal model—an important cell population that is responsible for the breakdown of the intestinal epithelial barrier during inflammation.

REFERENCES

1. OWEN, R.L. & A.L. JONES. 1974. Epithelial cell specialization within human Peyer's patches: an ultrastructural study of intestinal lymphoid follicles. Gastroenterology **66:** 189–203.
2. WOLF, J.L. & W.A. BYE. 1984. The membraneous epithelial (M) cell and the mucosal immune system. Annu. Rev. Med. **35:** 95–112.
3. NEUTRA, M.R., E. PRINGAULT & J.P. KRAEHENBUHL. 1996. Antigen sampling across epithelial barriers and induction of mucosal immune responses. Annu. Rev. Immunol. **14:** 275–300.
4. NEUTRA, M.R., P.J. GIANNASCA, K.T. GIANNASCA *et al.* 1995. M cells and microbial pathogens. *In* Infections of the Gastrointestinal Tract. M.J. Blaser, P.D. Smith, J.I. Ravdin, J.B. Greenberg & R.L. Guerrant, Eds.: 163–178. Raven Press. New York.
5. TRIER, J.S. 1991. Structure and function of intestinal M cells. Mucosal immunology I: basic principles. Gastroenterol. Clin. North Am. **20** (3)**:** 531–547.
6. SCHMITZ, H., C. BARMEYER, M. FROMM *et al.* 1999. Altered tight junction contributes to the impaired epithelial barrier function in ulcerative colitis. Gastroenterology **116:** 301–309.
7. STRÄTER, J., I. WELLISCH, S. RIEDL, *et al.* 1997. CD95 (APO-1/Fas)-mediated apoptosis in colon epithelial cells: a possible role in ulcerative colitis. Gastroenterology **113:** 160–167.

8. BYE, W.A., C.H. ALAN & J.S. TRIER. 1984. Structure, distribution and origin of M cells in Peyer's patches of mouse ileum. Gastroenterology **86:** 789–801.
9. MAURY, J., C. NICOLETTI, I. GUZZO-CHAMBRAUD & S. MAROUX. 1995. The filamentous brush border glycocalyx, a mucin-like marker of enterocyte hyper-polarization. Eur. J. Biochem. **228:** 323–331.
10. KRAEHENBUHL, J.P. & M.R. NEUTRA. 1992. Molecular and cellular basis of immune protection of mucosal surfaces. Physiol Rev. **72:** 853–879.
11. ITO, S. 1974. Form and function of the glycocalyx on free cell surfaces. Philos. Trans. R. Soc. Lond. **268:** 55–66.
12. PIMPLICAR, S.W. & K. SIMONS. 1993. Role of hetrodimeric G protein in polarized membrane transport. J. Cell Sci. Suppl. **17:** 27–32.
13. NEUTRA, M.R. 1999. M cells in antigen sampling in mucosal tissues. Curr. Topics Microbiol. Immunol. **236:** 17–32.
14. RAUTENBERG, K., C. CICHON, G. HEYER et al. 1996. Immunocytochemical characterization of the follicle-associated epithelium of Peyer's patches: anti-cytokeratin 8 antibody (clone 4.1.18) as a molecular marker for rat M cells. Eur. J. Cell Biol. **71:** 363–370.
15. GEBERT, A., H.J. ROTHKÖTTER & R. PABST. 1994. Cytokeratin 18 is an M cell marker in porcine Peyer's patches. Cell Tissue Res. **276:** 213–221.
16. KUCHARZIK, T., N. LÜGERING, K.W. SCHMID et al. 1998. Human intestinal M cells exhibit enterocyte-like intermediate filaments. Gut **42:** 54–62.
17. OWEN, R.L. & D.K. BHALLA. 1983. Cytochemical analysis of alkaline phosphatase and esterase activities and of lectin-binding and anionic sites in rat and mouse Peyer's patch M cells. Am. J. Anat. **168:** 199–212.
18. KERNEIS, S., A. sBOGDANOVA, E. COLUCCI-GUYON et al. 1996. Cytosolic distribution of villin in M cells from mouse Peyer's patches with the absence of a brush border. Gastroenterology **110:** 515–521.
19. GIANNASCA, P.J., K.T. GIANNASCA, P. FALK et al. 1994. Regional differences in glycoconjugates of intestinal M cells in mice: potential targets for mucosal vaccines. J. Physiol. **267:** G1108–1121.
20. GEBERT, A. & G. HACH. 1993. Differential binding of lectins to M cells and enterocytes in the rabbit cecum. Gastroenterology **105:** 1350–1361.
21. AMERONGEN, H.M., G.A.R. WILSON, B.N. FIELDS & M.R. NEUTRA. 1994. Proteolytic processing of reovirus is required for adherence to intestinal M cells. J. Virol. **68:** 8428–8432.
22. FREY, A., W.I. LENCER, R. WELTZIN et al. 1996. Role of the glycocalix in regulating access of microparticles to apical plasma membranes of intestinal epithelial cells: implications for microbial attachment and oral vaccine targeting. J. Exp. Med. **184:** 1045–1060.
23. CUVELIER, C.A., J. QUATACKER, H. MIELANTS et al. 1994. M cells are damaged and increased in number in inflamed human ileal mucosa. Histopathology **24:** 417-426.
24. FUJIMURA, Y., M. HOSOBE & T. KIHARA. 1992. Ultrastructural study of M cells from colonic lymphoid nodules obtained by colonic biopsy. Dig. Dis. Sci. **37:** 1089–1098.
25. SAVIDGE, T.C. & M.W. SMITH. 1995. Evidence that membraneous (M) cell genesis is immunoregulated. In Advances in Mucosal Immunology. J. Mestecky, Ed.: 239–241.
26. SMITH, M.W., P.S. JAMES & D.R. TIVEY. 1987. M cell number increase after transfer of SPF mice to a normal animal house environment. Am. J. Pathol. **128:** 385–389.
27. KERNÉIS, S., A. BODGDANOVA, J.P. KRAEHENBUHL et al. 1997. Conversion by Peyer's patch lymphocytes of human enterocytes into M cells that transport bacteria. Science **277:** 949–952.
28. SARTOR, R.B. 1991. Animal models of intestinal inflammation: relevance to inflammatory bowel disease. In Inflammatory Bowel Disease. R.P. MacDermott & W.F. Stenson, Eds.: 337–353. Elsevier. New York.
29. BECK, W.S., H.T. SCHNEIDER, K. DIETZEL et al. 1990. Gastrointestinal ulcerations induced by anti-inflammatory drugs in rats. Arch. Toxicol. **64:** 210–217.
30. YAMADA, T., E. DEITCH, R.D. SPECIAN et al. 1993. Mechanisms of acute and chronic intestinal inflammation induced by indomethacin. Inflammation **6:** 641–662.
31. ROBERT, A. & T. ASANO. 1977. Resistance of germfree rats to indomethacin-induced intestinal lesions. Prostaglandins **14:** 333–341.

32. SATOH, H., P. GUTH & M.I. GROSSMAN. 1983. Role of bacteria in gastric ulceration produced by indomethacin in the rat: cytoprotective action of antibiotics. Gastroenterology **84:** 483–489.
33. KENT, T.H., R.M. CARDELLI & F.W. STAMLER. 1969. Small intestinal ulcers and intestinal flora in rats given indomethacin. Am. J. Pathol. **54:** 237–245.
34. SMITH, M.W. & M.A. PEACOCK. 1980. M cell distribution in follicle-associated epithelium of mouse Peyer's patch. Am. J. Anat. **159:** 167–175.
35. BYE, W.A., C.H. ALLAN & J.S. TRIER. 1984. Structure, distribution, and origin of M cells in Peyer's patches of mouse ileum. Gastroenterology **86:** 789–801.
36. SICINSKY, P., J. ROWINSKY, J.B. WARCHOL *et al.* 1986. Morphometric evidence against lymphocyte-induced differentiation of M cells from absorptive cells in mouse Peyer's patches. Gastroenterology **90:** 609–616.
37. SAVIDGE, T.C., M.W. SMITH, P.S. JAMES *et al.* 1991. Salmonella-induced M cell formation in germ-free mouse Peyer's patch tissue. Am. J. Pathol. **139:** 177–184.
38. SMITH, M.W. & M.A. PEACOCK. 1992. Microvillus growth and M cell formation in mouse Peyer's patch follicle-associated epithelial tissue. Exp. Physiol. **77:** 389–392.
39. BHALLA, D.K. & R.L. OWEN. 1982. Cell renewal and migration in lymphoid follicles of Peyer's patches and cecum—an autoradiographic study in mice. Gastroenterology **82:** 232–242.
40. KUCHARZIK, T., A. LÜGERING, N. LÜGERING *et al.* 2000. Characterization of M cell development during indomethacin-induced ileitis in rats. Aliment. Pharmacol. Ther. **14:** 247–256.
41. LOCKHART MUMMERY, H.E. & B.C. MORSON. 1960. Crohn's disease (regional enteritis) of the large intestine and its distinction from ulcerative colitis. Gut **1:** 87–105.
42. FUJIMARA, Y., R. KAMOI & M. IIDA. 1996. Pathogenesis of aphthoid ulcers in Crohn's disease: correlative findings by magnifying colonoscopy, electron microscopy, and immunohistochemistry. Gut **38:** 724–732.
43. IWAMOTO, M., T. KOJ, K. MAKIYAMA *et al.* 1996. Apoptosis of crypt epithelial cells in ulcerative colitis. J. Pathol. **180:** 152–159.

Nutrient Transporter Function Studied in Heterologous Expression Systems

HANNELORE DANIEL[a]

Institute of Nutritional Sciences, Technical University of Munich,
85350 Freising-Weihenstephan, Germany

ABSTRACT: Although a large number of plasma cell nutrient transport proteins has been cloned in the last couple of years, much remains to be learned about their structure–function relationships, membrane topology, posttranslational regulation, and bioenergetics of transport. Major progress in the study of the human and animal transporters has come from heterologous expression systems, which offer the benefits of ease of genetic selection and manipulation, short generation time of the organisms in which transporters are expressed, and comparatively high levels of expression of the recombinant proteins. Because our main focus is mammalian peptide transporters, the intestinal peptide transporter, PEPT1, and its renal counterpart, PEPT2, will serve here as models for the analysis of their structure and function when they are heterologously expressed in different cell systems.

PHYSIOLOGY OF MAMMALIAN PEPTIDE TRANSPORTERS

Dipeptides, tripeptides, as well as a large number of peptide-like drugs are rapidly taken up into intestinal and renal epithelial cells by specific apical peptide transporters encoded by the *Pept1* and *Pept2* genes. The cDNAs of intestinal and renal transporters of different species have been cloned from cDNA libraries.[1-6] The PEPT1 and PEPT2 proteins show significant homologies to peptide transporters isolated from bacteria, fungi, and plants that all belong to the PTR family of proton-dependent peptide transporters.[7] Although at the amino acid level, the sequence identity of PEPT1 and PEPT2 is 47%, there are major differences in the functional phenotype, including the pH dependency, substrate affinity and specificity, as well as voltage dependence of transport function.[8-11]

The physiological role of PEPT1 lies in the absorption of peptide-bound amino acids from the intestinal tract after peptide release by enzymatic breakdown of dietary or endogenous proteins. In addition, the high availability of orally active peptide-based drugs, such as almost all aminocephalosporin antibiotics, ACE inhibitors like captopril, or peptidase inhibitors like bestatin, results from their active transport mediated by PEPT1.[1-3,12-14] The PEPT2 transporters contribute to renal amino acid homeostasis and the efficiency of conservation of amino acid nitrogen by reabsorption of peptides from the tubular fluids.[11,15-18] PEPT2-mediated transport, like that of PEPT1, occurs also by flux coupling to movement of protons down an electro-

[a]Address for correspondence: Hannelore Daniel, Institute of Nutritional Sciences, Technical University of Munich, Hochfeldweg 2, 85350 Freising-Weihenstephan, Germany. Voice: +49-8161-713400; fax: +49-8161-713999.
daniel@weihenstephan.de

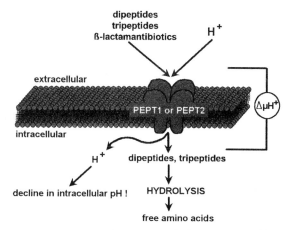

FIGURE 1. Peptide transporter function in apical membranes of intestinal and renal epithelial cells.

chemical proton gradient, causing the peptide transporter to operate as an acid loader in epithelial cells.[19] In addition to transport of di- and tripeptides from the tubular lumen into the cell, the transporters play a role in the renal handling of xenobiotics that possess a peptide backbone.[17] We recently showed that PEPT1 and PEPT2 are also involved in transmembrane transport of delta-aminolevulinic acid, which serves as a precursor in phorphyrin synthesis, suggesting that the peptide transporters may also have a role in interorgan metabolism of metabolites that are not structurally similar to peptides.[20]

By using isolated brush border membrane vesicles (BBMV) it was demonstrated almost two decades ago that in the apical membrane of intestinal and renal tubular cells, specific electrogenic peptide transporters are present. Functional analysis of peptide uptake into renal and intestinal BBMV established that peptide transport is Na^+ independent but energized by an inwardly directed transmembrane proton gradient and dependent on the inward negative membrane potential.[10,11] So far the epithelial peptide transporters remain the only well-characterized and cloned systems in mammals that have this unique capability of electrogenic uphill transport of organic solutes energized by a proton motive force. Another unique feature of the peptide transporters is the sequence-independent transport of thousands of different dipeptides, tripeptides, and peptidomimetics. FIGURE 1 summarizes the function of peptide transporters.

EXPRESSION OF PEPT1 AND PEPT2 IN *XENOPUS LAEVIS* OOCYTES FOR THE ELECTROPHYSIOLOGICAL ANALYSIS OF TRANSPORTER FUNCTION

The *Xenopus* oocyte is a robust and convenient system for the transient expression of many different animal and plant proteins and have been demonstrated to translate, process, and target most proteins properly. This expression system can also be used to clone genes, characterize protein function, and study posttranslational processing of proteins. We and others employed *Xenopus* oocytes for the analysis of PEPT1 and PEPT2 transporter function by conventional flux studies using radiolabeled peptides and by the two-electrode voltage clamp technique.[1,8,9,21–23] The latter allows a detailed characterization of inward currents generated by peptide/proton cotransport as a function of substrate and proton concentrations as well as the voltage dependence of currents.

The advantage of the oocyte expression system is that the cell is very large, with a diameter of 1 to 1.5 mm, enabling injection of the cRNA into the cell directly in defined and even large amounts. Another advantage of cell size is that multiple electrodes (potential, current, pH) can be impaled simultaneously for measuring membrane voltage, currents, and intracellular pH. The expression level may vary depending on the quality of the oocytes and the cRNA injected, but transport rates can exceed those in control oocytes up to 50-fold.

Injection of the peptide transporter's corresponding cRNAs into oocytes induced transport activities that resembled the characteristics of peptide transport obtained in intestinal and renal membrane vesicles and tissue preparations in all aspects. Both transporters are capable of translocating di- and tripeptides and aminocephalosporin antibiotics. Their activity depends on the membrane potential, and transport is rheogenic as a consequence of peptide flux coupled to proton cotransport that occurs with a 1:1 stoichiometry independent of the substrate's net charge.[8,9] However, there are distinct differences between the two transporter forms with respect not only to substrate affinity but also to their dependence on external pH and membrane potential.[8,9]

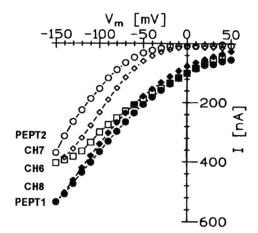

FIGURE 2. I-V relationships in oocytes expressing the different transporters and chimeras. Steady state current voltage (I-V) relationships were measured by the two-electrode voltage clamp in oocytes superfused at buffer pH 6.0 (2.5 mM Gly-Gln), and membrane potential was held at −60 mV. Then membrane potential was stepped symmetrically to the test potentials shown, and substrate-dependent currents were obtained as the difference measured in the absence and the presence of 2.5 mM of Gly-Gln.

To be able to relate transport function of PEPT1 and PEPT2 to certain protein domains, we studied a variety of chimeric transporters in which small stretches of PEPT2 were incorporated into the PEPT1 backbone replacing the homologous regions in the wild-type PEPT1. All regions replaced were derived from the aminoterminal region of PEPT2 that had been shown previously to be important for the PEPT2 phenotype.[24,25] Whereas most of the typical characteristics of PEPT1, such as substrate affinity and substrate specificity, remained unchanged in chimeras designated CH6, CH7, and CH8, significant changes in the voltage dependence of inward currents where obtained for CH7. FIGURE 2 shows the I-V relationships of the wild-type and chimeric transporters registered under identical experimental conditions, that is, pH_{out} 6.0, 2.5 mM glycyl-glutamine (Gly-Gln) as substrate. As demonstrated in FIGURE 2, all transporters induced significant inward currents when the dipeptide-containing solution was perfused. Substrate-evoked inward currents increased with increasing inside negative membrane potential in oocytes expressing PEPT1 and the chimeras CH 6 and CH8. By contrast, currents generated by PEPT2 and CH7 showed a significant response to membrane hyperpolarization only at membrane potentials > −60 mV. Because the chimera CH7 contained only a 20-residue-long stretch of PEPT2, replacing the corresponding region in PEPT1, its pronounced change in the I-V response towards a PEPT2 phenotype suggests that this protein region plays a crucial role in the voltage dependence of the transporter protein. We then deleted a 9–amino acid stretch within the replaced region to obtain CH8, which changed, as shown in FIGURE 2, the PEPT2 phenotype in the I-V relationship back to a PEPT1 type. This strongly suggests that we hit an important protein domain that somehow determines the characteristic voltage dependence difference in transport function of PEPT1 and PEPT2.

EXPRESSION OF PEPT1 IN THE YEAST *PICHIA PASTORIS* FOR DETERMINING THE BASIS OF THE SUBSTRATE SPECIFICITY

Preliminary studies on the substrate specificity of PEPT1 and PEPT2 indicated that the transporters—like their nonmammalian counterparts—transport almost all possible dipeptides, tripeptides, and numerous peptidomimetics. Free amino acids and tetrapeptides appeared not to be accepted as substrates. Although transport occurs generally in a stereospecific manner,[2,3,13] the transporters discriminate possible substrates only by differences in affinity for binding and/or maximal transport capacity. Based on the naturally occurring amino acids provided either as L- or D-amino enantiomers and by the huge number of different peptide-like xenobiotics, almost 1×10^6 potential substrates can be identified. Considering the wide distribution of this novel class of solute transporters throughout nature and their nutritional as well as pharmacological importance, identification of the minimal structural determinants of substrates affecting their affinity and capability for transport would greatly advance our understanding of these carrier proteins. This, however, requires a test system that allows high throughput screening for potential substrates and for the determination of their affinity for interaction with the substrate binding sites.

For this purpose we generated a series of transgenic yeast (*Pichia pastoris*) strains expressing the various mammalian peptide transporters. Construction of *Pichia* strains expressing either the rabbit PEPT1 or rabbit PEPT2 proteins have been described previously.[26,27] In the transgenic yeast strain GS-PEPT1, the PEPT1-cDNA was placed under the transcriptional control of the AOX1 (alcohol oxidase) promotor, which is inducible by methanol. Alternatively PEPT1 or PEPT2 expression can be placed under control of a strong GAPDH promotor, but here expression is dependent on the carbon source provided in the culture medium.[28] Nevertheless, using the right substrate in the growth medium drives expression levels above that obtained under control of the AOX promotor. PEPT1 and PEPT2 in *Pichia* currently represent the only mammalian transport proteins successfully expressed with full function in this yeast. As shown previously, PEPT1 and PEPT2 in the transgenic *Pichia* cells retain all functional characteristics known for the proteins.

The main advantage of the yeast system is that it provides high functional expression levels with uptake rates for dipeptides exceeding those in wild-type yeast cells by more than 100-fold. In addition, large quantities of cells can be grown in inexpensive media, allowing large-scale applications for functional analysis as well as for isolation of the expressed proteins.

We established a microwell plate–based substrate competition assay[26] as one of the applications of the yeast system. A fluorescent dipeptide derivative served as the transported substrate, and a large variety of potential competitors have been screened for interaction with the substrate binding site of PEPT1. We identified omega amino fatty acids (AFA) as one of the classes of potent competitors. Uptake of the fluorescent dipeptide conjugate into PEPT1-expressing *Pichia pastoris* cells was dose dependently inhibited by AFA when the compounds contained four or more CH_2 units in the backbone. In the case of 4-aminobutanoic acid (4-ABA = GABA), there was no noticable affinity. Elongation of the backbone by one additional CH_2 group to yield 5-aminopentanoic acid (5-APA) increased affinity dramatically to EC_{50} values similar to those obtained for di- and tripeptides when assayed under identical experimental conditions.[5] Inasmuch as competition with a dipeptide for binding at the substrate binding site does not establish that a compound is indeed transported, we employed a radiolabeled 6-aminohexanoic acid (^{14}C-6-AHA) as a representative structure to assess its transport characteristics in *Pichia* cells mediated by PEPT1. This compound was taken up into the transgenic yeast cells but not into the control cells, and uptake displayed saturation kinetics mediated by a single transport site with an apparent $K_{0.5}$ value of 0.92 ± 0.07 mM. To demonstrate that the transport of AFA by PEPT1 is independent of the expression system, we additionally employed *Xenopus* oocytes that expressed PEPT1. When these oocytes were perfused with AFA, positive inward currents as high as those induced by Gly-Gln were obtained for all AFA, except for 4-ABA (FIG. 3a). Moreover, 6-AHA acid and 8-aminooctanoic acid (8-AOA) also produced the same I-V relationships as the dipeptide Gly-Gln, suggesting that PEPT1 handles these substrates in the same way that it handles dipeptides (FIG. 3c).

As shown in the competition assay as well as by electrophysiology, PEPT1 obviously sharply discriminates in binding and transport between 4-ABA and 5-APA, and the longer chain AFA. We therefore employed computational model analysis to predict the compound's possible conformations based on energy minimization calculations to relate the conformation to transport capability. From these studies it became obvious that a >500 pm <630 pm distance between the two charged centers (carboxylic carbon and amino nitrogen) in AFA was sufficient for substrate recogni-

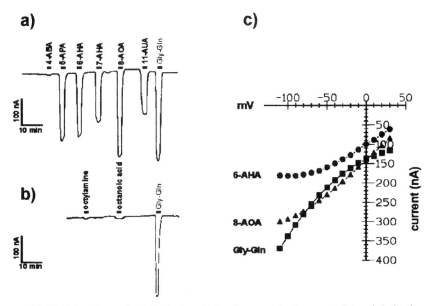

FIGURE 3. Electrophysiological analysis of transport of omega-AFA and derivatives in *Xenopus* oocytes expressing mammalian PEPT1. (**a**) Representative inward currents in oocytes expressing PEPT1 evoked by superfusion with AFA of increasing chain length. Three days postinjection of 5 ng of PEPT1-cRNA, oocytes were clamped to −60 mV and superfused with 2.5 mM of the AFA or the dipeptide Gly-Gln at pH 6.5. (**b**) Assessment of inward currents in oocytes expressing mammalian PEPT1 when octylamine, octanoic acid, or the dipeptide Gly-Gln served as substrates. Oocytes were clamped to −60 mV and superfused with 2.5 mM of the AFA or the dipeptide Gly-Gln at pH 6.5. (**c**) Recordings of current-voltage (I-V) relationships in oocytes superfused with 6-aminohexanoic acid (AHA), 8-aminooctanoic acid (AOA), or the dipeptide Gly-Gln. The membrane potential was stepped symmetrically to test potentials between −150 and +50 mV in steps of 10 mV and current responses as the difference between the presence or the absence of 2.5 mM of the substrates at pH 6.5.

tion and transport. Removal of either the amino group or the carboxy group in AFA to obtain octanoic acid or octylamine maintained the compound's affinity for interaction with the transporter in the competition experiment but abolished the capability for electrogenic transport (see FIG. 3b).

In summary, by using omega-AFA as model compounds, we show for the first time that substrate recognition and transport by the mammalian intestinal peptide transporter requires as a minimum only two ionized groups (i.e., amino and carboxy group) separated by an intramolecular spacer of four CH_2 units. Conformational analysis based on energy minimization calculations predicts the centers of the functional groups to be separated on a distance of > 500 < 635 pm to enable binding to PEPT1 followed by transport.[29] This provides a primary basis for understanding the versatility of di-/tripeptide carriers and their independent sequence transport of substrates. Regarding PEPT1, this information could also allow a more rational drug de-

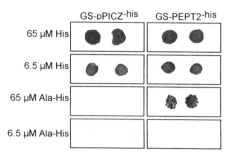

FIGURE 4. Growth properties of histidine-auxotrophic *Pichia pastoris* strains express-ing mammalian PEPT2. Cell growth of the histidine-auxotrophic strains GS-PEPT2-his or GS-pPICZ-his (wild type = control) in methanol medium containing either L-histidine (His) or L-alanyl-L-histidine (Ala-His) at concentrations of 65 μM or 6.5 μM. Three hours after induction of PEPT2 expression, 10 mL of cells (3.5×10^5 cells) from the strains GS-PEPT2-his (PEPT2) and GS-pPICZ-his (control) were dropped on methanol plates containing His or Ala-His and grown for 4 days.

sign for both peptide as well as nonpeptide mimetics, to enable their interaction with PEPT1 and consequently increase their intestinal absorption.

In addition to the fluorescence-based competition assay, we developed a new *Pichia* strain (GS-PEPT2[-his]) on the basis of a histidine-auxotroph phenotype. The characterization of PEPT2 function in this strain also showed the preserved PEPT2-phenotype with saturable transport activity for the hydrolysis-resistant dipeptide D-Phe-Ala and the driving force of the carrier provided by an inwardly directed proton gradient. Moreover, other functional characteristics, such as substrate specificity and affinity, were also very similar to the characteristics of PEPT2 as found in apical membrane vesicles of kidney tubular cells or in other systems, such as oocytes or mammalian cells, in which PEPT2 has been expressed. Cells of this *Pichia* strain ex-pressing PEPT2 can grow only when histidine is provided in the culture medium. However, the clone expressing PEPT2 should, in contrast to the histidine-aux-otrophic control strain, be able to grow also when histidine is provided as a histidyl-dipeptide, such as Ala-His. As shown in FIGURE 4, by using a simple growth test on agar plates, the functional expression of PEPT2 in the GS-PEPT2[-his] strain can be monitored directly in media when histidine is provided in dipeptide form; whereas the histidine-auxotroph control strain, GS-pPICZ[-his], grows only in media contain-ing free histidine. When the Ala-His concentration in the medium was lowered to 6.5 μM, the GS-PEPT2[-his] could not grow, suggesting that this concentration was too low to deliver sufficient histidine for cell growth.

This very simple and fast growth test can now be used for a variety of applica-tions, such as identification of compounds that compete with the Ala-His for uptake via the peptide transporter and consequently reduce growth. However, more impor-tantly, the growth test should be valuable for screening purposes and for employing cells in which the transporter has been submitted to random mutagenesis. Peptide transporters with altered phenotypes, as a consequence of mutations, can be identi-fied easily by the changed growth properties in Ala-His containing media with re-spect to changes in substrate affinity or the pH dependence of transport.

SUMMARY

As shown here with mammalian peptide transporters normally expressed in intestinal and renal epithelial cells at low protein levels, heterologous expression systems are very useful in characterizing the structure and function of the carrier proteins. However, different expression systems provide different advantages with respect to ease of handling, functionality, and costs. Whereas the oocyte as a large cell can be easily impaled with microcapillaries and electrodes for injection of cRNA or for determining the electrophysiological characteristics of rheogenic transporters, its use is limited by the necessity of handling individual cells. Other expression systems, such as the yeast system described here, possess the advantage of obtaining large quantities of transgenic cells cheaply and rapidly. The yeast system can be used for functional analysis of the transporters as well as for production of larger amounts of the carrier proteins. Moreover, this system may be useful for studies on structure–function analysis by employing yeast genetics in combination with rapid screening for altered transporter phenotypes based on the growth test described above.

REFERENCES

1. FEI, Y.J., Y. KANAI, S. NUSSBERGER et al. 1994. Expression cloning of a mammalian proton-coupled oligopeptide transporter. Nature 368(6471): 563–566.
2. BOLL, M., D. MARKOVICH, W-M. WEBER et al. 1994. Expression cloning of a cDNA from rabbit small intestine related to proton-coupled transport of peptides, β-lactam antibiotics and ACE-inhibitors. Pflügers Archiv. 429: 146–149.
3. BOLL, M., M. HERGET, M. WAGENER et al. 1996. Expression cloning and functional characterization of the kidney cortex high affinity proton-coupled peptide transporter. Proc. Natl. Acad. Sci. USA 93: 284–289.
4. LIANG, R., Y.J. FEI, P.D. PRASAD et al. 1995. Human intestinal H$^+$/peptide cotransporter. Cloning, functional expression, and chromosomal localization. J. Biol. Chem. 270(12): 6456–6463.
5. LIU, W., R. LIANG, S. RAMAMOORTHY et al. 1995. Molecular cloning of PEPT 2, a new member of the H$^+$/peptide cotransporter family, from human kidney. Biochim. Biophys. Acta 1235(2): 461–466.
6. SAITO, H., T. TERADA, M. OKUDA et al. 1996. Molecular cloning and tissue distribution of rat peptide transporter PEPT2. Biochim. Biophys. Acta 1280(2): 173–177.
7. STEINER, H.Y., F. NAIDER & J.M. BECKER. 1995. The PTR family: a new group of peptide transporters. Mol. Microbiol. 16(5): 825–834.
8. AMASHEH, S., U. WENZEL, M. BOLL et al. 1997. Transport of charged dipeptides by the intestinal H$^+$/peptide symporter PepT1 expressed in Xenopus laevis oocytes. J. Membr. Biol. 155: 247–256.
9. AMASHEH, S., U. WENZEL, W-M. WEBER et al. 1997. Electrophysiological analysis of the renal peptide transporter expressed in Xenopus laevis oocytes. J. Physiol. (Lond.) 504.1: 169–174.
10. FEI, Y.J., V. GANAPATHY & F.H. LEIBACH. 1998. Molecular and structural features of the proton-coupled oligopeptide transporter superfamily. Prog. Nucleic Acid Res. Mol. Biol. 58: 239–261.
11. LEIBACH, F.H. & V. GANAPATHY. 1996. Peptide transporters in the intestine and the kidney. Annu. Rev. Nutr. 16: 99–119.
12. RIES, M., U. WENZEL & H. DANIEL. 1994. Transport of cefadroxil in kidney brush border membranes is mediated by two electrogenic 1994-coupled transport systems. J. Pharmacol. Exp. Ther. 271: 1327–1333.
13. WENZEL, U., D.T. THWAITES & H. DANIEL. 1995. Stereoselective uptake of beta-lactam antibiotics by the intestinal peptide transporter. Br. J. Pharmacol. 116: 3021–3027.

14. WENZEL, U., H. GEBERT, I. WEINTRAUT et al. 1996. Transport characteristics of differently charged cephalosporin antibiotics in oocytes expressing the cloned peptide transporter PepT1 and in human intestinal Caco-2 cells. J. Pharmacol. Exp. Ther. 277: 831–839.
15. DANIEL, H. 1996. Function and molecular structure of brush border membrane peptide/H^+ symporters. J. Membr. Biol. 154: 197–203.
16. DANIEL, H. 1997. First insights into the operational mode of epithelial peptide transporters. J. Physiol. (Lond.) 498: 561.
17. DANIEL, H. & M. HERGET. 1997. Cellular and molecular mechanisms of renal peptide transport. Am. J. Physiol. 273: F1–F8.
18. DANIEL, H., F. DÖRING, M. HERGET & U. WENZEL. 1998. Mechanisms of renal peptide transport. Nova Acta Leopoldina 78: 195–200.
19. WENZEL, U., D. DIEHL, M. HERGET & H. DANIEL. 1998. LLC-PK1 cells express the renal high-affinity H^+/peptide cotransporter: its kinetic characterization and role as an acid loader. Am. J. Physiol. 275: C1573–C1579.
20. DÖRING, F., J. WALTER, J. WILL et al. 1998. Delta-aminolevulinic acid transport by intestinal and renal peptide transporters and its physiological and clinical implications. J. Clin. Invest. 101: 2761–2767.
21. MACKENZIE, B., D.D. LOO, Y. FEI et al. 1996. Mechanisms of the human intestinal H^+-coupled oligopeptide transporter hPEPT1. J. Biol. Chem. 271(10): 5430–5437.
22. STEEL, A., S. NUSSBERGER, M.F. ROMERO et al. 1997. Stoichiometry and pH dependence of the rabbit proton-dependent oligopeptide transporter PepT1. J. Physiol. (Lond.) 498 (Pt. 3): 563–569.
23. NUSSBERGER, S., A. STEEL, D. TROTTI et al. 1997. Symmetry of H^+ binding to the intra- and extracellular side of the H^+-coupled oligopeptide cotransporter PepT1. J. Biol. Chem. 272(12):7777–7785.
24. DÖRING, F., D. DORN, U. BACHFISCHER et al. 1996. Functional analysis of a chimeric mammalian peptide transporter derived from the intestinal and renal isoforms. J. Physiol. (Lond.) 497: 773–779.
25. DÖRING, F., J. WALTER, M. FÖCKING et al. 1998. The aminoterminal region of the renal peptide transporter PEPT2 determines its high substrate affinity. Nova Acta Leopoldina 306: 269–274.
26. DÖRING, F., S. THEIS & H. DANIEL. 1997. Expression and functional characterization of the mammalian intestinal peptide transporter PepT1 in the methylotrophic yeast Pichia pastoris. Biochem. Biophys. Res. Commun. 232: 656–662.
27. DÖRING, F., T. MICHEL, A. RÖSEL et al. 1998. Expression of the mammalian renal peptide transporter PEPT2 in the yeast Pichia pastoris and applications of the yeast system for functional analysis. Mol. Membr. Biol. 15: 79–88.
28. DÖRING, F., M. KLAPPER, S. THEIS et al. 1998. Use of the glyceraldehyde-3-phosphate dehydrogenase promoter for production of functional mammalian membrane transport proteins in the yeast Pichia pastoris. Biochem. Biophys. Res. Commun. 250: 531–535.
29. DÖRING, F., J. WILL, S. AMASHEH et al. 1998. Minimal molecular determinants of substrates for recognition by the intestinal peptide transporter. J. Biol. Chem. 273: 23211–23218.

Epithelial Barrier Defects in HT-29/B6 Colonic Cell Monolayers Induced by Tumor Necrosis Factor–α

ALFRED H. GITTER,[a,b] KERSTIN BENDFELDT,[a] HEINZ SCHMITZ,[c] JÖRG-DIETER SCHULZKE,[c] CARL J. BENTZEL,[d] AND MICHAEL FROMM[a]

[a]Institut für Klinische Physiologie and [c]Medizinische Klinik I, Gastroenterologie und Infektiologie, Universitätsklinikum Benjamin Franklin, Freie Universität Berlin, 12200 Berlin, Germany

[d]Department of Medicine, East Carolina University, Greenville North Carolina 27858, USA

ABSTRACT: The barrier function of intestinal epithelia relies upon the continuity of the enterocyte monolayer and intact tight junctions. After incubation with tumor necrosis factor–α TNF-α, however, the number of strands that form the tight junctions decreases, and apoptosis is induced in intestinal epithelial cells. These morphological changes lead to a rise of transepithelial ion permeability, because the paracellular ion permeability increases and leaks associated with sites of apoptosis increase by number and magnitude. Thus apoptosis and degradation of tight junctions contribute to the increased permeability observed after exposure to TNF-α. These mechanisms explain clinical manifestations in the inflamed intestinal wall containing cytokine-secreting macrophages—for example, leak flux diarrhea and invasion of bacterial enterotoxins.

The epithelial monolayer that lines the luminal surface of the intestine separates the mucosal and the serosal side by sealing the intercellular space with tight junctions. Impairment of the barrier function of intestinal epithelia may be the predominant mechanism in the pathogenesis of diarrhea in intestinal bowel disease (IBD).[1] Secretory mechanisms as a cause for IBD-related diarrhea are unlikely, because the inflamed mucosa shows decreased, rather than increased, rheogenic transport.[2]

During the inflammation, cytokine-secreting macrophages are recruited, and thus the intestinal concentration of tumor necrosis factor–α (TNF-α) rises.[3–5] Up to 10 ng/mL of bioactive TNF-α were measured in the circulation of patients with intestinal bowel disease.[6,7] Hence, 100 ng/mL or more may be reached in the relatively small volume of interstitial fluid space at the site of inflammation. Exposure to TNF-α leads to increased ion permeability of epithelia *in vitro*—for example, in LLC-PK1

[b]Address for correspondence: Dr. Alfred H. Gitter, Institut für Klinische Physiologie, Universitätsklinikum Benjamin Franklin, 12200 Berlin, Germany. Voice: +49-30-8445-2789; fax: +49-30-8445-4239.

gitter@medizin.fu-berlin.de

kidney cells,[8] endothelial cells,[9] HT-29 clone 19A[10], and T84 intestinal cells.[11] In addition, however, there is evidence that macrophages induce apoptosis of enterocytes.[12]

In order to investigate the putative role of TNF-α in epithelial barrier defects, we looked for morphological changes in cultured monolayers of human intestinal epithelial cells (HT-29/B6).[13] Employing the conductance scanning technique,[14] we determined the contribution of apoptosis to the overall conductivity of the epithelium.

METHODS

Cell Culture

HT-29/B6, a subclone of the human colorectal cancer cell line HT-29, grows as highly differentiated, polarized epithelium with properties of Cl$^-$ and mucus-secreting crypt cells. In order to elicit apoptosis, the serosal side of HT-29/B6 cells was incubated in culture medium containing 100 ng/mL of TNF-α for 7 hours. Recombinant human TNF-α (10^7 units/mg) was provided by Schering (Berlin).

Electrophysiology

Confluent monolayers were mounted horizontally between the two halves of the conductance scanning chamber described previously.[14] During electrophysiological recordings, mucosal and serosal surface of the epithelium were superfused with (concentrations in mmol/l) 113.6 NaCl, 2.4 Na$_2$HPO$_4$, 0.6 NaH$_2$PO$_4$, 21 NaHCO$_3$, 5.4 KCl, 1.2 CaCl$_2$, 1.2 MgCl$_2$, 10 D(+)-glucose, 2.5 L-glutamine, D(+)-mannose, and β-hydroxybutyric acid. The solution was gassed with carbogen (95% O$_2$, 5% CO$_2$), which set the pH to 7.4. The temperature was maintained at 37°C.

Ussing techniques and flux measurements were performed as described previously. Samples for determination of the serosal-to-mucosal flux of ^{22}Na$^+$ were taken at regular 30-minute intervals.

Conductance scanning experiments were performed under bright-field light microscopical control. Two pattern types could be discriminated in the epithelial monolayer: nonapoptotic areas with approximately hexagonal symmetry and apoptotic spots that where surrounded by neighboring cells positioned in a distinctive rosette-like arrangement.

The local conductivity (conductance referred to gross tissue area) was probed by measurement of the local electric field, generated by sinusoidal transepithelial current (AC, 0.3 mA/cm^2, 24 Hz) in the bath solution on the the mucosal side of the epithelium. By multiplication of the field strength measured locally and the specific resistivity of the bath solution, the current density was calculated.

In nonapoptotic areas the distribution of transepithelial current was homogeneous, and the conductivity equaled the ratio of current density to transepithelial voltage. Near apoptotic rosettes the transepithelial current was inhomogeneously elevated. Here the current associated with a single rosette was computed by numerical integration of the current density exceeding the current density of nonapoptotic areas. From rosette current and transepithelial voltage, the conductance of a single rosette was derived.

All values are given as mean ± SEM for 20 apoptotic sites in 9 control monolayers or 21 apoptoses in 11 monolayers treated with TNF-α.

Histochemistry

In order to prove that the rosettes observed with intravital light microscopy indeed indicate apoptosis in the center of the rosette, we demonstrated apoptosis using different histological techniques. Monolayers were fixed in formaldehyde at room temperature and embedded in paraffin. Serial sections, made with a thickness of 3 μm, were stained with hematoxylin and eosin or dewaxed for immunofluorescence localization of apoptoses.

Cellular DNA was either stained with 4,6-diamidino-2-phenylindole-2-HCl (DAPI) or inspected with a TUNEL (TdT-mediated X-dUTP nick end labeling) assay. In the latter, the blunt ends of double-stranded DNA exposed by strand breaks were, according to the description of the manufacturer (Boehringer Mannheim, Germany), visualized by means of enzymatic labeling (using terminal deoxynucleotidyl transferase) of the free 3′-OH termini with modified nucleotides (fluorescein-dUTP) and fluorescence microscopy.

Electron Microscopy

Mounted in the *in vitro* setup, HT-29/B6 monolayers were fixed at room temperature with phosphate-buffered 2% glutaraldehyde. For scanning electron microscopy the tissue was frozen in Freon 22 and liquid nitrogen ($-100°C$) and fractured with a double replica device. Freeze fractures were shadowed with platinum and carbon and examined in a Phillips 200 electron microscope.

Morphometric analysis was performed using coded prints of freeze-fracture electron micrographs ($\times 60,000$ magnification) on all tight junction regions with a delimited meshwork of strands in the tight junction. Vertical grid lines perpendicular to the most apical strand were drawn at intervals of 167 nm. Counting the intersections of strands with a grid line yielded its number of strands. The distance between the most apical and the most basal strands defined the depth of the tight junction. Aberrant strands were not observed, either in controls or after TNF-α exposure. In controls, 280 grid lines from 22 different tight junction regions were analyzed; in monolayers exposed to TNF-α, 320 grid lines from 22 tight junctions were analyzed.

Ultrathin sections for transmission electron microscopy were prepared following standard procedures.

RESULTS

Overall transepithelial ion permeability of HT-29/B6 monolayers was evaluated by measurement of total epithelial conductivity (or its reciprocal, transepithelial resistance). After incubation with TNF-α (100 ng/mL), the transepithelial conductivity increased from 3.24 ± 0.07 mS $\cdot$ cm^{-2} ($309\ \Omega \cdot$ cm^2) to 14.65 ± 0.31 mS $\cdot$ cm^{-2} ($68\ \Omega \cdot$cm^2). The short circuit current did not change under TNF-α; it was 0.1 ± 0.01 μmol $\cdot$ h$^{-1} \cdot$cm^{-2} ($n = 6$) under control conditions and with the cytokine.

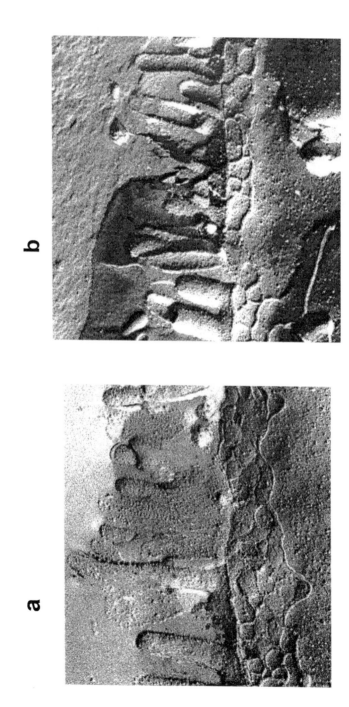

FIGURE 1. The junctional region of HT-29/B6 monolayers comprised a delimited meshwork of strands in the tight junction. The freeze-fracture electron micrographs were made (**a**) under control conditions and (**b**) after incubation with 100 ng/mL TNF-α for 24 hours. The depth of the tight junction decreased from 265 ± 17 (control, $n = 22$) to 200 ± 14 nm (TNF-α, $n = 22$; $p < 0.01$).

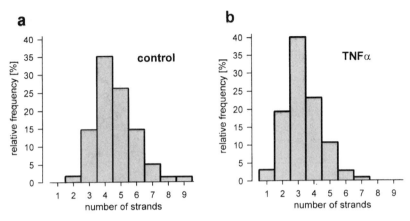

FIGURE 2. Histograms showing the relative frequency of the number of horizontally oriented junctional strands for all tight junction sections analyzed. (**a**) Control monolayers ($n = 280$); (**b**) exposed to TNF-α ($n = 320$).

Necrosis could be excluded, because the percentage of LDH released into the bath solution was equal in controls and monolayers with 24 h of exposure to TNF-α ($4.0 \pm 0.3\%$ versus $4.2 \pm 0.3\%$, n.s.). TNF-α was not effective after it was boiled (100°C, 30 min), excluding the possibility of artifacts due to endotoxin contamination.

Degradation of Tight Junctions

Freeze-fracture replicas of tight junctions in regular Ringer's showed the typical branching and anastomosing network of strands that appeared as ridges or furrows in the plane of the cell membrane (FIG. 1a). In the 280 tight junction sections inspected, the number of horizontally oriented strands was between 2 and 9, with a mean of 4.7 ± 0.2 (FIG. 2a). The mean depth of the tight junction was 267 ± 18 nm (22 cells).

The network of strands was also observed after incubation with TNF-α (FIG. 1b); the mean number of strands had decreased to 3.3 ± 0.2 ($p < 0.001$ vs. control) (FIG. 2b), and the depth of tight junctions to 200 ± 14 nm ($p < 0.01$). In contrast to control cells, TNF-α–treated cells showed a significant amount of tight junctional regions with only 1 or 2 strands (FIG. 2).

With and without TNF-α, the histogram of strand numbers appeared Gaussian, but tight junctional regions without horizontal strands cannot be excluded because the method does not assess those. These findings suggest that the TNF-α–induced decrease in the the number of strands affects the majority of the tight junctions in the same way, probably by interference with assembly/disassembly mechanisms.

Paracellular Permeability

In order to assess paracellular permeability, we measured the serosal-to-mucosal flux of $^{22}Na^+$. Under control conditions, it was $1.6 \pm 0.1 \ \mu mol \cdot h^{-1} \cdot cm^{-2}$ ($n = 6$). After 4.5 hours of preincubation with TNF-α, the flux rose to $10.2 \pm 1.0 \ \mu mol \cdot h^{-1} \cdot cm^{-2}$ ($n = 6$). Thus, the paracellular ion permeability increased under TNF-α.

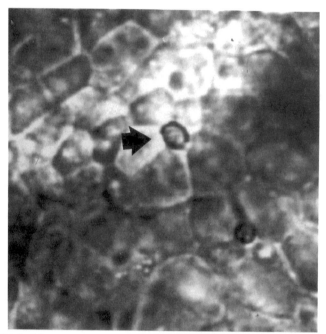

FIGURE 3. Monolayers of living HT-29/B6 cells in the conductance scanning chamber, viewed by light microscopy with bright-field illumination, show clusters suggestive of a rosette, composed of radially elongated cells around a central apoptotic cell (*arrow*).

Evidence of Apoptosis

In the chamber used for conductance scanning experiments, the living monolayer was observed with an ×40 water immersion object lens. Within the otherwise inconspicuous cell sheet, we found rosette-like arrangements composed of radially elongated cells with a small cell (if any) in the center (FIG. 3). The frequency of these clusters, $3,700 \pm 300$ cm^{-2} under control conditions, increased to $11,100 \pm 500$ cm^{-2} after TNF-α treatment.

The rosettes resembled the tissue remodeling during TNF-α–induced apoptosis in LLC-PK1 renal epithelial cells.[16] Hence we performed several histological tests in order to prove apoptosis. Apoptotic fragments of the nucleus in the center of the rosette were visualized by staining with hematoxylin-eosin or the fluorochrome DAPI, or by fluorescent DNA end labeling (TUNEL).

Transmission electron micrographs showed apoptotic cells and bodies in the center of rosettes and fragments engulfed by adjacent cells (FIG. 5). In some sections there was a hole in the center of the rosette, suggesting a superficial lacuna due to incomplete closure of the space that had been occupied by the apoptotic cell (FIG. 4).

Conductance Scanning

In order to determine the local conductance associated with single apoptoses, we scanned the monolayer with a conductance probe.[14] In nonapoptotic areas, a con-

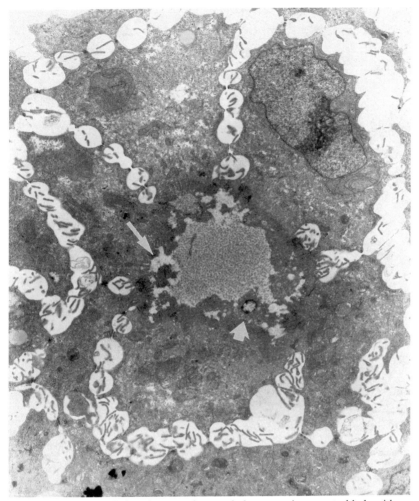

FIGURE 4. Transmission electron micrograph demonstrating a central hole with apoptotic bodies (*short arrow*) and cytoplasma vacuoles (*long arrow*) in a section through the apical part of intestinal epithelial cells forming an apoptotic rosette. This finding suggests incomplete closure of the space that had been occupied by the apoptotic cell. The mean cell diameter was 6 μm.

stant value reflecting the basic epithelial conductivity was recorded. Above apoptotic rosettes, however, the local conductivity increased along a line between the area with homogeneous conductivity and the rosette's center. Integration yielded the conductance associated with a single apoptotic rosette.

The morphological appearance of a rosette correlated with its conductance. Three types were distinguished, as illustrated in the cartoon (FIG. 6): (a) a shrunk, sharply contoured cell with almost no conductance (except the basic epithelial conductivity);

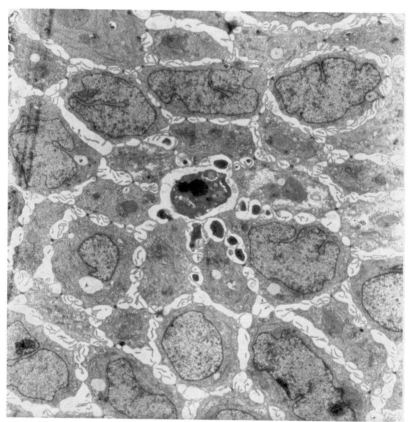

FIGURE 5. Transmission electron micrograph of a section through the basal part of intestinal epithelial cells forming an apoptotic rosette. In this plane, a single apoptic cell is surrounded by normal neighbors (rosette). Hence, there is no hole of cellular dimension in the epithelium, but the relatively wide intercellular space with few interdigitating membrane foldings suggests the possibility of disconnection between the apoptic cell and its neighbors.

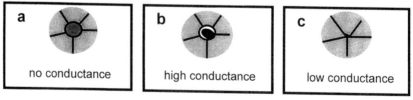

FIGURE 6. Three types of apoptotic rosettes observed with intravital light microscopy were correlated with apoptotic conductance: (**a**) a shrunk, sharply contoured cell, with no conductance (except the basic epithelial conductivity); (**b**) a transparent center and apoptotic fragments, with low or high conductance; (**c**) no central cell and the surrounding cells closing in, with no or low conductance.

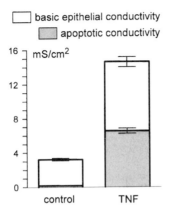

FIGURE 7. Histogram showing epithelial conductivity and its components, basic epithelial conductivity and apoptotic conductivity, in HT-29/B6 monolayers under control conditions (*left*) and with TNF-α (*right*). After exposure to TNF-α, the epithelial conductivity rose dramatically, because both basic epithelial conductivity as well as frequency and conductance of apoptotic rosettes increased.

(b) a transparent center and apoptotic fragments, with low or high conductance; and (c) no central cell and the surrounding cells closing in, with no or low conductance.

Incubation with TNF-α increased the mean conductance of single apoptoses from 48 nS to 597 nS. Together with the three times higher frequency of apoptoses under TNF-α, the contribution of apoptoses to the total epithelial conductivity increased from 0.18 ± 0.02 mS · cm^{-2} (5.5%) under control conditions to 6.57 ± 0.30 mS · cm^{-2} (45% of the total epithelial conductivity). However, the basic epithelial conductivity also increased under TNF-α, from 3.06 ± 0.07 mS · cm^{-2} to 8.08 ± 0.05 mS · cm^{-2} (FIG. 7).

DISCUSSION

The present report documents two morphological alterations of the intestinal epithelial cells treated with TNF-α: degradation of tight junctions and apoptosis. With the conductance scanning technique the contribution of both effects to the increase in epithelial ion permeability was measured.

TNF-α, applied at 100 ng/mL increased the basic epithelial conductivity and led to apoptosis associated with pronounced local barrier defects. Measurement of serosal-to-mucosal flux of ^{22}Na$^+$ indicated that the increase in basic epithelial conductivity was caused by an increase in paracellular conductivity. The latter can be attributed to a reduction in complexity of the network of strands that form the tight junction, as demonstrated by freeze-fracture analysis of TNF-α–treated monolayers. Absence of changes in short-circuit current in monolayers treated with TNF-α indicate that transcellular ion permeability plays a minor role. Chloride secretion induced by TNF-α via prostaglandins is short-lived and not accompanied by a significant change in the transepithelial resistance.[17,18]

Our data demonstrate leaks in an epithelial monolayer caused by spontaneous apoptosis and, with more pronounced effect, by TNF-α–induced apoptosis. The higher rate and leakiness of apoptoses, contrasting TNF-α–treated with control epithelia, may be related to the degradation of tight junctions observed in freeze fractures, because cell–cell contacts can be involved in the control of apoptosis.[16] Thus, epithelial integrity may be compromised during TNF-α–mediated inflammatory processes, causing a clinical manifestation of leakiness in the intestinal wall — diarrhea, for example. Invasion of bacterial enterotoxins may start a vicious cycle.

ACKNOWLEDGMENTS

We thank D. Sorgenfrei and S. Lüderitz for electrifying and cultivating collaboration. The work was supported by the Deutsche Forschungsgemeinschaft and by Freie Universität Medical Faculty funds.

REFERENCES

1. SCHULZKE, J.D., M. FROMM, H. SCHMITZ, C. BARMEYER & E.O. RIECKEN. 1995. Barrier impairment in gastrointestinal diseases: ulcerative colitis and celiac sprue. Z. Gastroenterol. **33:** 571–572.
2. SANDLE, G.I., N. HIGGS, P. CROWE et al. 1990. Cellular basis for defective transport in inflamed human colon. Gastroenterology **99:** 97–105.
3. SARTOR, R.B. 1994. Cytokines in intestinal inflammation. Pathophysiological and clinical considerations. Gastroenterology **106:** 533–539.
4. BREESE, E.J., C.A. MICHIE, S.W. NICHOLLS et al. 1994. Tumor-necrosis factor-α producing cells in the intestinal mucosa of children with inflammatory bowel disease. Gastroenterology **106:** 1455–1466.
5. MCKAY, D.M., K. CROITORU & M.H. PERDUE. 1996. T cell-monocyte interactions regulate epithelial physiology in a coculture model of inflammation. Am. J. Physiol. **270:** C418–C428.
6. SATEGNA-GUIDETTI, C., R. PULITANO, L. FENOGLIO et al. 1993. Tumor necrosis factor/cachectin in Crohn's disease. Relation of serum concentration to disease activity. Rec. Prog. Med. **84:** 93–99.
7. MARANO C.W., S.A. LEWIS, L.A. GARULACAN et al. 1998. Tumor necrosis factor-alpha increases sodium and chloride conductance across the tight junction of CACO-2 BBE, a human intestinal epithelial cell line. J. Membr. Biol. **161:** 263–274.
8. MULLIN, J.M., K.V. LAUGHLIN, C.W. MARANO et al. 1992. Modulation of tumor necrosis factor-induced increase in renal (LLC-PK$_1$) transepithelial permeability. Am. J. Physiol. **263:** F915–F924.
9. BURKE-GAFFNEY, A. & A.K. KEENAN. 1993. Modulation of human endothelial cell permeability by combinations of the cytokines interleukin-1 α/β, tumor necrosis factor-α and interferon-γ. Immunopharmacology **25:** 1–9.
10. RODRIGUEZ, P., M. HEYMAN, C. CANDALH et al. 1995. Tumor necrosis factor-α induces morphological and functional alterations of intestinal HT-29 cl.19A cell monolayers. Cytokine **7:** 441–448.
11. MADARA, J.L. & J. STAFFORD. 1989. Interferon-γ directly affects barrier function of cultured intestinal epithelial monolayers. J. Clin. Invest. **83:** 724–727.
12. IWANAGA, T., O. HOSHI, H. HAN et al. 1994. Lamina propria macrophages involved in cell death (apoptosis) of enterocytes in the small intestine of rats. Arch. Histol. Cytol. **57:** 267–276.
13. KREUSEL, K.M., M. FROMM, J.D. SCHULZKE & U. HEGEL. 1991. Cl⁻ secretion in epithelial monolayers of mucus-forming human colon cells (HT-29/B6). Am. J. Physiol. **261:** C574–582.

14. GITTER, A.H., M. BERTOG, J.D. SCHULZKE & M. FROMM. 1997. Measurement of paracellular epithelial conductivity by conductance scanning. Pflügers Arch. **434:** 830–840.
15. SCHULZKE, J.D., M. FROMM, C.J. BENTZEL *et al.* 1992. Epithelial ion transport in the experimental short bowel syndrome of the rat. Gastroenterology **102:** 497–504.
16. PERALTA SOLER, A., J.M. MULLIN, K.A. KNUDSEN & C.W. MARANO. 1996. Tissue remodeling during tumor necrosis factor-induced apoptosis in LLC-PK1 renal epithelial cells. Am. J. Physiol. **270:** F869–879.
17. SCHMITZ, H., M. FROMM, C.J. BENTZEL *et al.* 1996. Tumor necrosis factor alpha induces Cl⁻ and K⁺ secretion in human distal colon driven by prostaglandin E$_2$. Am. J. Physiol. **271:** G669–G674.
18. KANDIL, H.M., H.M. BERSCHNEIDER & R.A. ARGENZIO. 1994. Tumour necrosis factor α changes porcine intestinal ion transport through a paracrine mechanism involving prostaglandins. Gut **35:** 934–940.
19. HEYMAN M., N. DARMON, C. DUPONT *et al.* 1994. Mononuclear cells from infants allergic to cow's milk secrete tumor necrosis factor-α, altering intestinal function. Gastroenterology **106:** 1514–1523.
20. RUSSELL, R.G., M. O'DONNOGHUE, D.C. BLAKE JR. *et al.* 1993. Early colonic damage and invasion of *Campylobacter jejuni* in experimentally challenged infant *Macaca mulatta*. J. Infect. Dis. **168:** 210–215.
21. MCKAY, D.M. & P.K. SINGH. 1997. Superantigen activation of immune cells evokes epithelial (T84) transport and barrier abnormalities via IFN-γ and TNFα: inhibition of increased permeability, but not diminished secretory responses by TGF-β$_2$. J. Immunol. **159:** 2382–2390.

Enhanced Expression of iNOS in Inflamed Colons of IL-2–Deficient Mice Does Not Impair Colonic Epithelial Barrier Function

M. HARREN,[a,e] C. BARMEYER,[b,e] H. SCHMITZ,[b] M. FROMM,[c] I. HORAK,[d] A. DIGNASS,[a] M. JOHN,[b] B. WIEDENMANN,[a,f] AND J.-D. SCHULZKE[b]

[a]Universitätsklinikum Charité der Humboldt Universität zu Berlin, Campus Virchow-Klinikum, Medical Clinic, Department of Hepatology and Gastroenterology, Berlin, Germany

[b]Freie Universität Berlin, Universitätsklinikum Benjamin Franklin, Medical Clinic I, Department of Gastroenterology and Infectiology, Berlin, Germany

[c]Freie Universität Berlin, Universitätsklinikum Benjamin Franklin, Department of Clinical Physiology, Berlin, Germany

[d]Forschungsinstitut für Molekulare Pharmakologie, Department of Molecular Genetics, Berlin, Germany

ABSTRACT: On the basis of recently observed high levels of iNOS expression that correlated with intestinal inflammation in interleukin-2–deficient [IL-2(−/−)] mice, it was postulated that nitric oxide may damage colonic epithelial cells or impair intestinal epithelial barrier function. This damage may result in an increased permeability of the colonic epithelium leading to high antigenic exposure of the intestinal immune system, which may perpetuate chronic inflammation. Our data demonstrate that high expression of iNOS in IL-2(−/−) mice is correlated with the length/weight ratio (L/W ratio), a widely accepted marker for intestinal inflammation. However, no reduction of epithelial resistance was observed, as would be expected in case of a damaged, leaky epithelium. Our results suggest that enhanced formation of NO in IL-2(−/−) mice does not cause impairment of epithelial barrier function.

INTRODUCTION

Mice bearing targeted deletions for IL-2, IL-10, MHC class II, or TCR genes develop intestinal inflammation.[1,2] These findings stimulated investigations about regulation of immune responses in the intestinal epithelium.[3] Although the presence of luminal bacteria or their products seem to be critical for the development of intestinal inflammation, the pathophysiological mechanisms of bowel inflammation in these genetically altered animals remain unclear. Disturbances in the balance of im-

[e] M. Harren and C. Barmeyer contributed equally to this paper.

[f]Address for correspondence: Bertram Wiedenmann, M.D., Charité–Campus Virchow Klinikum, Med. Klinik mit Schwerpunkt Hepato- und Gastroenterologie, Augustenburger Platz 1, 13353 Berlin, Germany. Voice: +49-30-450-53022; fax: +49-30-450-53902.
bertram.wiedenmann@charite.de.

mune regulatory lymphocyte subsets have been postulated to play an important role in the pathogenesis of inflammation in these mice.[4]

Interleukin-2–deficient mice [IL-2(–/–)] spontaneously develop colitis and other autoimmune defects when kept under conventional conditions or following immunization with a specific antigen (e.g., 2,4,6-trinitrophenol, coupled to keyhole limpet hemocyanin).[2] With increasing age, animals develop colitis and chronic diarrhea, usually between weeks 6–15.[3] The absence of IL-2 in mice leads to the development of activated autoreactive CD4 positive lymphocytes, which are considered pivotal for the development of colitis.[3] This notion is supported by transfusion of isolated CD4 lymphocytes from IL-2(–/–) mice into wild-type mice, which leads to colitis in these animals.[5] Characterization of these CD4-positive lymphocytes of IL-2(–/–) mice revealed a strong Th1-secretion pattern, which is characterized by high levels of IFN-γ in the presence of low IL-4 levels.[2]

IFN-γ is one of the most potent inducers of the macrophage-type nitric oxide synthase (inducible NO synthase, iNOS, NOS type II), which in turn generates high amounts of NO. Besides being a potent vasodilator and neurotransmitter, NO can also act as a cytotoxic agent.[6] Recent studies suggest that nitric oxide can disturb epithelial barrier function by damaging colonic epithelial cells. Thus, an altered epithelial barrier may lead to an increased exposure of the mucosal immune system with luminal antigens that may induce, enhance, or maintain chronic inflammation (FIG. 1).[7–9] On the basis of recently observed high iNOS expression in inflamed co-

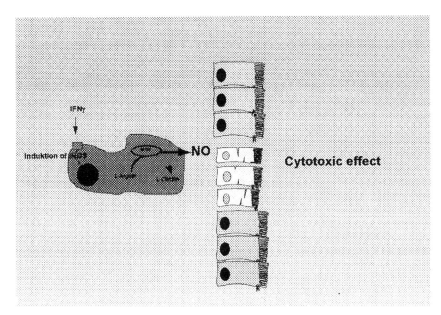

FIGURE 1. Nitric oxide may damage colonic epithelial cells and facilitate the permeation of antigens in the mucosa. A disturbance of the epithelial barrier function by NO-induced damage of colonic epithelial cells has been postulated. A leaky epithelial barrier may lead to an increased antigenic exposure of the mucosal immune system, resulting in a chronic inflammatory response.

lonic segments of interleukin-2–deficient mice,[1] we assessed the effects of NO on intestinal epithelial barrier function. We performed electrophysiological studies in iNOS-expressing colonic tissues and analyzed the conductivity of the colonic epithelium. Furthermore, we determined the location of iNOS by means of immunofluorescence microscopy.

METHODS AND MATERIAL

Mice

Mice with a C57BL6 genetic background and targeted disruption of the IL-2 gene (described previously[3]) were used. The animals were kept under standard conditions and were monitored for the development of loose bowel movements and wasting. Severely ill animals were sacrificed after ether anesthesia, the colon was dissected from the abdomen, flushed free of contents, and cut into three segments (ascending colon, a.c.; transverse colon, t.c.; descending colon, d.c.).

Morphological Analysis

To investigate altered crypt architecture, single crypt diameter and the distance between two crypts, represented by the shortest distance between the crypt margins, were determined on 4% formaldehyde–fixed native mucosal surface preparations. Measurements were performed on two IL-2(–/–) mice and two wild-type mice by using light microscopy (Zeiss Axioplan, Germany) combined with a charged coupled device (CCD) camera (Hamamatsu, Japan). Images were processed on a screen by an image-processing device (Argus-10, Hamamatsu, Japan). Parameters were calculated from images.

To correlate our previous findings of an increased iNOS expression in inflamed colonic segments in IL-2(–/–) mice,[1] we determined the length/weight ratio (L/W ratio) in the ascending colon, the transverse colon, and the descending colon of 9 IL-2(–/–) and 7 wild-type mice as an additional marker for inflammation.

Immunofluorescence Microscopy

Indirect immunofluorescence microscopy was performed using a rabbit polyclonal antibody against iNOS as a primary antibody (Biomol, Plymouth). The secondary antibody was Texas red conjugated goat antirabbit antibody.

Electrophysiological Studies

For electrophysiological studies ascending colon was mounted as unstripped preparation into Ussing-type chambers. The exposed tissue area was 0.28 cm^2. Alternating current (AC) impedance analysis was applied as described by Schmitz et al.[12] This technique permits discrimination between the epithelial (G^e) and the subepithelial (G^{sub}) component of the transmural conductivity (G^t). Briefly, the voltage responses after transepithelial application of 35 μA/cm^2 eff. sine-wave alternat-

A. wild type mouse

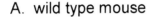

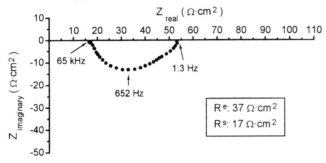

B. IL-2-/--mouse

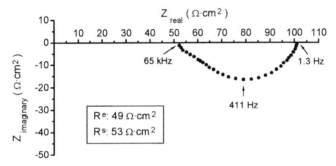

FIGURE 2. Impedance analysis. Original impedance locus plots of wild-type mouse colon (**A**) and IL-2(−/−) mouse colon (**B**). Z_{real} is the ohmic component, and $Z_{imaginary}$ is the reactive component of the complex impedance. Intersections between the semicircle and x-axis at low and high frequencies represent transmural (R^t) and subepithelial resistance (R^{sub}), respectively. Epithelial resistance (R^e) can be calculated as $R^e = R^t − R^{sub}$. The conductivity G as given in TABLE 2 is the reciprocal value of the resistance R.

ing current of 48 discrete frequencies in a range from 1 to 65 kHz were detected by phase-sensitive amplifiers (Mod. 1250 Frequency Response Analyzer and Mod. 1286 Electrochemical Interface; Solartron Schlumberger, Farnborough Hampshire, UK). Complex impedance values were calculated and corrected for the conductivity of the bathing solution and the frequency behavior of the measuring setup for each frequency. Then, for each tissue the impedance locus was plotted in a Nyquist diagram (FIG. 2), and a circle segment was fitted by least square analysis. From this circle segment, three variables of an electric equivalent circuit were obtained, which consisted of a resistor and a capacitor in parallel, representing the epithelium and a resistor in series to this unit representing the subepithelium. Because the electrical characteristics of the capacitor are frequency dependent, G^t was obtained at low frequencies, whereas G^{sub} was obtained at high frequencies. The epithelial conductivity G^e was calculated from $G^{e-1} = G^{t-1} − G^{sub-1}$.

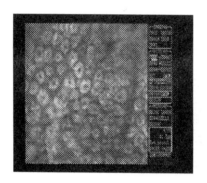 Wildtype

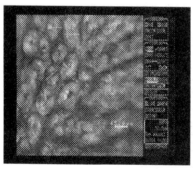 IL-2 (-/-)

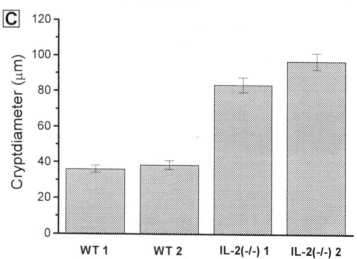

FIGURE 3. *See following for caption.*

TABLE 1. Length/weight (L/W) ratio of IL-2–deficient mice[a]

	Ascending colon (a.c.) (mg/cm)	Transverse colon (t.c) (mg/cm)	Descending Colon (d.c.) (mg/cm)	n
Control	46.3 ± 4.2	29.9 ± 3.5	27.2 ± 2.6	8
IL-2(–/–)	79.6 ± 9.1*	65.2 ± 13.6*	35.4 ± 6.0[n.s.]	7

[a]L/W ratio of the ascending, transverse, and and descending colon of controls and IL-2(–/–) mice. All values are given as means ± SEM. * p <0.05 versus control mice; n.s., not significant versus control mice.

RESULTS

Morphometric Analysis

Microscopic analysis of the intestinal mucosal surface of wild-type mice shows a regular arrangement of crypts (FIG. 3, Panel A), with diameter varying in a small range from 25 μm to 49 μm, (mean 37.4 μm, SD 7.5 μm) (FIG. 3, Panel C, column WT 1 and WT 2). The distance between two crypts varies from 2 μm to 12.6 μm (mean 8.2 μm, SD 1.7 μm).

By contrast, the intestinal mucosal surface of IL-2(–/–) mice show an irregular arrangement of crypts and a diffuse infiltration and enlargement of the mucosal layer leading to a pronounced thickening of the bowel wall (FIG. 3, Panel B). The crypt diameter shows a wide variation ranging from normal to significantly increased diameters of 60 μm to 122 μm (mean 90.25 μm, SD 22.25 μm) (FIG. 3, Panel C, column IL-2(–/–) 1 and IL-2(–/–) 2); the distance between two crypts is increased from 5 μm to 45 μm (mean 24 μm, SD 4.9 μm).

L/W Ratio

The L/W ratio in colonic segments of IL-2(–/–) mice is increased (FIG. 4, TABLE 1). Sections of the ascending colon show a mean L/W ratio of 79.6 mg/cm (FIG. 4, IL-2(–/–), row a.c.), the transverse colon a mean of 65.2 mg/cm (FIG. 4, IL-2(–/–), row t.c.), and the descending colon a mean of 35.4 mg/cm (FIG. 4, IL-2(–/–), row d.c.).

In comparison, the colonic segments of wild-type mice show lower mean L/W-ratio values especially in the ascending colon: a.c. 46.3 mg/cm; t.c. 29.8 mg/cm; d.c. 27.15 mg/cm (FIG. 4, wild-type mice, row a.c., t.c., and d.c.).

To identify iNOS-expressing cells, we assessed iNOS expression with an immunofluorescence technique in 4 IL-2(–/–) and 4 wild-type mice (FIG. 5). Immunoreactivity for iNOS was detectable in the epithelium and in the lamina propria in inflamed areas of colonic sections of IL-2(–/–) mice. Strong immunoreactivity could

FIGURE 3. Crypt diameter in IL-2(–/–) and wild-type mice. Light microscopy (magnification ×20) of the mucosal surface of wild-type mice showed a regular arrangement of crypts (**A**). Mean crypt diameters of two wild-type mice are given in **C**, column WT 1 and WT 2. By contrast, the mucosal surface in IL-2(–/–) mice showed an irregular arrangement of crypts and a diffuse infiltration and enlargement of the mucosal layer, which led to a pronounced thickening of the bowel wall (**B**). The mean crypt diameter was increased [**C**, column IL-2(–/–) 1 and IL- 2(–/–) 2].

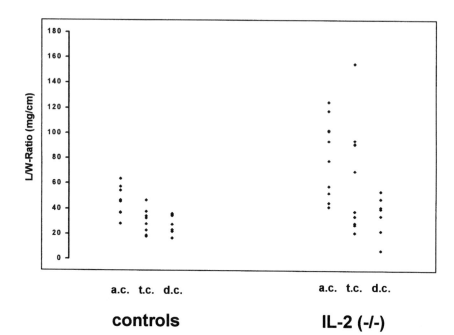

FIGURE 4. Length/weight ratio (L/W ratio) of colonic segments in IL-2(−/−) and wild-type mice. Compared to IL-2(−/−) mice, colonic segments of wild-type mice showed lower L/W ratios in the ascending colon (a.c.) and transverse colon (t.c.), whereas in the descending colon (d.c.) this tendency did not reach statistical significance.

TABLE 2. Transmural, epithelial, and subepithelial electrical conductivity[a]

	G^t (mS/cm^2)	G^e (mS/cm^2)	G^{sub} (mS/cm^2)	n
Control	18.3 ± 2.0	27.5 ± 4.4	59.5 ± 3.9	6
IL-2(−/−)	11.7 ± 1.4*	20.4 ± 1.3[n.s.]	32.0 ± 7.0**	5

[a]Transmural (G^t), epitheial (G^e), and subepitheial (G^{sub}) conductance as determined by alternating current impedance analysis. All values are given as means ± SEM. ** p <0.05 versus control mice; n.s., not significant versus control mice.

especially be observed in the ascending colon of IL-2(−/−) mice. In microscopically normal tissues of IL-2(−/−) mice, iNOS expression was detected only in the lamina propria. wild-type mice showed only a sporadic, weak iNOS immunoreactivity in the lamina propria.

Alternating Current Impedance Analysis

The data and statistical evaluation of the intestinal barrier function are summarized in TABLE 2. In control mice, G^t was 18.28 ± 1.95 mS/cm^2. In IL-2(−/−) mice, G^t was decreased to 11.71 ± 1.44 mS/cm^2. By using impedance analysis the conduc-

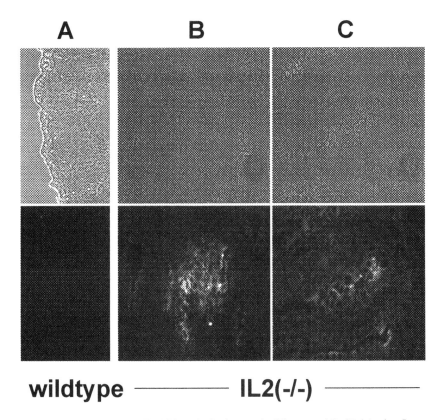

wildtype ———————— IL2(-/-) ————————

FIGURE 5. Location of iNOS in colonic tissues of wild-type and IL-2(−/−) mice. Immunoreactivity for iNOS was detectable in the epithelium and in the lamina propria in inflamed areas of colonic sections of IL-2(−/−) mice (**B** and **C**). Especially in the ascending colon of IL-2(−/−) mice, a strong immunoreactivity could be observed. In microscopically normal tissues of IL-2(−/−) mice, iNOS expression was only detected in the lamina propria. Wild-type mice showed only weak iNOS immunoreactivity in the lamina propria sporadically (**A**).

tivity of the epithelium was assessed. Epithelial conductivity of IL-2(−/−) mice was 26% lower compared with control mice (20.39 ± 1.26 versus 27.49 ± 4.42 mS/cm^2). Thus, an impairment of mucosal barrier function—at least for ionic permeability—was not observed in IL-2(−/−) mice. These findings are in contrast to recent studies in patients with ulcerative colitis (UC), where an increase in epithelial conductivity was detected in inflamed colonic segments indicating a pronounced disturbance of the epithelial barrier function in ulcerative colitis.[12] By contrast, G^{sub} was altered in IL-2(−/−) mice. G^{sub} decreased from 59.48 ± 3.89 mS/cm^2 in controls to 32.03 ± 7.03 in IL-2−/− mice. This increase in subepithelial tightness may reflect the inflammatory process within the intestinal wall, which includes edema and infiltration with inflammatory cells.

DISCUSSION

iNOS and Epithelial Conductivity

Attention has recently been focused on the enhanced production of nitric oxide as a key player in the pathogenesis of inflammatory bowel disease (IBD). This notion is supported by the fact that NO levels have been shown to be elevated in rectal dialysates of patients with ulcerative colitis.[13] In addition to being a potent vasodilator and neurotransmitter, NO has been observed to have cytotoxic effects at high concentrations. A possible mechanism of NO-induced epithelial cell damage may be its interaction with superoxide to produce peroxynitrite, which interacts with tyrosine to form nitrotyrosine in cellular proteins[8] (FIG. 1). It has been suggested that the increased NO production may cause epithelial cell damage, disturbing the epithelial barrier function, thereby leading to an increased antigenic exposure of the mucosal immune system resulting in a chronic inflammatory response (FIG.1).[7–9]

Until now, however, no information has been available about the epithelial barrier function in inflamed colonic tissues of IL-2(–/–) mice, with increased expression of iNOS.

Our results clearly demonstrate that elevated iNOS expression, which correlates with the histological grade of inflammation, also correlates with an increased L/W ratio, but is not associated with an increase in epithelial conductivity as would be expected in case of a damaged, leaky epithelium. Our results suggest that at least in IL-2(–/–) mice, which develop some key features of IBD, the proposed mechanism of NO-induced alteration of the epithelial barrier function is not likely.

The absence of IL-2 in mice leads to the development of activated autoreactive CD4-positive lymphocytes, which are considered pivotal for the development of colitis.[3] CD4-positive lymphocytes in IL-2(–/–) mice show a strong Th1-secretion pattern, characterized by very low levels of IL-4[2] in the presence of high levels of IFN-γ, a potent inductor of iNOS. This results in high iNOS expression, which correlates with intestinal inflammation in IL-2(–/–) mice.[1] Recent publications about the immunoregulatory properties of nitric oxide emphasize the possible role of nitric oxide in the pathogenesis of colitis in IL-2(–/–) mice. Huang et al. showed that NO is capable of inhibiting Th1 cell development.[14] Tarrant et al. demonstrated that suppression of a Th1-mediated autoimmune disease, experimental autoimmune uveitis, by IL-12 is NO dependent.[15]

Data presented in the recent literature combined with our finding that a high expression of iNOS is not associated with an increased epithelial conductivity, may indicate that NO suppresses the Th1 immune response, which may play a crucial role in the pathogenesis of colitis in IL-2(–/–) mice. An enhanced Th1 cell response in iNOS-deficient mice underscores this hypothesis. Furthermore, for the first time we were able to quantitate alterations of the crypt architecture in IL-2(–/–) mice by morphometric studies. However, the etiologic factors that causes epithelial intestinal hyperplasia in IL-2(–/–) mice have not been identified.

In conclusion, we have shown that enhanced iNOS expression in inflamed colonic tissues in an experimental model of colitis is not associated with alterations of the epithelial barrier function. Further studies are necessary to evaluate the functional relevance of increased levels of NO and iNOS expression in patients with experimental colitis and IBD.

ACKNOWLEDGMENTS

The authors wish to thank Ines Eichhorn, Nura Sayed Suleiman, and Sieglinde Lüderitz for their skillful technical assistance and our electronic engineer Detlef Sorgenfrei.

REFERENCES

1. HARREN, M. *et al.* 1998. High expression of inducible nitric oxide synthase correlates with intestinal inflammation of interleukin-2-deficient mice. Ann. N.Y. Acad. Sci. **859:** 210–215.
2. EHRHARDT, R.O. *et al.* 1997. Induction and prevention of colonic inflammation in IL-2-deficient mice. J. Immunol. **158:** 566–573.
3. SADLACK, B. *et al.* 1993. Ulcerative colitis-like disease in mice with a disrupted inter-leukin-2 gene. Cell **75:** 253–261.
4. HORAK, I. *et al.* 1995. Interleukin-2 deficient mice: a new model to study autoimmmu-nitiy and self-tolerance. Immun. Rev. **148:** 35–44.
5. LUDVIKSSON, B.R. *et al.* 1997. Dysregulated intrathymic development in the IL-2-defi-cient mouse leads to colitis-inducing thymocytes. J. Immunol. **158:** 104–111.
6. NATHAN, C. & Q. Xie. 1994. Nitric oxide synthases: roles, tolls, and controls. Cell **78:** 915–918.
7. MCKENZIE, S.J. *et al.* 1996. Evidence of oxidant-induced injury to epithelial cells dur-ing inflammatory bowel disease. J. Clin. Invest. **98:** 136–141.
8. SINGER, I.I. *et al.* 1996. Expression of inducible nitric oxide synthase and nitrotyrosine in colonic epithelium in inflammatory bowel disease. Gastroenterology **111:** 871–885.
9. KUBES, P. *et al.* 1992. Nitric oxide modulates epithelial permeability in the feline small intestine. Am. J. Physiol. **262:** G1138–G1142.
10. EPPLE, H.J. *et al.* 1995. Enzyme- and mineralocorticoid receptor-controlled electro-genic Na$^+$ absorption in human rectum *in vitro*. Am. J. Physiol. **269** (Gastrointest. Liver Physiol. **32**)**:** G42–G48.
11. TAI, Y.H. *et al.* 1981. The conventional short-circuiting technique under-short-circuits most epithelia. J. Membr. Biol. **59:** 173–177.
12. SCHMITZ, H. *et al.* 1999. Altered tight junction structure contributes to the impaired epithelial barrier function in ulcerative colitis. Gastroenterology **116:** 301–309.
13. BOUGHTON-SMITH, N.K. *et al.* 1993. Nitric oxide activity in ulcerative colitis and Crohn's disease. Lancet **342:** 338–340.
14. HUANG, F.P. *et al.* 1998. Nitric oxide regulates Th1 cell development through the inhi-bition of IL-12 synthesis by macrophages. Eur. J. Immunol. **28** (12): 4062–4070.
15. TARRANT, T.K. *et al.* 1999. Interleukin-12 protects from a T helper type 1–mediated autoimmune disease, experimental autoimmune uveitis, through a mechanism involv-ing interferon gamma, nitric oxide, and apoptosis. J. Exp. Med. **189** (2): 219–230.

Regulation of Intercellular Tight Junctions by Zonula Occludens Toxin and Its Eukaryotic Analogue Zonulin

ALESSIO FASANO[a]

Division of Pediatric Gastroenterology and Nutrition, Gastrointestinal Pathophysiology Section, Center for Vaccine Development, and Department of Physiology, University of Maryland School of Medicine, Baltimore, Maryland 21201, USA

ABSTRACT: The intestinal epithelium represents the largest interface between the external environment and the internal host milieu and constitutes the major barrier through which molecules can either be absorbed or secreted. There is now substantial evidence that tight junctions (tj) play a major role in regulating epithelial permeability by influencing paracellular flow of fluid and solutes. Tj are one of the hallmarks of absorptive and secretory epithelia. Evidence now exists that tj are dynamic rather than static structures and readily adapt to a variety of developmental, physiological, and pathological circumstances. These adaptive mechanisms are still incompletely understood. Activation of PKC either by Zonula occludens toxin (Zot) or by phorbol esters increases paracellular permeability. Alteration of epithelial tj is a recently described property for infectious agents. *Clostridium difficile* toxin A and B and influenza and vesicular stomatitis viruses have been shown to loosen tj in tissue culture monolayers. Unlike what occurs after the Zot stimulus, these changes appear to be irreversible and are associated with destruction of the tj complex. On the basis of this observation, we postulated that Zot may mimic the effect of a functionally and immunologically related endogenous modulator of epithelial tj. We were able to identify an intestinal Zot analogue, which we named *zonulin*. It is conceivable that the zonulins participate in the physiological regulation of intercellular tj not only in the small intestine, but also throughout a wide range of extraintestinal epithelia as well as the ubiquitous vascular endothelium, including the blood-brain barrier. Disregulation of this hypothetical zonulin model may contribute to disease states that involve disordered intercellular communication, including developmental and intestinal disorders, tissue inflammation, malignant transformation, and metastasis.

The intestinal epithelium represents the largest interface (more then 2,000,000 cm^2) between the external environment and the internal host milieu and constitutes the major barrier through which molecules can either be absorbed or secreted. There is

[a]Address for correspondence: Alessio Fasano, M.D., Division of Pediatric Gastroenterology and Nutrition, University of Maryland School of Medicine, 685 W. Baltimore St., HSF Building, Room 465, Baltimore, MD 21201. Voice: 410-328-0812; fax 410-328-1072.

afasano@umaryland.edu

now substantial evidence that tight junctions (tj) play a major role in regulating epithelial permeability by influencing paracellular flow of fluid and solutes. Tj is one of the hallmarks of absorptive and secretory epithelia. As a barrier between apical and basolateral compartments, it selectively controls the passive diffusion of ions and small water-soluble solutes through the paracellular pathway, thereby counterregulating any gradients generated by transcellular pathways.[1] Variations in transepithelial conductance can usually be attributed to changes in the permeability of the paracellular pathway, since the resistance of the enterocyte plasma membrane is relatively high.[2] The tj represent the major barrier within this paracellular pathway, and the electrical resistance of epithelial tissues seems to depend on the number of transmembrane protein strands and their complexity within the tj as observed by freeze-fracture electron microscopy.[3] Evidence now exists that tj, once regarded as static structures, are in fact dynamic and readily adapt to a variety of developmental,[4–6] physiological,[7–10] and pathological[11–13] circumstances. These adaptive mechanisms are still incompletely understood.

In the presence of Ca^{2+}, tj assembly is the result of a complex cascade of biochemical events that ultimately lead to the formation of an organized network of tj elements, the composition of which has been only partially characterized. Two candidates for the transmembrane protein strands, occludin and clandins, have recently been identified[14] (FIG. 1). Several proteins have been identified in a cytoplasmic submembraneous plaque underlying membrane contacts (FIG.1), but their function remains to be established. ZO-1 and ZO-2 each exists as a heterodimer[15] in a detergent-stable complex with an uncharacterized 130-kDa protein (ZO-3). Most immunoelectron microscopic studies have localized ZO-1 to precisely beneath membrane contacts.[16] Both ZO-1 and ZO-2 belong to the membrane-associated guanylate kinase (MAGUK) family of proteins.[17] Several other peripheral membrane proteins have been localized to the tj, including cingulin,[18] 7H6,[19] rab 13,[20] $G\alpha_{i-2}$,[21,22] and PKC.[22] Recently, a novel protein (symplekin) has been described that not only associates with tj but can also be localized to the nucleus.[23] Similar to ZO-1, symplekin is also expressed by cells that do not form tj, where it appears to be only in the nucleus. ZO-1 also can be localized to the nucleus, but unlike symplekin, only in growing but not in differentiated epithelial cells.[24] This dual localization for these tj components suggests that tj might also be involved in the regulation of gene expression, cell growth, and differentiation.[25] Beside rab 13, other small GTP-binding proteins are known to regulate the cortical cytoskeleton; Rho regulates actin polymerization and focal adhesion formation.[26] In polarized epithelial cells, Rho also regulates tj organization and permeability.[27] Other proteins, such as Rac and focal adhesion kinase (FAK), play a role in plasma membrane ruffling and focal adhesion formation.[28] Whether these molecules also participate in tj regulation is unknown.[29] On the basis of the bidirectional signaling that is transduced across focal adhesions[30] and the zonula adherens,[31] it is conceivable that tj-associated proteins are similarly involved in transducing signals in one or more directions across the cell membrane and in regulating links to the cortical actin cytoskeleton. In eukaryotic cells, junctional complex proteins, actin filaments, microtubules, and intermediate filaments interact to form the cytoskeleton network involved in determination of cell architecture, intracellular transport, modulation of surface receptors, paracellular permeability, mitosis, cell motility, and differentiation.[32]

Intestinal lumen

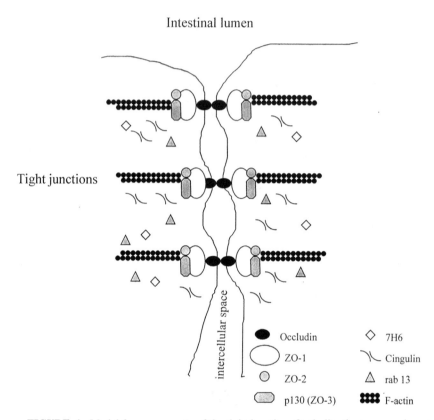

FIGURE 1. Model for components of the tight junction. Occludin, the transmembrane protein strand, is anatomically and functionally connected with the cell cytoskeleton via the junctional complex. This complex comprises a series of proteins, including ZO-1, ZO-2, and p130 (ZO-3). Other proteins, such as cingulin, 7H6, rab13, rho, and ras, are located further from the cell membrane. However, they also seem involved in the regulation of tight junction permeability.

There is now a large body of evidence that structural and functional linkage exists between the actin cytoskeleton and the tj complex of absorptive cells.[33–35] The actin cytoskeleton is composed of a complicated meshwork of microfilaments the precise geometry of which is regulated by a large cadre of actin binding proteins. The architecture of the actin cytoskeleton appears to be critical for tj function. Most of the actin is positioned under the apical junctional complex where myosin II and several actin binding proteins, including α-catenin, vinculin, and radixin, have been identified.[29] Myosin movement along actin filaments is regulated by ATP and phosphorylation of the regulatory light chain by Ca^{2+}-calmodulin–activated myosin light-chain kinase.[36] In several systems, increases in intracellular Ca^{2+} can affect phosphorylation of myosin regulatory light chain contraction of perijunctional actin and cause increased paracellular permeability.[37] We have recently demonstrated PKC-dependent actin polymerization associated with increments in paracellular permeability.[38]

A variety of intracellular mediators have been shown to alter tj function and/or structure. Tj within the amphibian gallbladder[39] and goldfish[40] and flounder[41] intestine display enhanced resistance to passive ion flow as intracellular cAMP is elevated. In addition, exposure of amphibian gallbladder to Ca^{2+} ionophore appears to enhance tj resistance and induce alterations in tj structure.[42] Last, activation of PKC either by Zonula occludens toxin (Zot)[38] or by phorbol esters[43–45] increases paracellular permeability. Alteration of epithelial tj is a recently described property of infectious agents. *Clostridium difficile* toxin A[46] and B[47] and influenza and vesicular stomatitis viruses[48] have been shown to loosen tj in tissue culture monolayers. However, unlike what occurs after the Zot stimulus, these changes appear to be irreversible and are associated with destruction of the tj complex.[46,47]

To meet the many diverse physiological challenges to which both epithelia and endothelia are subjected, tj must be capable of rapid and coordinated responses. This requires the presence of a complex regulatory system that orchestrates the state of assembly of the tj multiprotein complex. While it is well accepted that tj are dynamic structures, surprisingly little is known about their regulation. The discovery of Zot, a protein elaborated by *Vibrio cholerae*, has shed some light on the intricate mechanisms involved in the modulation of the intestinal paracellular pathway. Zot action is mediated by a cascade of intracellular events that lead to a protein kinase C (PKC)α–dependent polymerization of actin microfilaments strategically localized to regulate the paracellular pathway.[38] These changes are a prerequisite to opening of tj and are evident at a toxin concentration as low as 1.1×10^{-13} M.[49] The toxin exerts its effect by interacting with the surface of enteric cells. By using immunofluorescence binding studies, we have shown that Zot binding varies within the intestine, being more detectable in the jejunum and distal ileum, and decreasing along the villous-crypt axis.[49] This binding distribution coincides with the regional effect of Zot on intestinal permeability[49] and with the preferential F-actin redistribution induced by Zot in the mature cells of the villi,[38] suggesting that the regional effect of Zot is associated to its binding of a surface receptor (r) the distribution of which varies within the intestine and along the villous-crypt axis. These data showed that Zot regulates tj in a rapid, reversible, and reproducible fashion and probably activates intracellular signals, which are operative during the physiologic modulation of the paracellular pathway (FIG. 2).

On the basis of this observation, we postulated that Zot may mimic the effect of a functionally and immunologically related endogenous modulator of epithelial tj. The combination of affinity-purified anti-Zot antibodies and the Ussing chamber assay allowed us to identify an intestinal Zot analogue, which we named *zonulin*.[50] When zonulin was studied in a nonhuman primate model, it reversibly opened intestinal tj. Since *V. cholerae* infections are strictly confined to the gastrointestinal tract, we anticipated an exclusively intraintestinal role for zonulin. Much to our surprise, this protein was detected in a wide range of extra-intestinal tissues and has now been purified from human intestine, heart, and brain. We have provided evidence that the zonulins comprise a family of tissue-specific regulators of tight junctions.[50] Each family member has an MW of ~47 kDa, a distinct N-terminal receptor binding motif that confers tissue specificity, and a C-terminal tau-like domain probably involved in the cytoskeleton rearrangement. Amino acid substitution within the N-terminal binding motif identified three amino acid residues that dictate tissue specificity, allowing local autocrine/paracrine regulation in response to local requirements.[50] The

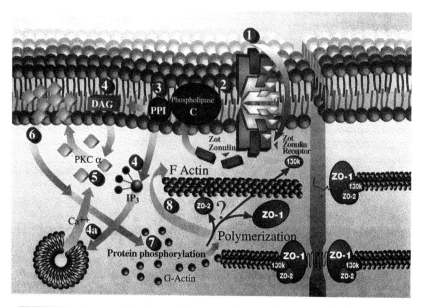

FIGURE 2. Proposed Zot intracellular signaling leading to the opening of intestinal tight junctions. Zot interacts with a specific surface receptor (**1**) whose distribution within the intestine varies. The protein is then internalized and activates phospholipase C (**2**), which hydrolyzes phosphatidyl inositol (**3**) to release inositol 1,4,5-tris phosphate (PPI-3) and diacylglycerol (DAG) (**4**). PKCα is then activated (**5**), either directly (via DAG) (**4**) or through the release of intracellular Ca^{++} (via PPI 3) (**4a**). PKCα catalyzes the phosphorylation of target protein(s), with subsequent polymerization of soluble G actin in F actin (**7**). This polymerization causes the rearrangement of the filaments of actin and the subsequent displacement of proteins (including ZO-1) from the junctional complex (**8**). As a result, intestinal tight junctions become looser.

physiological function of the zonulins remains to be established; however, it is likely that they are involved in tj regulation responsible for the movement of fluid, macromolecules, and leukocytes between body compartments. It is likely that zonulin is also involved in pathological conditions, since tj dysfunction occurs in a variety of clinical conditions affecting the gastrointestinal tract, including food allergies,[51] *V. cholerae* infection,[52] malabsorption syndromes,[53,54] and inflammatory bowel diseases.[55] It is not surprising that tj structure may be affected when the physiological state of absorptive cells is dramatically changed as it is in many of these disease states. In cholera, jejunal biopsies obtained during the acute phase of the disease showed a marked widening of the lateral intercellular spaces that was present only in the upper third of the villi and was maximal at the villous tips, gradually decreasing towards the middle of the villus.[52] In symptomatic celiac patients, small intestinal barrier dysfunction also has been ascribed to disorganization of the tj complex.[53,54] These changes are more pronounced in villous enterocytes, and they resolve once the patients are treated with a gluten-free diet.[54] The tj derangement that occurs with cholera and CD coincides with Zot/zonulin receptor distribution along the gastrointestinal tract.[38] These findings, together with our observation that zonulin is overexpressed during the acute phase of CD,[56] suggest that this protein contributes to CD pathogenesis by increasing tj permeability typical of the early stages of this clinical condition. During intestinal inflammatory states, transient loss of the tj barrier (as revealed by diminished tissue resistance and increased permeability to macromolecules) also has been implicated in transepithelial migration of granulocytes.[57]

Although tumor cell invasion and metastasis are complex processes, a key step in both is physical disengagement from the primary tumor.[58] This step involves disruption of normal cell–cell adhesion in epithelial tissues. Although much of the work in this area has focused on the adherence junction,[59] evidence now exists that tj elements are similarly involved. Willott and coworkers have recently demonstrated that the tj protein ZO-1 is homologous to the *Drosophila* discs large tumor suppressor protein of septate junctions and that *dlg* mutations result in loss of apical-basolateral epithelial cell polarity and in neoplastic growth.[60] These data suggest that tj dysfunction may be involved in carcinogenesis.

In conclusion, it is conceivable that the zonulins participate in the physiological regulation of intercellular tj not only in the small intestine, but also throughout a wide range of extraintestinal epithelia (e.g., the tracheobronchial tree and the renal tubule) as well as the ubiquitous vascular endothelium, including the blood-brain barrier. Disregulation of this hypothetical zonulin model may contribute to disease states that involve disordered intercellular communication, including developmental and intestinal disorders, tissue inflammation, malignant transformation, and metastasis.

REFERENCES

1. DIAMOND, J.M. 1977. The epithelial junction: bridge, gate and fence. Physiologist **20:** 10–18.
2. MADARA, J.L. 1989. Loosing tight functions lessons from the intestine. J. Clin. Invest. **83:** 1089–1094.
3. MADARA, J.L. & K. DHARMSATHAPHORN. 1985. Occluding junction structure-function relationships in a cultured epithelial monolayer. J. Cell Biol. **101:** 2124–2133.
4. MAGNUSON, T., J.B. JACOBSON & C.W. STACKPOLE. 1978. Relationship between intercellular permeability and junction organization in the preimplantation mouse embryo. Dev. Biol. **67:** 214–224.
5. REVEL, J.P. & S.S. BROWN. 1976. Cell junctions in development with particular reference to the neural tube. Cold Spring Harbor Symp. Quant. Biol. **40:** 443–455.
6. SCHNEEBERGER, E.E., D.V. WALTERS & R.E. OLIVER. 1978. Development of intercellular junctions in the pulmonary epithelium of the foetal lamb. J. Cell Sci. **32:** 307–324.
7. GILULA, N.B., D.W. FAWCETT & A. AOKI. 1976. The sertoli cell occluding junctions and gap junctions in mature and developing mammalian testis. Dev. Biol. **50:** 142–168.
8. MADARA, J.L. & J.R. PAPPENHEIMER. 1987. Structural basis for physiological regulations of paracellular pathways in intestinal epithelia. J. Membr. Biol. **100:** 149–164.
9. MAZARIEGOS, M.R., L.W. TICE & A.R. HAND. 1984. Alteration of tight junctional permeability in the rat parotid gland after isoproteranol stimulation. J. Cell Biol. **98:** 1865–1877.
10. SARDET, C., M. PISAM & J. MAETZ. 1979. The surface epithelium of teleostean fish gills. Cellular and junctional adaptations of the chloride cell in relation to salt adaptation. J. Cell Biol. **80:** 96–117.
11. MILKS, L.C., G.P. CONYERS & E.B. CRAMER. 1986. The effect of neutrophil migration on epithelial permeability. J. Cell Biol. **103:** 2729–2738.

12. NASH. S., J. STAFFORD & J.L. MADARA. 1988. The selective and superoxide-independent disruption of intestinal epithelial tight junctions during leukocyte transmigration. Lab. Invest. **59**: 531–537.
13. SHASBY, D.M., M. WINTER & S.S. SHASBY. 1988. Oxidants and conductance of cultured epithelial cell monolayers: inositol phospholipid hydrolysis. Am. J. Physiol. **255** (Cell Physiol. **24**): C781–C788.
14. TSUKITA, S. & M. FURUSE. 1999. Occludin and clandins in tight-junction strands: leading or supporting players. Trends Cell Biol. **9**: 268–273.
15. GUMBINER, B., T. LOWENKOPF & D. APATIRA. 1991. Identification of 160-kDa polypeptide that binds to the tight junction protein ZO-1. Proc. Natl. Acad. Sci. USA **88**: 3460–3464.
16. STEVENSON, B.R., J.M. ANDERSON & S. BULLIVANT. 1988. The epithelial tight junction: structure, function and preliminary biochemical characterization. Molec. Cell Biochem. **83**: 129–145.
17. ANDERSON, J.M. 1996. Cell signalling: MAGUK magic. Curr. Biol. **6**: 382–384.
18. CITI, S., H. SABANNAY, R. JAKES et al. 1988. Cingulin, a new peripheral component of tight junctions. Nature (London) **333**: 272–275.
19. ZHONG, Y., T. SAITOH, T. MINASE et al. 1993. Monoclonal antibody 7H6 reacts with a nevel tight junction-associated protein distinct from ZO-1, cingulin and ZO-2. J. Cell Biol. **120**: 477–483.
20. ZAHRAOUI, A., G. JOBERTY, M. ARPIN et al. 1994. A small rab GTPase is distributed in cytoplasmic vesicles in nonpolarized cells but colocalized with the tight junction marker ZO-1 in polarized eipithelial cells. J. Cell Biol. **124**: 101–115.
21. DENKER, B.M., C. SAHA, S. KHAWAJA & S.K. NIGAM. 1996. Involvement of a heterotrimeric G protein a subunit in tight junction biogenesis. J. Biol. Chem. **271**: 25750–25753.
22. DODANE, V. & B. KACHAR. 1996. Identification of isoforms of G proteins and PKC that colocalize with tight junctions. J. Membr. Biol. **149**: 199–209.
23. KEON, B.H., S. SCHAFER, C. KUHN et al. 1996. Symplekin, a novel type of tight junction plaque protein. J. Cell Biol. **134**: 1003–1018.
24. GOTTARDI, C.J., M. ARPIN, A.S. FANNING & D. LOUVARD. 1996. The junction-associated protein, zonula occludens-1, localizes to the nucleus before the maturation and during the remodeling of cell-cell contacts. Proc. Natl. Acad. Sci. USA **93**: 10779–10784.
25. BALDA, M.S. & K. MATTER. 1998. Tight junctions. J. Cell Sci. **111**: 541–547.
26. RIDLEY, A.J. & A. HALL. 1992. The small GTP-binding protein rho regulates the assembly of focal adhesions and actin stress fibers in response to growth factors. Cell **70**: 389–399.
27. NUSRAT, A., M. GIRY, J.R. TURNER et al. 1995. Rho protein regulates tight junctions and perijunctional actin organization in polarized epithelia. Proc. Natl. Acad. Sci. USA **92**: 10629–10633.
28. HANKS, S.K. & T.R. POLTE. 1997. Signaling through focal adhesion kinase. Bioessays **19**: 137–145.
29. DENKER, B.M. & S.K. NIGAM. 1998. Molecular structure and assembly of the tight junction. Am. J. Physiol. **274**: F1–9.
30. GUAN, J.L. & D. SHALLOWAY. 1992. Regulation of focal adhesion-activated protein tyrosin kinase by both cellular adhesion and onogenic transformation. Nature **358**: 690–692.
31. TSUKITA, S., M. ITOH, A. NAGAFUCHI et al. 1993. Submembranous junctional plaque proteins include potential tumor suppressor molecules. J. Cell Biol. **123**: 1049–1053.
32. MACRAE, T.H. 1992. Towards an understanding of microtubule function and cell organization: an overview. Biochem. Cell Biol. **70**: 835–841.
33. GUMBINER, B. 1987. Structure, biochemistry, and assembly of epithelial tight junctions. Am. J. Physiol. **253**: C749–C758.
34. MADARA, J.L., D. BARENBERG & S. CARLSON. 1986. Effects of cytochalasin D on occluding junctions of intestinal absorptive cells: further evidence that the cytoskeleton may influence paracellular permeability and junctional charge selectivity. J. Cell Biol. **102**: 2125–2136.

35. DRENCHAHN, D. & R. DERMIETZEL. 1988. Organization of the actin filament cytoskeleton in the intestinal brush border: a quantitative and qualitative immunoelectron microscope study. J. Cell Biol. **107:** 1037–1048.
36. HECHT, F. et al. 1996. Expression of the catalytic domain of myosin light chain kinase increases paracellular permeability. Am. J. Physiol. **271:** C1678–C1684.
37. TSUNEO, K., U. BRAUNEIS, Z. GATMAITAN & I. ARIAS. 1991. Extracellular ATP, intracellular calcium and canalicular contraction in rat hepatocye doublets. Hepatology **14:** 640–647.
38. FASANO, A., C. FIORENTINI, G. DONELLI et al. 1995. Zonula occludens toxin modulates tight junctions through protein kinase C–dependent actin reorganization, in vitro. J. Clin. Invest. **96:** 710–720.
39. DUFFEY, M.E., B. HAINAN, S. HO, C.J. BENTLEY. 1981. Regulations of epithelial tight junction permeability by cyclic AMP. Nature **204:** 451–452.
40. BAKKER, R. & J.A. GROOT. 1984. cAMP-mediated effects of quabain and theophylline on paracellular ion selectivity. Am. J. Physiol. **246:** G213–G217.
41. KRASNEY, E., MADARA, D. DIBONA & R. FRIZZELL. 1983. Cyclic AMP regulates tight junction permselectivity in flounder intestine. Fed. Proc. **42:** 1100. (Abstr.)
42. PALANT, C.E., M.E. DUFFEY, B.K. MOOKERJEE et al. 1983. Ca^{2+} regulation of tight junction permeability and structure in Necturus gallbladder. Am. J. Physiol. **245:** C203–C212.
43. THELEN, M., A. ROSEN, A.C. NAIM & A. ADEREM. 1991. Regulation by phospherylation of reversible association of a myristoylated protein kinase C substrate with the plasma membrane. Nature **351:** 320–322.
44. ELLIS, B., E.E. SCHNEEBERGER & C.A. RABITO. 1992. Cellular variability in the development of tight junctions after activation of protein kinase C. Am. J. Physiol. **263:** F293–F300.
45. STENSON, W.F., R.A. EASOM, T.E. RIEHL & J. TURK. 1993. Regulation of paracellular permeability in Caco-2 cell monolayers by protein kinase C. Am. J. Physiol. **265** (Gastrointest. Liver Physiol. **28**): G955–G962.
46. HECHT, C., C. POTHOULAKIS, J.T. LaMONT & J.L. MADARA. 1988. *Clostridium difficile* toxin A perturbs cytoskeletal structure and tight junction permeability of cultured human intestinal epithelial monolayers. J. Clin. Invest. **82:** 1516–1524.
47. FIORENTINI, C. & M. THELESTAM. 1991. *Clostridium difficile* toxin A and its effects on cells. Toxicon **29:** 543–567.
48. LOPEZ-VANCELL, R., G. BEATY, E. STEFANI et al. 1984. Changes in paracellular and cellular ionic permeabilities of monolayers of MDCK cells infected with influenza or vascular stomatitis viruses. J. Membr. Biol. **81:** 171–180.
49. FASANO, A., S. UZZAU, C. FIORE & K. MARGARETTEN. 1997. The enterotoxic effect of zonula occludens toxin (Zot) on rabbit small intestine involves the paracellular pathway. Gastroenterology **112:** 839–846.
50. WANG, W., S. UZZAU, S.E. GOLDBLUM & A. FASANO. Human zonulin, a potential modulator of intestinal tight junctions. J. Cell Sci. In press.
51. PAGANELLI, R., R.J. LEVINSKI, J. BROSTOFF & D.G. WRAITH. 1979. Immune complexes containing food proteins in normal and atopic subjects after oral challange and effect of sodium cromoglycate on antigen absorption. Lancet **i:** 1270–1272.
52. MATHAN, M.M., G. CHANDY & V.I. MATHAN. 1997. Ultrastructural changes in the upper small intestinal mucosa in patient with cholera. Gastroenterology **112:** 839–846.
53. MADARA, J.L. & J.S. TRIER. 1980. Structural abnormalities of jejunal epithelial cell membranes in celiac sprue. Lab. Invest. **43:** 254–261.
54. SCHULZKE, J.D., C.J. BENTZEL, I. SCHULZKE et al. 1998. Epithelial tight junction structure in the jejunum of children with acute and treated celiac sprue. Pediatr. Res. **43:** 435–441.
55. HOLLANDER, D. 1988. Crohn's disease: a permeability disorder of the tight junction. Gut **29:** 1621–1624.
56. FASANO, A., T. NOT, W. WANG, S. UZZAU, I. BERTI, A. TOMMASINI & S.E. GOLDBLUM. 2000. Zonulin, a newly discovered modulator of intestinal tight junctions, and its expression in coeliac disease. Lancet **355:** 1518–1519.

57. MADARA, J.L. 1990. Contributions of the paracellular pathway to secretion, absorption, and barrier function in the epithelium of the small intestine. *In* Textbook of Secretory Diarrhea. E. Lebenthal & M. Duffey, Eds.: 125–138. Raven Press. New York.
58. LIOTTA, L.A. 1991. Tumor invasion and metastasis: an imbalance of positive and negative regulation. Cancer Res. **51:** 5054S–5059S.
59. JIANG, W.G. 1996. E-chaderin and its associated protein catenins, cancer invasion and metastasis. Br. J. Surg. **83:** 437–446.
60. WILLOTT, E., M.S. BALDA, A.S. FANNING *et al.* 1993. The tight junction protein ZO-1 is homologous to the *Drosophila* discs-large tumor suppressor protein of septate junctions. Proc. Natl. Acad. Sci. USA **90:** 7834–7838.

Analysis of Low-Molecular-Weight GTP-Binding Proteins in Two Functionally Different Intestinal Epithelial Cell Lines

JÜRGEN STEIN,[a] RUTH BAUSKE, AND RALF GERHARD

Medizinische Klinik II, Johann-Wolfgang-Goethe-University,
Frankfurt am Main, 60950 Frankfurt, Germany

ABSTRACT: Low molecular weight GTP-binding proteins are molecular switches that are believed to play pivotal roles in cell growth, differentiation, cytoskeletal organization, and vesicular trafficking. In this study, for the first time, members of this family of proteins in two functionally different intestinal epithelial cell lines are identified and characterized. [α-^{32}P]GTP blot overlay assays of cytosolic and membranous fractions revealed the presence of specific GTP-binding proteins in the range of 20–30 kDa (small GTPases) in both fractions, with considerably higher amounts in the membranous insoluble fraction. Analysis by two-dimensional electrophoresis, immunoprecipitation using monoclonal and sequence-specific polyclonal antibodies, and C3 exoenzyme-mediated ADP ribosylation demonstrated the presence of Ras, Rap, Rho, Rac, Rab, and several other small GTPases. The pattern of small GTP-binding proteins corresponded to the characteristics of the cell lines. Caco-2 cells showed a Rab5 protein that is known to be involved in endocytosis but was not found in T$_{84}$ cells. On the contrary Rab3 has been shown to participate in secretory processes. It is highly expressed in T$_{84}$ cells (sixfold compared to Caco-2 cells).

INTRODUCTION

Low-molecular-weight GTP-binding proteins (small GTPases) are molecular switches that regulate diverse cellular processes, such as growth, differentiation, signal transduction, organization of the actin cytoskeleton, and intracellular transport.[1–3] More than fifty members of the Ras superfamily of small GTPases have been reported; these proteins are divided into five groups based on sequence homology—namely, the Ras, Rho, Arf, Rab, and Ran subfamilies.[4–6] In normal cells these proteins cycle between the GDP- (inactive) and GTP- (active) bound forms to regulate various essential biological processes. This cycling is positively regulated by guanine nucleotide exchange factors, which promote formation of active GTP-bound proteins, and negatively by GTPase-activating proteins (GAPs), thus promoting formation of inactive GDP-bound proteins.[1] The Ras superfamily proteins behave as molecular switches and regulate a diverse spectrum of intracellular processes, including cellular proliferation and differentiation (Ras, Rap), vesicular trafficking (Rab, Arf), and cytosk-

[a]Address for correspondence: J. Stein, M.D., Ph.D., Medizinische Klinik II, Johann-Wolfgang-Goethe-University, Theodor-Stern-Kai 7, 60950 Frankfurt, Germany. Voice: +49 69-6301 6899; fax: +49 69-6301 6448.
j.stein@em.uni-frankfurt.de

eletal control (Rho, Rac). The mammalian Rho subfamily comprises ~10 distinct proteins: Rho (A, B, and C), Rac (1 and 2), Cdc42 (also known as G25K), RhoE, RhoG, RhoD, and TC10.[7] Functional aberrations involving some of these regulatory proteins have been shown to underlie various pathological conditions.[8,9]

To study the distribution of low-molecular-weight GTP-binding proteins in two functionally different intestinal epithelial cell lines, an [α-^{32}P]GTP overlay assay was used, in combination with a new two-dimensional isoelectric focusing (IEF) and SDS/PAGE mapping technique.

METHODS

Materials

GTP, GDP, and ATP were purchased from Sigma Chemical Co. (St. Louis, MO, USA). C3-exoenzyme of *Clostridium botulinum* was supplied by Calbiochem (San Diego, CA, USA). [α-^{32}P]GTP (3000 Ci mmol) and [^{32}P]NAD (1000 Ci mmol) were obtained from DuPont NEN (Boston, MA, USA) and Amersham (Arlington Heights, IL, USA), respectively. The monoclonal and polyclonal antibodies (sequence specific) were purchased from Santa Cruz Biotechnology, Inc. (Santa Cruz, CA, USA). All antibodies have been confirmed as specific for the respective GTP-binding proteins by this commercial source. *C. botulinum* C3 ADP-ribosyl transferase (C3) was obtained from List Biological Laboratories, Inc. (Campbell, CA).

Cell Culture

Both cell lines (T$_{84}$ and Caco-2) obtained from the American Type Cell Collection (ATCC) were cultured in a humidified incubator at 37°C in 5% CO$_2$-95% air and grown in Dulbecco's modified Eagle's medium (DMEM)/F12 (Gibco) supplemented with 5% neonatal calf serum, penicillin (100 U/mL), and streptomycin (100 μg/mL). Caco-2 cells were held under the same conditions and maintained in DMEM (Gibco) supplemented with 10% fetal calf serum, penicillin (100 U/mL), and streptomycin (100 μg/mL). Cells were seeded in culture tissue plastic flasks, and medium was changed every other day. Confluence was reached after seven days, controlled by contrast phase microscopy.

Cell Fractioning and In Vitro ADP Ribosylation

ADP-ribosylation assays were done by the method described by Chong *et al.*[15] Cytosolic and membrane fractions were prepared as recently described.[10] Cells were homogenized in cold 50 mM Tris buffer, pH 7.5, containing 0.5 mM EDTA, 0.5 mM PMSF, and 2 mM 2-mercaptoethanol, followed by centrifugation at 35000 *g* for 60 min at 4°C. The supernatant obtained after this step was the soluble fraction, and the insoluble pellet was rehomogenized in the same buffer and centrifuged for 30 min at 35000 *g*. The resulting pellet was used as a membranous insoluble fraction. Protein was determined by the method of Bradford.[11] Membrane or cytosolic fractions (40 μg) were incubated in the presence of 10 μCi/mL [^{32}P]NAD (Amerham Corp.), and 5 μg/mL C3 toxin. ADP-ribosylated proteins were separated by SDS-PAGE and visualized by autoradiography using Kodak X-omat film (Eastman kodak Co., Rochester, NY).

Two-Dimensional SDS/PAGE and Transfer to Nitrocellulose for
[α-^{32}P]GTP Overlay and Immunoblotting

A combination of isoelectric focusing (IEF) and SDS/PAGE was used to resolve proteins in two dimensions, as described previously.[6] For SDS-PAGE analysis, [α-^{32}P]GTP overlay assays, and immunoblotting, suitable aliquots of soluble and insoluble fractions were dissolved in 65 mm Tris buffer, pH 7.4, containing 2% SDS and 2% 2-mercaptoethanol and separated on 12.5% acrylamide gels according to Laemmli.[12] Transfer of proteins from SDS-PAGE gels to nitrocellulose was carried out in Tris-glycine buffer using a Bio-Rad transfer unit. For two-dimensional electrophoresis, soluble and insoluble fractions were prepared in 9 M urea containing 4% Nonidet P-40 (NP-40), 100 mm DTT, and 2% ampholytes (pH range 3.5 to 10.0). For the first-dimensional IEF, gel strips (11×0.5 cm) were used with a linear, immobile pH gradient (Pharmacia, Sweden). Samples were applied and IEF was run at 15°C with 20,000 Vh applied. Gel strips were equilibrated for 2×20 min at 50°C under constant gentle shaking and loaded on a 12% polyacrylamide gel to resolve the proteins into the second dimension. For transfer to nitrocellulose membranes, gels were initially washed with 50 mm Tris-HCl, pH 7.5, containing 20% glycerol, and electrophoretically transferred to nitrocellulose paper in 10 mM NaHCO$_3$/3 mM NaCO$_3$ (pH 9.8), as described by Huber *et al.*[6]

GTP Blot Overlay Assays

[α-^{32}P]GTP binding to proteins on nitrocellulose was carried out as described by Lapetina and Reep[13] and Huber *et al.*[6] Transfer blots from SDS-PAGE were rinsed 15 min with GTP-binding buffer, consisting of Tris, pH 7.5, 0.2% Tween 20, 2 mM MgCl$_2$, 1 mM EGTA, 1 mM DTT, and 4 μM ATP. The blots were then incubated with [α-^{32}P]GTP (1 μCi mL^{-1}: specific activity, 3000 Ci mmol^{-1}; 1 Ci = 37 Gbq) in binding buffer for 90 minutes. The blots were then washed extensively with several changes of binding buffer. All these incubations were carried out at room temperature. Finally blots were air dried and subjected to autoradiography (2–24 h at −80°C). To test the specificity of [α-^{32}P]GTP binding, blots were preincubated with GTP, GDP, and ATP in binding buffer for 30 minutes. Transblots of two-dimensional gels were incubated with labeled GTP, as described by Huber *et al.*[6] Blots were initially washed for 30 min with binding buffer consisting of 50 mm NaH$_2$PO$_4$, pH 7.5, 10 mM MgCl$_2$, 2 mM DTT, 0.2% Tween 20, 4 μM ATP, and 0.1 mM EGTA. The rest of the binding procedure was similar to that used for transblots of SDS-PAGE gels.

RESULTS AND DISCUSSION

Members of the Ras-related family of small GTP-binding proteins are involved in regulating many different biological functions.[1–3] As recently shown by Huber *et al.*,[6] high-resolution two-dimensional gel electrophoresis (2DGE) provides a convenient and sensitive method for analyzing small GTP-binding proteins of a given cell,

TABLE 1. Distribution of low molecular weight GTP-binding proteins in Caco-2 cells

	Homogenate	BLM	BBM
smg p28	+	+/−	+
smg p27	+	+/−	+
smg p25	+	+	+
smg p23	+	−	++
smg p21	+	−	+

NOTE: +, strong labeling; ±, weak labeling; −, no labeling.

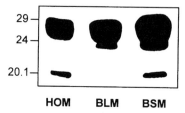

FIGURE 1. Binding of [^{32}P]GTP to GTP-binding proteins of homogenate, basolateral, and apical membrane (HOM, BLM, BBM). BLM and BBM show different patterns of these proteins. The 28-, 27-, and 25-kDa GTP-binding proteins were detectable in both fractions, whereas the 21-kDa and the highly enriched 25-kDa proteins were located only on the brushborder membrane. X-ray film was exposed for 16 hours.

tissue, or subcellular fraction. The authors have been able to identify 28 small GTP-binding proteins in MDCK cells, BHK-21 cells, and rat hippocampal neurons. This assay has been used previously to analyze these proteins from several other tissues. The technique relies on the ability of proteins to bind GTP ligand following separation by SDS-PAGE and blotting to nitrocellulose membrane.[5,6,13] To study the distribution of Ras-related GTP-binding proteins in the two functionally different intestinal cell lines—that is, Caco-2 and T$_{84}$—we have modified this mapping technique by providing the IEF with the Rotophor® system using a pH gradient from 3.5–10.

FIGURE 1 illustrates the [α-^{32}P]GTP-binding proteins from Caco-2 cells as detected by using their ability to bind [α^{32}P]GTP. Equal amounts of protein (40 µg) were subjected to the assay, after having been separated electrophoretically by SDS-PAGE and transferred to nitrocellulose. The specificity of GTP binding by these proteins was tested by preincubating blots with unlabeled GTP, GDP, or ATP. Although 0.2–1.0 µM concentration of GTP or GDP completely inhibited the binding of labeled GTP, 4 µM ATP had no effect (data not shown). Using this technique, five distinct smg proteins with molecular masses of 28, 27, 25, 23, and 21 kDa were identified in the homogenate of Caco-2 cells. In both fractions, BBM and BLM, smg proteins of 28, 27, and 25 kDa were detectable. Furthermore, proteins with molecular masses of 25 and 23 kDa were highly enriched in the BBM fraction, whereas no 23- and 21-kDa proteins were detectable in the BLM fraction (TABLE 1). These results show a different distribution of smg proteins between BBM and BLM and in-

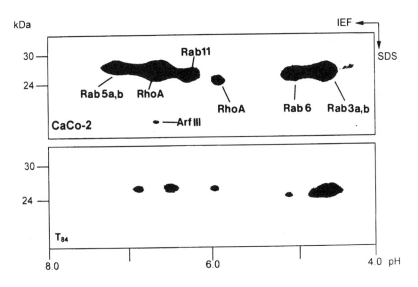

FIGURE 2. Image of [^{32}P]GTP binding of small GTP-binding proteins from Caco-2 and T$_{84}$ cells after high-resolution two-dimensional PAGE. The area shown represents the 20–25 kDa range after SDS/12.5% PAGE. The pH range is shown at the bottom. The directions of SDS and IEF are indicated.

dicate a role for smg proteins associated with the apical membrane in receptor-mediated processes.

To identify new and study already known GTP-binding proteins, we combined the GTP-overlay method with the analysis by two-dimensional PAGE. The protein blots were checked for GTP-binding proteins using the GTP overlay assay. FIGURE 2 illustrates the detection of different small GTPases (20–30 kDa), with pH ranging from 3.5 to 10. Whereas one-dimensional SDS-PAGE-GTP overlay analysis showed the presence of both the 20- to 25-kDa and 27- to 29-kDa GTP-binding proteins, the 2D electrophoresis-GTP overlay assay failed to demonstrate the 33- to 45-kDa group. The reasons for this are not clear, but it could conceivably be due to the duration of the electrophoresis, differences in the transfer of proteins to the membrane, or decreased efficiency of renaturation. To confirm that these 27- to 29-kDa GTP-binding proteins do not arise as a result of incomplete dissociation, we compared the GTP-binding protein profiles of samples that were solubilized in Laemmli buffer with those solubilized in the 9-M urea containing 4% NP-40 and 100 mM DTT used in preparing samples for two-dimensional analysis. The two preparations were directly loaded onto a one-dimensional SDS/PAGE and showed identical patterns of GTP-binding proteins, including both 20- to 25- and 27- to 29-kDa proteins.

All GTP-binding proteins mapped here are listed in TABLE 2, with molecular mass and isoelectric point (IEP). One unidentified protein spot with an apparent molecular mass of 21 kDa was found at identical electrophoretic positions in all cell types studied and may serve as an internal position marker.

TABLE 2. Summary of the GTP-binding proteins found by two-dimensional electrophoresis

IEP	Caco-2	T₈₄	Predicted GTPase
pH 3.7	16.0	5.0	Rab protein
pH 4.7–4.8	5.2	31	Rab 3a,b
pH 5.3	3.8	16.0	Rab6
pH 6.1	18.6	13.7	RhoA[a]
pH 6.5	22.1	15.1	Rab11
pH 6.8	19.0	10.8	RhoA[a]
pH 8.0	10.8		Rab5

[a]Mapped by antibodies; values are given as percent of total [32P]GTP binding.

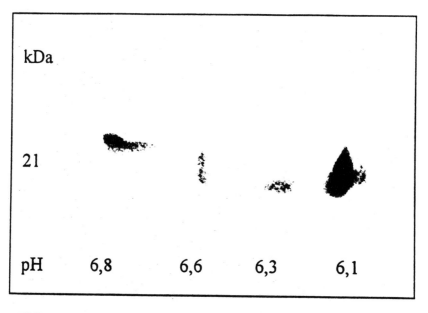

FIGURE 3. At least 4 Rho proteins are the target for C3 toxin-mediated ADP ribosylation in T₈₄ cells. Cells were ADP ribosylated by C3 *in vitro*, using [32P]NAD as substrate, and samples were analyzed by 2DGE (pH range 3.5–10.0) followed by autoradiography. The positions of RhoA (pI 6.1), RhoC (pI 6.8), and RhoD (pI 6.6 or 6.3) are shown.

Among the over 50 members of the family of small G proteins, only Rho proteins (RhoA, B, and C) are ADP ribosylated by C3 toxin. Rac proteins are very poor substrates for this enzyme *in vitro*.[14,16] To identify the 21-kDa proteins shown in FIGURE 3, C3-ADP-ribosylation-2DGE was used. Autoradiography of 2DGE revealed four different spots with pI values identical to RhoC (pI 6.8), RhoD (pI 6.6/6.3), and Rho A (pI 6.1).

The identity of the C3-labeled band was also confirmed by Western blotting with the anti-RhoA-specific monoclonal antibody (data not shown). Our results differ from those obtained by Santos *et al.*[7] showing that in IEC6 cells C3 specifically ADP ribosylated only one protein, with a PI corresponding to that of RhoA.

Although both cell lines are intestinal colon carcinoma cell lines, they have different functions. Caco-2 cells are distinguished as more absorptive, T_{84} as a more secretory cell line. The pattern of small GTP-binding proteins corresponded to the characteristics of the cell lines; for example, Caco-2 cells showed a Rab5 protein that is known to be involved in endocytosis but was not found in T_{84} cells. On the contrary, Rab3 has been shown to participate in secretory processes. It is highly expressed in T_{84} cells (sixfold compared to Caco-2 cells).

Much work remains to be done to place all the known and unknown small GTPases into their cellular context and learn about their specific roles. The Rab and Arf proteins provide molecular and functional specificity to membrane traffic between different organelles. The Rho and Rac proteins are involved in regulating cytoskeletal functions and respiratory burst. It is obvious from previous studies that the expression of many of these proteins will vary with the cell type; therefore, the pattern of smg proteins could help to identify and characterize a more secretory or more absorptive cell line.

ACKNOWLEDGMENT

This work is dedicated to Prof. Dr. W. F. Caspary on the occasion of his sixtieth birthday.

REFERENCES

1. BOGUSKI, M.S. & F. McCORMICK. 1993. Proteins regulating Ras and its relatives. Nature **366:** 643–654.
2. HALL, A. 1994. Small GTP-binding proteins and the regulation of the actin cytoskeleton. Annu. Rev. Cell Biol. **10:** 31–54.
3. NUOFFER, C. & W.E. BALCH. 1994. GTPases: multifunctional molecular switches regulating vesicular traffic. Annu. Rev. Biochem. **63:** 949–990.
4. CHARDIN, P. 1991. Small GTP-binding proteins of the Ras family: a conserved functional mechanism? Cancer Cells **3:** 117–126.
5. GROMOV, P. S. & J.E. CELIS. 1994. Several small GTP-binding proteins are strongly down-regulated in simian virus 40 (SV40) transformed human keratinocytes and may be required for the maintenance of the normal phenotype. Electrophoresis **15:** 474–81.
6. HUBER, L.A., O. ULLRICH, Y. TAKAL *et al.* 1994. Mapping of Ras-related GTP-binding proteins by GTP overlay following two-dimensional gel electrophoresis. Proc. Natl. Acad. Sci. USA **91:** 7874–7878.
7. SANTOS, M.F., S.A. McCORMACK, Z. GUO *et al.* 1997. Rho proteins play a critical role in cell migration during the early phase of mucosal restitution. J. Clin. Invest. **100:** 216–225.
8. BOS, J. L. 1989. *Ras* oncogenes in human cancer: a review. Cancer Res. **49:** 4682–4689.
9. HENKEMEYER, M., D.J. ROSSI, D.P. HOLMYARD *et al.* 1995. Vascular system defects and neuronal apoptosis in mice lacking Ras GTPase-activating protein. Nature **377:** 695–701.
10. STEIN, J., R. GERHARD, S. ZEUZEM & W.F. CASPARY. 1995. Cellular distribution of smg-binding proteins in CaCo-2 cells. Eur. J. Clin. Invest. **25:** 793–795.

11. BRADFORD, M.M. 1976. A rapid and sensitive method for the quantification of microgram quantities of protein utilizing the principle of protein-dye binding. Anal. Biochem. **72:** 248–254.
12. LAEMMLI, U.K. 1970. Cleavage of structural proteins during the assembly of the head of bacteriophage T4. Nature **227:** 680–685.
13. LAPETINA, E. G. & B.R. REEP. 1987. Specific binding of $[\alpha\text{-}^{32}P]$ GTP to cytosolic and membrane-bound proteins of human platelets correlates with the activation of phospholipase C. Proc. Natl. Acad. Sci. USA **84:** 2261–2265.
14. AKTORIUS, K. & I. JUST. 1995. *In vitro*-ribosylation of Rho by bacterial ADP-ribosyl transferases. Methods Enzymol. **256:** 184–195.
15. CHONG, L.D., A. TRAYNOR-KAPLAN, G.M. BOKOCH & M.A. SCHWARTZ. 1994. The small GTP-binding protein Rho regulates a phosphatidylinositol 4-phosphate 5-kinase in mammalian cells. Cell **79:** 507–513.
16. HILL, C.S., J. WYNNE & R. TREISMAN. 1995. The Rho family GTPases RhoA, Rac1, and CDC42 regulate transcriptional activation by SRF. Cell **80:** 1159–1170.

Increased Tight Junction Permeability Can Result from Protein Kinase C Activation/Translocation and Act as a Tumor Promotional Event in Epithelial Cancers

JAMES M. MULLIN,[a] KATHLEEN V. LAUGHLIN, NICOLE GINANNI,
COLLEEN W. MARANO, HILARY M. CLARKE, AND
ALEJANDRO PERALTA SOLER

*The Lankenau Medical Research Center, 100 Lancaster Avenue,
Wynnewood, Pennsylvania 19096, USA*

ABSTRACT: Exposure of LLC-PK$_1$ epithelial cell cultures to phorbol ester tumor promoters causes immediate translocation of protein kinase C–α (PKC-α) from cytosolic to membrane-associated compartments. With a very similar time course, a dramatic and sustained increase in tight junctional (paracellular) permeability occurs. This increased permeability extends not only to salts and sugars but macromolecules as well. Fortyfold increases of transepithelial fluxes of biologically active EGF and insulin occur. Recovery of tight junction barrier function coincides with proteasomal downregulation of PKC-α. The failure to downregulate activated membrane-associated PKC-α has correlated with the appearance of multilayered cell growth and persistent leakiness of tight junctions. Accelerated downregulation of PKC-α results in only a partial and transient increase in tight junction permeability. Transfection of a dominant/negative PKC-α results in a slower increase in tight junction permeability in response to phorbol esters. In a separate study using rat colon, dimethylhydrazine (DMH)-induced colon carcinogenesis has been preceded by linear increases in both the number of aberrant crypts and transepithelial permeability, as a function of weeks of DMH treatment. Adenocarcinomas of both rat and human colon have been found to have uniformly leaky tight junctions. Whereas most human colon hyperplastic and adenomatous polyps contain nonleaky tight junctions, adenomatous polyps with dysplastic changes did possess leaky tight junctions. Our overall hypothesis is that tight junctional leakiness is a late event in epithelial carcinogenesis but will allow for growth factors in luminal fluid compartments to enter the intercellular and interstitial fluid spaces for the first time, binding to receptors that are located on only the basal-lateral cell surface, and causing changes in epithelial cell kinetics. Tight junctional leakiness is therefore a promotional event that would be unique to epithelial cancers.

Our group is studying the potential importance of increased tight junction (TJ) permeability in the development of epithelial cancers. Our initial work in epithelial cell cultures, principally the LLC-PK$_1$ cell line, has now been extended to epithelial tis-

---

[a]Address for correspondence: Dr. James M. Mullin, The Lankenau Medical Research Center, 100 Lancaster Avenue, Wynnewood, PA 19096. Voice: 610-645-2703; fax: 610-645-2205. mullinj@mlhs.org

sue, specifically rat and human colon. Toward this end we use phorbol esters as not only protein kinase C (PKC) activators,[1,2] but also as tumor promoting agents.[3,4] Their action as tumor promoters is likely mediated through PKC, but the current intense focus in the biomedical literature on signal transduction has largely obscured the fact that as tumor promoters, phorbol esters can act synergistically with carcinogens, such as DMBA, MMNG, or DMH, to increase the number and size of tumors.[4,5] Tumor promoters cause epigenetic events that follow upon the genetic events of the primary carcinogens to confer a growth advantage to the initiated population of cells. Phorbol esters, as PKC activators, have already enabled us to acquire unique insights into the signal transduction aspects of TJ regulation. However, we hope also to demonstrate that the increased TJ permeability engendered by these PKC-activating tumor promoters is able to play a causal role in the process of epithelial carcinogenesis.

Using the MDCK cell line, Ojakian[6] was the first to demonstrate that phorbol esters can increase transepithelial permeability. Using the LLC-PK$_1$ cell line[7] we then showed that not only is transepithelial electrical resistance (R_t) decreased immediately by phorbol esters, but paracellular fluxes of mannitol (MW 182) and polyethylene glycol (MW 4000) are dramatically increased, suggesting again that specifically TJ permeability was increased. Further work by our group then showed that diacylglycerols produced effects similar to phorbol esters,[8] as did nonphorbol ester, protein kinase C–activating tumor promoters such as aplysiatoxin and teleocidin.[9] These observations, along with the ability of protein kinase C inhibitors, such as bisindolylmaleimide[10] to block the effects of phorbol esters on TJ permeability, provided support that these effects were mediated by protein kinase C.

Using the electron-dense dye ruthenium red, we observed that even though TJs of phorbol ester–treated cell sheets appeared normal in transmission electron micrographs, these same junctions were *uniformly* leaky to this compound.[11,12] Therefore, phorbol esters were causing an increase in transepithelial permeability not by eliminating TJs but simply by increasing their permeability. Unlike R_t measurements or monitoring paracellular flux of ^{14}C-D-mannitol, use of electron dense dyes allowed us to examine the epithelial cell sheet not as a summed average of paracellular permeabilities, but rather on a cell-by-cell basis. This then permitted us to determine that, whereas an acute exposure (< 24 h) to phorbol esters renders all TJs leaky, prolonged exposure to phorbol esters (> 3 days) allows for discrete regions of cells to appear, the TJs of which have recovered their barrier function and will not allow transit of ruthenium red.[12]

The fact that *chronic* phorbol ester exposure was actually allowing for the reversal of *acute* phorbol ester exposure on TJ permeability was initially shown by recovery of R_t and diminution of paracellular ^{14}C-D-mannitol flux after 4 days of continuous exposure.[12,13] Interestingly, these parameters returned almost to the levels of time-matched controls, but never attained those levels. The use of electron-dense dyes revealed two populations of epithelia in chronically phorbol ester–treated cell sheets. Whereas most of the epithelium was the normal one-cell-layer thickness and possessed of TJs that would not allow ruthenium red to pass, there were also random, polyp-like regions, all of the TJs of which were leaky to ruthenium red.[12] This indicated that increased TJ permeability and aberrant cell growth were associated.

With the rapid increase in TJ permeability following acute phorbol ester exposure, an equally rapid translocation of total PKC *activity* (histone type III phosphorylation) from cytosolic to membrane-associated compartments occurred.[11] With cells *chronically* exposed to phorbol esters, PKC activity remained near absent in the cytosolic fraction but also declined in membrane-associated fractions as well. At this point, we decided to focus upon one isoform, PKC-α, as it had been reported to be a major PKC isoform in LLC-PK$_1$ epithelia.[14] Western immunoblots of this isoform mirrored the data previously obtained by activity assays. First, there was very rapid translocation of PKC-α from cytosolic to membrane-associated fractions after phorbol ester exposure. However, when phorbol ester exposure was continued from hours to days and even weeks, the amount of membrane-associated PKC-α now decreased but still showed a significant presence (even though cytosolic PKC-α remained absent).[13] As was true for the use of electron-dense dyes regarding permeability, immunofluorescent detection of PKC-α allowed for discrimination of different cell populations regarding this isoform's distribution within the phorbol ester chronically treated culture. These results showed that the one cell layer–thick regions of the cell sheet (which manifested ruthenium red impermeable TJs) had nearly completely downregulated PKC-α, whereas the multilayered polyp-like regions (the TJs of which were near uniformly permeable to ruthenium red) exhibited PKC-α levels as high or higher than those of control cultures never exposed to phorbol esters.[12,13] Failure to downregulate activated and translocated PKC-α was therefore correlating with tight junctional leakiness and aberrant cell growth.

Guided by the work of Lee *et al.*[15,16] we assumed for the phorbol ester–treated LLC-PK$_1$ cell sheets that PKC-α would also be downregulated through proteasomes. We have very recently observed that the PKC activator bryostatin causes a dramatically faster downregulation of PKC-α than phorbol esters. Interestingly, the decrease in R_t caused by bryostatin is less than half that caused by phorbol esters and reverses in less than 6 hours in the continued presence of bryostatin (compared to the 4 days required in the presence of phorbol esters). Along with the fact that the proteasome inhibitor MG-132 both blocks PKC-α downregulation in the presence of bryostatin and prevents spontaneous recovery of Rt, all of these data provide further evidence that PKC-α does play a regulatory role in TJ permeability and that its rapid downregulation following its activation is essential to maintaining epithelial barrier function.[32] This does not mean that other PKC isoforms are not equally important in the regulation of TJ permeability, as published literature in this and other cell types suggests.[17,18]

In analyzing the permeability change itself, our group has been increasingly occupied by the question of what types of solutes are permeable across the TJs of phorbol ester–treated cell sheets, inasmuch as the biomedical significance of the leakiness lies in the answer to this question. Our early studies showed dramatic, over 20-*fold* increases in the transepithelial paracellular fluxes of the relatively small molecules D-mannitol and polyethylene glycol.[7] More recently we have observed that phorbol ester–treated cell sheets manifest increased TJ permeability even to dextrans of 2 million molecular weight, with cationic ferritin being the only solute tested, the paracellular permeability of which was not increased by phorbol esters.[19] This signifies that TJs of phorbol ester–treated cells would be leaky not only to salts and sugars but also to many types of growth factor proteins as well. Indeed, the transepithelial flux of epidermal growth factor (EGF) and insulin increased over 20-fold in the presence of phorbol esters, coming across the cell sheet in a biologically active form.[20,21] When one considers that growth factor proteins can exist at extremely high levels in the luminal fluids of many epithelial tissues[22-24] and that growth factor receptors are almost uniformly polarized on epithelial cells to the basal-lateral or antiluminal cell surface,[25-27] the onset of *chronic* tight junctional leakiness to luminal growth factors during epithelial carcinogenesis might provide a growth advantage to a transformed focus of epithelia. It would, in effect, be an epigenetic change characteristic of the tumor promotion phase of epithelial carcinogenesis.

In order to test the *in vivo* relevance of many of the above cell culture findings, we are using colon tissue because of its ease of use in Ussing chamber studies, the large published literature on its transport properties, and its applicability as a source of large numbers of epithelial cancers. In initial studies we observed the action of phorbol esters on colon TJ permeability. Unlike the dramatic and sustained effects of phorbol esters on MDCK and LLC-PK$_1$ barrier function, we found only minor and transient increases in TJ permeability for both rat and human colon exposed to either TPA or PDBU at concentrations as high as 5×10^{-6}M (Ref. 28 and unpublished data). Although we have not as yet examined the possible role of rapid proteasomal PKC downregulation in such a weak response, we have identified high serine/threonine phosphatase activity as a contributing factor (unpublished results). Although the nature of the mechanisms responsible for a weak effect of phorbol esters on colon barrier function are not known as yet, we would argue teleologically that this may very well be beneficial for the tissue and the organism. A property as basic as barrier function should not be significantly compromised whenever PKC isoforms are activated in an epithelial cell layer, an event that may need to occur frequently and may play key roles in a wide range of unrelated physiological responses. Therefore, the nature of these compensatory mechanisms that prevent dramatic changes of barrier function are pivotal to understand, as the failure of such mechanisms may be central to many disease processes.

In our studies using rat and human colon to examine the state of TJ permeability in adenocarcinomas and preneoplastic foci, as would be predicted from the work of Martinez-Palomo[29] and Saito,[30] we found that the TJs of epithelia within adenocarcinomas of rat and human colon are leaky electrically and to the electron-dense dye ruthenium red.[31] However, the TJs between epithelia of human hyperplastic polyps and most adenomatous polyps are not leaky. Only adenomatous polyps of high atypia possess TJs leaky to ruthenium red. Moreover, rats treated weekly with the colon carcinogen dimethylhydrazine exhibit a linear increase in the number of aberrant crypts and a linear decrease in R_t as a function of weeks of exposure before any adenocarcinomas are present. A pattern of increased TJ permeability *before* the onset of epithelial cancer appears to be in evidence. A goal in future work will be to investigate increased permeability of growth and motility factors such as EGF and trefoil peptide across the preneoplastic colon, and to study the potential contributory role of such luminal growth factors in promoting the onset of colon tumors. Only in so doing can the TJ permeability findings derived from the cell culture models be seen as *causally* contributing to carcinogenesis in epithelial tissues *in vivo*.

REFERENCES

1. NISHIZUKA, Y. 1984. The role of protein kinase C in cell surface signal transduction and tumor promotion. Nature **308:** 693–698.
2. SHARKEY, N.A., K.L. LEACH & P.M. BLUMBERG. 1984. Competitive inhibition by diacylglycerol of specific phorbol ester binding. Proc. Natl. Acad. Sci. USA **81:** 607–610.
3. BLUMBERG, P.M., G. ACS, L.B. ARECES et al. 1994. Protein kinase C in signal transduction and carcinogenesis. Prog. Clin. Biol. Res. **387:** 3–19.
4. DIAMOND, L., T.G. O'BRIEN & W.M. BAIRD. 1980. Tumor promoters and the mechanism of tumor promotion. Adv. Cancer Res. **32:** 1–63.
5. BOUTWELL, R.K. 1974. The function and mechanism of promoters of carcinogenesis. CRC Crit. Rev. Toxicol. **2:** 419–443.
6. OJAKIAN, G.K. 1981. Tumor promoter–induced changes in the permeability of epithelial cell tight junctions. Cell **23:** 95–103.
7. MULLIN, J.M. & T.G. O'BRIEN. 1986. Effects of tumor promoters on LLC-PK1 renal epithelial tight junctions and transepithelial fluxes. Am. J. Physiol. **251:** C597–C602.
8. MULLIN, J.M. & M.T. MCGINN. 1988. Effects of diacylglycerols on LLC-PK1 renal epithelia: similarity to phorbol ester tumor promoters. J. Cell. Physiol. **134:** 357–366.
9. MULLIN, J.M., M.T. MCGINN, K.V. SNOCK & S. IMAIZUMI. 1990. The effects of teleocidin and aplysiatoxin tumor promoters on epithelial tight junctions and transepithelial permeability: comparison to phorbol esters. Carcinogenesis **11:** 377–385.
10. MARANO, C.W., K.V. LAUGHLIN, L.M. RUSSO & J.M. MULLIN. 1995. The protein kinase C inhibitor, bisindolylmaleimide, inhibits the TPA-induced but not the TNF-induced increase in LLC-PK1 transepithelial permeability. Biochem. Biophys. Res. Commun. **209:** 669–676.
11. MULLIN, J.M., K.V. SNOCK, R.D. SHURINA et al. 1992. Effects of acute vs. chronic phorbol ester exposure on transepithelial permeability and epithelial morphology. J. Cell. Physiol. **152:** 35–47.
12. MULLIN, J.M., A.P. SOLER, K.V. LAUGHLIN et al. 1996. Chronic exposure of LLC-PK1 epithelia to the phorbol ester TPA produces polyp-like foci with leaky tight junctions and altered protein kinase C–α expression and localization. Exp. Cell Res. **227:** 12–22.
13. MULLIN, J.M. J.A. KAMPHERSTEIN, K.V. LAUGHLIN et al. 1997. Transepithelial paracellular leakiness induced by chronic phorbol ester exposure correlates with polyp-like foci and redistribution of protein kinase C–α. Carcinogenesis **18:** 2339–2345.
14. WARTMANN, M., D.A. JANS, P.J. PARKER et al. 1992. Overexpression of the alpha-type protein kinase (PK) C in LLC-PK1 cells does not lead to a proportional increase in the induction of two 12-o-tertradecanoylphorbol-13-acetate–inducible genes. Cell Regul. **2:** 491–502.
15. LEE, H.W., L. SMITH, G. R. PETTIT et al. 1996. Dephosphorylation of activated protein kinase C contributes to downregulation by bryostatin. Am. J. Physiol. Cell Physiol. **271:** C304–C311.
16. LEE, H.W., L. SMITH, G. R. PETTIT et al. 1996. Ubiquitination of protein kinase C–α and degradation by the proteasome. J. Biol. Chem. **271:** 20973–20976.
17. DODANE, V. & B. KACHAR. 1996. Identification of isoforms of G proteins and PKC that colocalize with tight junctions. J. Membr. Biol. **149:** 199–209.
18. LUM, H. & A.B. MALIK. 1994. Regulation of vascular endothelial barrier function. Am. J. Physiol. Lung Cell. Mol. Physiol. **267:** L223–L241.
19. MULLIN, J.M., C.W. MARANO, K.V. LAUGHLIN et al. 1997. Different size limitations for increased transepithelial paracellular solute flux across phorbol ester and tumor necrosis factor–treated epithelial cell sheets. J. Cell. Physiol. **171:** 2 26–233.
20. MULLIN, J.M., N. GINANNI & K.V. LAUGHLIN. 1998. Protein kinase C activation increases transepithelial transport of biologically active insulin. Cancer Res. **58:** 1641–1645.
21. MULLIN, J.M. & M.T. MCGINN. 1987. The phorbol ester, TPA, increases transepithelial epidermal growth factor flux. FEBS Lett. **221:** 359–364.

22. DIGNASS, A., K. LYNCH-DEVANEY, H. KINDON *et al.* 1994. Trefoil peptides promote epithelial migration through a transforming growth factor beta–independent pathway. J. Clin. Invest. **94:** 376–383.
23. JORGENSEN, P.E., L. VINTER-JENSEN & E. NEXO. 1996. An immunoassay designed to quantitate different molecular forms of rat urinary epidermal growth factor with equimolar potency: application on fresh rat urine. Scand. J. Clin. Lab. Invest. **56:** 25–36.
24. MROCZKOWSKI, B. & M. REICH. 1993. Identification of biologically active epidermal growth factor in human fluids and secretions. Endocrinology **132:** 417–425.
25. BISHOP, W.P. & J.T. WEN. 1994. Regulation of Caco-2 cell proliferation by basolateral membrane epidermal growth factor receptors. Am. J. Physiol. Gastrointest. Liver Physiol. **267:** G892–G900.
26. PLAYFORD, R.J., A.M. HANBY, S. GSCHMEISSNER *et al.* 1996. The epidermal growth factor receptor (EGF-R) is present on the basolateral, but not the apical, surface of enterocytes in the human gastrointestinal tract. Gut **39:** 262–266.
27. SCHEVING, L.A., R.A. SHIURBA, T.D. NGUYEN *et al.* 1989. Epidermal growth factor receptor of the intestinal enterocyte. Localization to laterobasal but not brush border membrane. J. Biol. Chem. **264:** 1735–1741.
28. SIMONS, R.M., K.V. LAUGHLIN, J.A. KAMPHERSTEIN *et al.* 1998. Pentobarbital affects transepithelial electrophysiological parameters regulated by protein kinase C in rat distal colon. Dig. Dis. Sci. **43:** 632–640.
29. MARTINEZ-PALOMO, A. 1970. Ultrastructural modifications of intercellular junctions between tumor cells. In Vitro **6:** 15–20.
30. SAITO, T. 1984. Ultrastructural changes on the junctional complexes in the human urinary bladder carcinoma by thin sectioning and freeze fracture. J. Clin. Electron Micros. **17:** 201–209.
31. PERALTA SOLER, A., R.D. MILLER, K.V. LAUGHLIN *et al.* 1999. Increased tight junction permeability is associated with the development of colon cancer. Carcinogenesis **20:** 1425–1431.
32. CLARKE *et al.* 2000. Exp. Cell Res. In press.

Stress-Induced Decrease of the Intestinal Barrier Function

The Role of Muscarinic Receptor Activation

JACK GROOT,[a,b] PIETER BIJLSMA,[a] ANNETTE VAN KALKEREN,[a] AMANDA KILIAAN,[a] PAUL SAUNDERS,[c] AND MARY PERDUE[c]

[a]*Institute for Neurobiology, University of Amsterdam, Amsterdam, the Netherlands*
[c]*Intestinal Disease Research Program, McMaster University, Hamilton, Canada*

ABSTRACT: Recently the breakdown of the barrier function of the intestinal epithelium after application of an experimental psychological and physical stress protocol in rats has been observed. Not only did smaller molecules pass from the luminal to the serosal side, but so also did larger proteins with the dimensions of luminal antigens and toxins. The increased permeability for macromolecules is primarily due to a decrease of the tightness of the *zonula occludens*, but an increased endocytotic uptake indicates that transcytosis is increased also. From studies of model systems it can be concluded that activation of the intracellular protein kinase C route by muscarinic receptor activation or histamine receptor activation can be one of the underlying cellular pathways. The physical pathway relaying the stress from the brain to the intestinal tract appears to be the parasympathetic branch of the autonomic nervous system. The difference in reaction of different strains suggests that coping style is an important determinant of the response of the intestinal barrier to stress.

INTRODUCTION

Reports in this volume indicate the large increase in scientific interest in the regulation of the barrier function of the intestinal epithelial layer. In this review we want to draw attention to a field in intestinal physiology that to date has been studied far less, namely, the effect of stress on the loss of barrier function of the epithelium for large, possibly allergenic or toxic macromolecules. The word *stress* is used here to indicate the psychological or physical exogenous action on an individual animal, which may lead to a change in the activity of one or more physiological mechanisms. It may be that an observed change in activity is adaptational and part of the specific coping activity to maintain homeostasis, or it may be that the change is due to failure of the regulatory mechanism as its limits are exceeded.

Although studies on effects of stress on motor function are relatively numerous, studies of the mucosa are restricted to gastric ulceration or secretory activity.

The strong relation between occurrence of stressful events and gastrointestinal diseases has long been recognized. In particular, a relation between personality pro-

[b]Address for correspondence: J.A. Groot, Institute for Neurobiology, Kruislaan 320, 1098 SM Amsterdam, the Netherlands. Voice: +31-20-5257728; fax: +31-20-6659125.
groot@bio.uva.nl

file and intestinal diseases has been proposed,[1,2] although, obviously, disease affects the behavior, and therefore a causative relation is not clear. Moreover, the possible mechanisms relating psychological events and somatic diseases are only slowly being unraveled. In this aspect it is important to recognize that psycho-immuno-neuroendocrinology as a new field of biomedical research is developing as a promising approach in understanding psychosomatic interactions. This integrative part of physiology studies the interplay between neuronal, immunological, and endocrinological signals in relation to psychological state. It is now well recognized that physical and/or emotional stress has an impact on the immune system. Therefore stressful events may cause the dysregulation of immunological functions and may lead to relapse of inflammatory diseases. However, the activation of the immune reaction leading to inflammation usually requires the triggering of the immune system by allergens or haptens from the outside. In other words, the foreign substances must be presented to the immune system; they have to pass the epithelial barrier.

In the reported studies of intestinal epithelial barrier function, it is often not possible to answer the question whether permeability was increased before onset of the disease or whether the relapse of the disease precedes the increase of permeability. Moreover, the methodology to study permeability *in vivo* has not been standardized. The data often are ambiguous and do not allow one to draw conclusions about defects in the barrier function for immunoactive compounds, which are usually much larger than the permeability probes used. Furthermore, even if barrier function for macromolecules has been lost, it cannot be decided from *in vivo* studies whether this is due to histological observable lesion or to morphological undetectable changes in the transepithelial transport routes.

In animal studies, the *in vitro* approach with isolated intestine mounted in Ussing type chambers has shown that a number of endogenous signals can increase the transepithelial transport of macromolecules. We will illustrate that some of these findings, especially the observed increase of permeability for macromolecules induced by the muscarinic receptor agonist carbachol, may well be related to the mechanism involved in stress-induced changes of mucosal function.

Interesting and promising studies have been initiated with the use of human biopsy materials. In particular, the development of techniques wherein small forceps biopsies can be used have shown that, in most aspects related to the present paper, the effect of added drugs is similar to that in intestinal preparations of laboratory animals.[3]

Although the data are far from complete, the findings may be helpful in unraveling the causative relation between stressful events, coping style of the individual, and development or relapse of gastrointestinal diseases.

In the first section we will describe the effects of muscarinic receptor activation on transepithelial transport of macromolecules through isolated intestinal sheets.

In the second section we will develop the working hypothesis that in individuals with a coping style that involves a relatively strong increase of parasympathetic activity upon stressful events, the intestinal barrier may become impaired by physical or emotional stress. We propose that this may lead to the influx of allergenic materials, which in these individuals may lead to the development of immune reactions or other unwanted reactions triggered by the molecules that have gained access to the wrong side of the epithelium. In the case of an existing immune sensitization, a local anaphylactic response may be induced by a new challenge, but this requires that the allergen be able to penetrate in an immunological intact state through the epithelium. Stressful events may be instrumental in increasing the permeability. However, recent findings suggest that in the case of sensitization, the epithelial cells can already recognize the allergen, leading to augmented transcytosis and activation of sensitized mast cells.[4] It remains to be studied whether stressors can affect this activated endocytotic-transcytotic route. However, the finding that stress can increase the endosomal uptake[5] suggests that this is not an unreasonable possibility.

MUSCARINIC RECEPTOR ACTIVATION AND TRANSEPITHELIAL TRANSPORT OF MACROMOLECULES

Isolated intestinal sheets from small intestine of rats or guinea pigs were stripped from the muscle layers and mounted to separate the two compartments of a diffusion chamber, usually equipped with electrodes to measure the potential difference between the two compartments and to measure the resistance of the epithelium. Radiolabeled mannitol (MW 180 D) and PEG 4000 (MW 4 kD) were used as permeability probes, and horseradish peroxidase (HRP, MW 40 kD) was used as a model for the transport of macromolecules. The probes were added in the same concentration to the apical side, and samples were drawn every 15 minutes from the serosal side. HRP activity was measured enzymatically so that the flux of HRP indicated the flux of the intact molecule. The presence of HRP in the epithelial cells and underlying tissue was determined histologically.[6] Predictably, HRP flux was much smaller than PEG 4000, which was smaller than the mannitol flux (J_{man} : $J_{PEG4000}$: J_{HRP} = 40 : 7 : 1). The flux of HRP reached a more or less constant level after about 90 minutes. When at this time carbachol (10^{-5} M) was added to the serosal side, the flux of all probes increased, reaching a new semiconstant value 60 minutes after carbachol addition. Carbachol is a stable muscarinic receptor agonist. The muscarinic receptor antagonist atropine, when added before carbachol, totally prevented the increase of the fluxes, indicating that muscarinic receptor activation was involved in the increase of the permeability.

The relative increase of the fluxes (ΔJ_{man} : $\Delta J_{PEG4000}$: ΔJ_{HRP} = 18 : 2.5 : 1) was not different than the predicted relative diffusion coefficients based on the square root of the molecular weights: D_{man} : $D_{PEG4000}$: D_{HRP} = 18 : 3.2 : 1. This strongly suggests that the carbachol-induced increase of the fluxes was due to opening of pores, which allow diffusion of the three compounds. Because the increased flux of HRP represents the flux of the intact molecule, evidently this HRP escaped cellular breakdown, and this suggests that the increased flux was via the paracellular pathway. This suggestion was corroborated by the finding in electron micrographs that some tight junctions were heavily stained with the product of the HRP reaction and that the lateral intercellular spaces were filled with HRP product.

Carbachol activates the intracellular Ca^{2+} and PKC pathways that lead to Cl secretion. However, the increased permeability has no relationship with Cl secretion, because the phenomenon occurs in the presence of bumetanide (which inhibits basolateral NaK_2Cl cotransport and thereby Cl uptake from the serosal side) and Ba^{2+} (which blocks basolateral K channels and thereby decreases the driving force for Cl efflux from the cells). Therefore, Cl secretion and increased permeability appear to be two different effects of muscarinic receptor activation. Activation of another secretory pathway, namely the cAMP pathway by forskolin, did not increase the fluxes, again indicating that there was no causal relationship between secretion of Cl and increased permeability.

Careful examination of the electron micrographs suggested that carbachol also increased the amount of HRP product in endosomes in the epithelial cells. A semi-quantitative visual analysis showed that carbachol induced a transient increase of the endosomal uptake of HRP in villi as well as in crypt cells. By contrast, forskolin decreased the uptake. Thus, two mechanisms appear to increase the transepithelial transport of HRP through isolated small intestine of the rat, increased diffusional permeability through the tight junctions, and increased transcytosis.

This phenomenon is not restricted to the rat. Recent unpublished observations (Verheul *et al.*) showed that carbachol can also increase the permeability of the small intestine of guinea pigs. This was observed earlier in isolated small intestine of cow's milk–sensitized guinea pigs.[3] Interestingly, in these animals the spontaneous permeability for HRP (8 pmole/cm^2·h) and the spontaneous I_{sc} (60 µA/cm^2) could be reduced significantly by application of atropine to 3.2 pmole/cm^2·h and 30 µA/cm^2. This suggests that the intestinal sheets of these guinea pigs had a spontaneous high cholinergic tone that led to increased permeability for macromolecules. Electron micrographs showed HRP products in the lateral intercellular spaces as well as increased amounts in the endosomes, as in rat. It remains to be studied whether this was caused by the sensitization, or whether stressful events related to the housing conditions were causing the high secretory tone and high permeability. More recent data indicate that under carefully controlled conditions guinea pig intestine of non-sensitized animals do not show the high permeability.

Human biopsies that were obtained during endoscopical examination and mounted in Ussing type chambers likewise responded to carbachol with an increased permeability for HRP.[3] This could only be observed when the permeability was not already spontaneously high (arbitrarily less than 15 pmole/cm^2·h). Paracellular staining with HRP product could be observed in the tissues with spontaneously high permeability and after carbachol application. Carbachol also increased the endosomal uptake of HRP, as in rats.

So, although this effect of carbachol on tight junction permeability occurs in the intestine of a number of species, it is also evident from the literature that muscarinic receptor activation can also increase the permeability of tight junctions in other tissues like pancreas, liver, and airway epithelium (for review, see Ref. 6). This suggests that muscarinic receptor activation triggers a mechanism that exists in a number of epithelia (and possibly also in endothelia[7] that modulate the tight junction permeability). However, one should be aware that the application of carbachol, even in the presence of TTX, does not guarantee that the effect of carbachol on isolated stripped intestinal sheets is due to activation of muscarinic receptors on the epithelial cells only. For instance, mast cells can be activated to release their contents, which can lead to increased permeability.[8]

Although the molecular mechanisms involved are far from clear, it is interesting that there are a number of reports relating PKC activation by phorbol esters to increased permeability[9] in cell lines as well as in native intestine.[10,11] Moreover other secretagogues, like histamine and bile salts that may share their intracellular pathway with muscarinic receptor activation, induce increased permeability. Also toxins, for example, ZOT,[12] which acts via the Ca-PKC route, induce increased permeability for macromolecules. Thus, it may be that activation of this intracellular route triggers a change in tight junction structure, thereby increasing the diameter of pores in the tight junctional strands.

However, it is our own experience that the effect of carbachol does not occur in every piece of intestine of the same animal, and sometimes all six tissues of one animal were nonresponsive. We therefore suggest that other factor(s) beside the activation of the Ca-PKC pathway are involved. This putative factor(s) may come from the lamina propria cells and could bring the epithelial cells in a potentiated state; for example, enterocytes may have to be preactivated by, for instance, cytokines (Oprins *et al.*, this volume). Another possibility for the variation in reactions comes from an intriguing report describing that the effect of phorbol ester on *in vitro* intestinal epithelium was only apparent when the animals were not anesthetized with pentobarbital.[10] Although this finding has not yet been confirmed, it deserves the attention of investigators using intestinal *in vitro* material from anesthetized animals.

STRESS AND INTESTINAL PERMEABILITY

The findings discussed above strongly suggest that release of acetylcholine in the intestine can lead to increased permeability for macromolecules. Because acetylcholine is the common transmitter from the vagus nerve, it would be interesting to know whether stressors that activate the vagus can increase intestinal permeability. If so, this could well be the link between stress and the relapse of inflammatory bowel diseases. Although effects of stress on the visceral system have not been studied in large detail, a number of groups[13–16] have studied the effect of stress on gastrointestinal motor activity, and it is apparent that stressors increase the gastric emptying time, increase small intestinal transit time (but increase the colonic motor activity, leading to shortened transit time), and also increase fecal output in rodents. Detailed studies of the involvement of the stress signal between hypothalamus and pituitary, corticotropine releasing factor/hormone (CRF/CRH), have shown that intracerebral application of CRF mimics the effects of stress on gastrointestinal motor function. This signal plays a role in the PVN-dorsal vagal motor center, which is involved in the regulation of gastric motor function, whereas the PVN-locus ceruleus is involved in regulation of the colonic motor activity via the parasympathetic sacral system. Thus, it may well be that the parasympathetic nerve conveys the stress-signal from the CNS to the gut.[17] The effect of stress on gastric ulceration is via activated vagal activity,[18] and cold stress, especially, strongly activates the vagus via TRH release in the dorsal motor nucleus.[19] In healthy human subjects, using a triple lumen perfusion technique, psychological stress (dichotomous listening) induced decreased water absorption from the jejunum and reversed the net NaCl absorption into secretion. This effect could be prevented by intravenous atropine infusion, suggesting the involvement of a parasympathetic cholinergic mechanism.[20] From studies with rats and dogs it appeared that cerebral TRH but not intravenous TRH stimulates water secretion in the small intestine. The effect was mediated by the vagus and muscarinic and VIP-ergic mechanism. The latter was more important. These findings, in conjunction with the finding that central vagal activation by TRH can activate mast cell degranulation, which in turn affects the mucosal transport and barrier function,[21] illustrates the importance of the vagus nerve in the regulation of the intestinal mucosal function. Earlier Greenwood and Mantle[22] showed that direct electrophysiological vagus stimulation could increase the blood to lumen flux of proteins. Thus, there is ample evidence that the vagus nerve affects not only the absorption/secretion balance of the intestine but also the barrier function. Therefore, it is not surprising that in a stress-susceptible rat strain, characterized by a low acetylcholinesterase (AchE) activity, cold restraint stress could induce increased permeability for macromolecules in the small and large intestine, which could be prevented by atropine.[5,23,24] The permeability was measured in isolated and stripped intestine, *in vitro*, two hours after finishing the cold restraint stress. The permeability for the probe macromolecule (HRP) of tissue from stressed rats was more than four times larger than in the controls, and the permeability for Cr-EDTA was only 1.3 times larger than in controls. This is comparable with the differing effects of carbachol on the flux of large and smaller molecules and suggests that most of the increased HRP flux is via the paracellular pathway. This was corroborated by electronmiscroscopy, which showed filling of paracellular space and tight junctions with the product of HRP reaction in epithelia from stressed rats. Such a filling was never observed in the controls and was also not detectable in atropine pretreated stressed rats. Similar to carbachol, the enterocytes of stressed rats showed a large increase of HRP product-filled endosomes. This indicates that part of the increased transepithelial flux of intact macromolecules can be ascribed to transcytosis. Again, treatment with atropine prevented the increase of transcytosis.

These are, as far as we know, the first direct indications that some types of stressors can induce an increased permeability for antigenic macromolecules. The consequence of the finding of an increased paracellular flux of macromolecules induced by stress may not only be in the understanding of effects of stress on the relapse of inflammation or induction of allergic reactions in sensitized individuals but may also be of importance in the understanding of processes leading to development of oral intolerance/sensitization. Induction of oral tolerance is considered to be dependent on antigen-presenting activity of the epithelial cells and CD8[+] antigen-nonspecific suppressor T cells. An increased load of antigens via the paracellular route may circumvent this safeguard and may activate the mechanisms leading to oral intolerance.[25]

It is important to know whether this phenomenon is specific for this strain of rats; in other words, is the effect of stress dependent on the coping style of the animal? This question has been touched on in a preliminary study by comparing two strongly related strains, namely, the Roman High Avoidance and the Roman Low Avoidance rats in a parallel study with the same stress protocol as in the Wistar Kyoto rats.[26] The Swiss sublines of Roman high- and low-avoidance (RHA/Verh and RLA/Verh) rats have been selected and bred for rapid (RHA/Verh) versus extremely poor (RLA/Verh) acquisition of two-way active avoidance. Behavioral and physiological measures of reactivity to stress appear to be prominent characteristics differentiating both rat lines. The Low Avoidance rats reacted to the cold restraint stress with an increased permeability of the small intestine for HRP that was 3.9 times larger than in the controls. Small intestinal sheets from the stressed High Avoidance rats showed no increase of the flux as compared with their controls. Injection with atropine before the cold restraint stress prevented the stress-induced increase of the flux in Low Avoidance rats. The results may be related to the fact that the two rat strains respond differently to infusion of CRF into the central nucleus of the amygdala, which has direct connections with autonomic regulatory nuclei in the brain stem.[27]

The RLA rats are considered to react with a parasympathetic stress response (J.M. Koolhaas, personal communication). The decrease of the intestinal barrier function in the RLA rats is another example of the differing coping style of these rats, and the absence of effect in the RHA rats shows that the intestinal permeability of some individuals can be totally unsusceptible to this type of stress. The strain comparison studies may reveal useful regulatory differences that may exist between individual coping styles with stress. It is an important issue for which to try to find behavioral characteristics, which may predict the possible breakdown of intestinal barrier function.

The difference in reaction of Wistar Kyoto and the parent Wistar strain was not as extreme as observed in the Roman lines. In the Wistar rats the increase of permeability was less compared with Wistar Kyoto rats but still significantly higher than in controls (P.R. Saunders, personal communication). An important difference between the two strains is their AchE activity, which is much lower in the Wistar Kyoto rats.[24] This, and the fact that methyl-nitrate-atropine (which is considered to act only in the periphery, because it cannot pass the blood brain barrier) can prevent the breakdown of the barrier, whereas hexamethonium was without effect, again suggests the involvement of a peripheral cholinergic muscarinic mechanism in the regulation of epithelial tightness. Quite important in this respect is the observation in Sprague-Dawley rats that cholinergic enzyme activities (choline acetyltransferase, ChAT, and AchE) in the gastrointestinal tract are affected by acute stress, probably via adrenal corticoids.[28] The AchE activity after cold-restraint stress was somewhat decreased in the ileum, but strongly increased in the duodenum and colon. ChAT activity was decreased in the duodenum, ileum, and colon. The relationship between these effects and the effects of stress on barrier function cannot be given; however, it may be that the changes in enzyme activity are related to adaptive mechanisms. In this respect it should be mentioned that the HPA axis via CRF and corticosterone might reduce the effect of stress in a trinitrobenzenesulfonic acid colitis rat model, again implying a protective role of glucocorticosteroids on stress effects. Although a large body of evidence suggests that muscarinic receptors are involved in the breakdown of the barrier function, this does not mean that the ultimate response comes from the muscarinic receptors on the epithelial cells only. The mast cells in the lamina propria can be triggered to release preformed and newly synthesized signals upon activation by the vagus nerve.

Histamine can increase paracellular leak in *in vitro* rat small intestine,[29] and it has been shown that Rat Mast Cell Protease II (RMCP II) can induce leakiness of the intestinal epithelium.[30] The importance of mast cells has been clearly shown by using mast cell stabilizers[31] and mast cell-deficient rats.[32] The data concerning the effects of CRF on intestinal motility, of course, triggered experiments where CRF effects (intracerebral and peripheral) were investigated. The application of peripheral CRF mimics the effect of stress on colonic permeability, and the CRF antagonist alpha helical CRF 9–41 prevented the effect of CRF. However, atropine, hexamethonium, and bretylium could also prevent the effect of CRF, indicating that the peptide can activate a multitude of peripheral signaling pathways. The release of mast cell signals may also play a role in the effect of peripheral CRF on permeability[31] and mucin release,[33] because mast cell stabilizers stop the CRF effect on permeability and mucin release, and CRF receptors are expressed on the mast cells.[34] Nerve–mast cell interaction has been illustrated functionally by showing the existence of a Pavlovian reflex of mast cells employing psychological conditioning.[35,36]

The type of stressors used and whether they are chronically applied or acute may also affect the response of the intestine. For instance, it appeared that isolated housing can increase intestinal permeability of the small intestine of mice (A. van Kalkeren *et al.*, unpublished observations), and preliminary evidence suggests that the same applies for guinea pigs. Application of subchronic mild acoustic stress (95 dB) during 7 days increased the HRP flux through isolated ileum of Wistar rats two times via increased endocytotic uptake.[3] There was no evidence for increased paracellular flux. Interestingly, an increased acoustic load (105 dB) was without effect on intestinal permeability.

This review underscores the large gaps in our knowledge of regulation of the barrier function of the intestine in the organism. Although usually suggested, it has not yet been proven that breakdown of barrier function is related to the induction of oral intolerance or the relapse of inflammatory diseases. To study this relationship, experiments relating oral sensitization and stress should be performed, and we should also have the possibility of maintaining the barrier function with a pharmacological tool during the stress protocol.

REFERENCES

1. ALMY, T., F. KERN, JR. & M. TULIN. 1949. Alterations in colonic function in man under stress: experimental production of sigmoid spasm in healthy persons. Gastroenterology **12:** 425–436.
2. SULLIVAN, A. 1935. Psychogenic factors in ulcerative colitis. Am. J. Dig. Dis. **2:** 651–656.
3. BIJLSMA, P.B. 1999. An aproach to intestinal permeability. Ph.D. thesis. University of Amsterdam, Amsterdam.
4. BERIN, M.C., A.J. KILIAAN, P.C. YANG *et al.* 1997. Rapid transepithelial antigen transport in rat jejunum: impact of sensitization and the hypersensitivity reaction. Gastroenterology **113:** 856–864.
5. KILIAAN, A.J., P.R. SAUNDERS, P.B. BIJLSMA *et al.* 1998. Stress stimulates transepithelial macromolecular uptake in rat jejunum. Am. J. Physiol. **275:** G1037–G1044.
6. BIJLSMA, P.B., A.J. KILIAAN, G. SCHOLTEN *et al.* 1996. Carbachol, but not forskolin, increases mucosal-to-serosal transport of intact protein in rat ileum *in vitro*. Am. J. Physiol. **271:** G147–G155.
7. FRIEDMAN, A., D. KAUFER, J. SHEMER *et al.* 1996. Pyridostigmine brain penetration under stress enhances neuronal excitability and induces early immediate transcriptional response. Nat. Med. **2:** 1382–1385.
8. RAMAGE, J.K., A. STANISZ, R. SCICCHITANO *et al.* 1988. Effect of immunologic reactions on rat intestinal epithelium: correlation of increased permeability to chromium 51-labeled ethylenediaminetetraacetic acid and ovalbumin during acute inflammation and anaphylaxis. Gastroenterology **94:** 1368–1375.
9. GROOT, J.A. 1998. Correlation between electrophysiological phenomena and transport of macromolecules in intestinal epithelium. Vet. Q. **20** (Suppl. 3): S45–S49.
10. SIMONS, R.M., K.V. LAUGHLIN, J.A. KAMPHERSTEIN *et al.* 1997. Pentobarbital affects transepithelial electrophysiological parameters regulated by protein kinase C in rat distal colon. Dig. Dis. Sci. **43:** 632–640.

11. PEREZ, M., A. BARBER & F. PONZ. 1997. Modulation of intestinal paracellular permeability by intracellular mediators and cytoskeleton. Can. J. Physiol. Pharmacol. **75:** 287–292.
12. FASANO, A., C. FIORENTINI, G. DONELLI *et al.* 1995. *Zonula occludens* toxin modulates tight junctions through protein kinase C-dependent actin reorganization, *in vitro*. J. Clin. Invest. **96:** 710–720.
13. MCRAE, S., K. YOUNGER, D.G. THOMPSON & D.L. WINGATE. 1982. Sustained mental stress alters human jejunal motor activity. Gut **23:** 404–409.
14. MONNIKES, H., B.G. SCHMIDT & Y. TACHE. 1993. Psychological stress-induced accelerated colonic transit in rats involves hypothalamic corticotropin-releasing factor. Gastroenterology **104:** 716–723.
15. BUENO, L., M. GUE & C. DELRIO. 1992. CNS vasopressin mediates emotional stress and CRH-induced colonic motor alterations in rats. Am. J. Physiol. **262:** G427–G431.
16. LEE, C. & S.K. SARNA. 1997. Central regulation of gastric emptying of solid nutrient meals by corticotropin releasing factor. Neurogastroenterol. Motil. **9:** 221–229.
17. TACHE, Y., V. MARTINEZ, M. MILLION & J. RIVIER. 1999. Corticotropin-releasing factor and the brain-gut motor response to stress. Can. J. Gastroenterol. **13** (Suppl. A): 18A–25A.
18. CHO, C.H., B.S. QUI & I.C. BRUCE. 1996. Vagal hyperactivity in stress induced gastric ulceration in rats. J. Gastroenterol. Hepatol. **11:** 125–128.
19. YANG, H., S.V. WU, T. ISHIKAWA & Y. TACHE. 1994. Cold exposure elevates thyrotropin-releasing hormone gene expression in medullary raphe nuclei: relationship with vagally mediated gastric erosions. Neuroscience **61:** 655–663.
20. BARCLAY, G.R. & L.A. TURNBERG. 1987. Effect of psychological stress on salt and water transport in the human jejunum. Gastroenterology **93:** 91–97.
21. SANTOS, J., E. SAPERAS, M. MOURELLE *et al.* 1996. Regulation of intestinal mast cells and luminal protein release by cerebral thyrotropin-releasing hormone in rats. Gastroenterology **111:** 1465–1473.
22. GREENWOOD, B. & M. MANTLE. 1992. Mucin and protein release in the rabbit jejunum: effects of bethanechol and vagal nerve stimulation. Gastroenterology **103:** 496–505.
23. SAUNDERS, P.R., U. KOSECKA, D.M. MCKAY & M.H. PERDUE. 1994. Acute stressors stimulate ion secretion and increase epithelial permeability in rat intestine. Am. J. Physiol. **267:** G794–G799.
24. SAUNDERS, P.R., N.P. HANSSEN & M.H. PERDUE. 1997. Cholinergic nerves mediate stress-induced intestinal transport abnormalities in Wistar-Kyoto rats. Am. J. Physiol. **273:** G486–G490.
25. MAYER, L. & D. EISENHARDT. 1990. Lack of induction of suppressor T cells by intestinal epithelial cells from patients with inflammatory bowel disease. J. Clin. Invest. **86:** 1255–1260.
26. BIJLSMA, P.B., P.R. SAUNDERS, P. DRISCOLL *et al.* 1998. Acute stress enhances intestinal permeability to intact protein in Roman low avoidance rats but not in Roman high avoidance rats [abstract]. Gastroenterology **114:** A352.
27. WIERSMA, A., J.P KONSMAN, S. KNOLLEMA *et al.* 1998. Differential effects of CRH infusion into the central nucleus of the amygdala in the Roman high-avoidance and low-avoidance rats. Psychoneuroendocrinology **23:** 261–274.
28. ORIAKU, E.T. & K.F. SOLIMAN. 1986. Effect of stress and glucocorticoids on the gastrointestinal cholinergic enzymes. Arch. Int. Pharmacodyn. Ther. **280:** 136–144.
29. BIJLSMA, P.B., M. ROTS, J.A. TAMINIAU & J.A. GROOT. 1993. Modulation of macromolecular permeability of rat ileum *in vitro* by histamine and NO [abstract]. Ger. J. Gastroenterol. **31:** 566.
30. SCUDAMORE, C.L., E.M. THORNTON, L. MCMILLAN *et al.* 1995. Release of the mucosal mast cell granule chymase, rat mast cell protease-II, during anaphylaxis is associated with the rapid development of paracellular permeability to macromolecules in rat jejunum. J. Exp. Med. **182:** 1871–1881.
31. SANTOS, J., D. YATES & M.H. PERDUE. 1999. Nerve-mast cell interactions regulate acute corticotropine releasing hormone (CRH) induced abnormalities in rat colonic epithelium [abstract]. AGA poster session G 2799.
32. SANTOS, J., M. BENJAMIN & M.H. PERDUE. 1999. Chronic stress induces mast cell dependent epithelial changes in the rat jejunum and colon [abstract]. AGA poster session G 4024.

33. CASTAGLIUOLO, I., J.T. LAMONT, B. QIU *et al.* 1996. Acute stress causes mucin release from rat colon: role of corticotropin releasing factor and mast cells. Am. J. Physiol. **271:** G884–G892.
34. SINGH, L.K., W. BOUCHER, X. PANG *et al.* 1999. Potent mast cell degranulation and vascular permeability triggered by urocortin through activation of corticotropin-releasing hormone receptors. J. Pharmacol. Exp. Ther. **288:** 1349–1356.
35. MACQUEEN, G., J. MARSHALL, M.H. PERDUE *et al.* 1989. Pavlovian conditioning of rat mucosal mast cells to secrete rat mast cell protease II. Science **243:** 83–85.
36. RUSSELL, M., K.A. DARK, R.W. CUMMINS *et al.* 1984. Learned histamine release. Science **225:** 733–734.

Immunologically Mediated Transport of Ions and Macromolecules

LINDA C. H. YU AND MARY H. PERDUE[a]

Intestinal Disease Research Program, Department of Pathology and Molecular Medicine, McMaster University, Hamilton, Ontario L6J3X6, Canada

ABSTRACT: There is increasing evidence supporting the involvement of immune cells and mediators in the control of intestinal physiology. Cell coculture systems and epithelial cell lines have provided convenient model systems for the study of immunomodulation of epithelial function. Abundant cytokines and immune mediators have been shown to directly or indirectly alter epithelial transport of ions and macromolecules. Animal models of hypersensitivity have shown that luminal antigen challenge in the intestine of sensitized rats induces a rapid ion secretory response due to enhanced transepithelial transport of antigen. Transport of ions and macromolecules is highly regulated and an important component of host defense. Dysregulation of epithelial function may play a role in several intestinal disorders, such as inflammatory bowel diseases and food allergy.

INTRODUCTION

The intestinal mucosa is composed of a single-celled epithelial layer covering the lamina propria. The epithelium separates the external environment from the internal milieu of the body and thus forms an important barrier restricting uptake of noxious substances from the intestinal lumen. The epithelium contains mainly transporting enterocytes and other cell types, such as goblet cells, enteroendocrine cells, and Paneth cells. M cells are specialized cells in the epithelium of Peyer's patches that have little cytoplasm, allowing facilitated transport of antigens into lymphoid regions for the induction of immune responses.

The main roles of the intestinal epithelium are (1) its transport function for nutrients, ions, and water; and (2) its barrier function. Enterocytes are derived from stem cells found in the crypt. While in the crypt, enterocytes have a secretory phenotype; but as they migrate onto the villus, they develop an absorptive phenotype. Secretion of water is important both to dissolve ingested material and to flush out toxins, pathogens, and other harmful materials. Tight junctions between adjacent epithelial cells prevent back diffusion of transported substances and produce a physiological barrier to the uptake of luminal antigens. Tight junctions are normally not permeable to macromolecules larger than 11.5 Å in Stokes radius.[1] Small quantities of macromolecules are transported across epithelial cells via an endocytic pathway. However,

[a]Address for correspondence: Dr. M.H. Perdue, Intestinal Disease Research Program, McMaster University, HSC-3N5C, 1200 Main Street West, Hamilton, Ontario, Canada, L8N 3Z5. Voice: 905-525-9140 ext. 22585; fax: 905-522-3454.

perdue@mcmaster.ca

most proteins are degraded by enzymes in lysosomes that fuse with endosomes. The physical barrier property of the intestine also involves mucus produced by goblet cells, peristalsis, and secretion of immunoglobulin (Ig) A, which binds to pathogens and prevents them from gaining access to the body. Paneth cells at the base of crypts synthesize cryptidins/defensins with antimicrobial activity.

Immune cells are present in the epithelium and lamina propria. Intraepithelial lymphocytes (IEL) are mainly CD8[+], with T cell receptors (TcR) of either the $\alpha\beta$ or the $\gamma\delta$ type. Underlying the intestinal epithelium, the lamina propria contains a wide variety of immune cells—for example, T and B lymphocytes, mast cells, and macrophages. It has become increasingly clear in recent years that the intestine is the largest lymphoid organ in the body, containing ~40% of the immune cells in the body.[2] The gut-associated lymphoid tissue (GALT) includes IELs, lymphocyte aggregates in Peyer's patches, and lymphocytes diffusely distributed throughout lamina propria. GALT not only serves as the front line of immune defense against luminal antigens/pathogens, but also is involved in regulating intestinal epithelial physiology.

EVIDENCE FOR IMMUNOPHYSIOLOGY

The involvement of immune system in the alteration of intestinal morphology and function was first suggested in studies of parasite infection. Rats infected with *Nippostrongylus brasiliensis* (Nb) or *Trichinella spiralis* (Ts) showed enterocyte detachment, partial or subtotal villous atrophy, and crypt hyperplasia. However, the expected enteropathy during infection was ablated in rats that were previously depleted of T cells by thymectomy and irradiation.[3,4] It was later shown that activated mucosal T cells were associated with crypt cell hyperplasia in an explant human small intestine culture system.[5] The recognition of immune regulation of intestinal function has resulted in the use of a new term, *immunophysiology*. In the last few years, studies in many systems ranging from *in vitro* cell culture to *in vivo* animal models, have confirmed the critical role for immune cells in controlling epithelial transport of ions and macromolecules.

Studies Using Cell Cultures

Coculture of Different Cell Types

To understand the role of immune factors in the control of intestinal physiology, cell culture systems have been used extensively. Culture systems offer a reductionist approach to study the effect of specific immune cells and mediators in altering the intestinal epithelial transport and permeability. In coculture models, intestinal epithelial cell monolayers are grown on semipermeable filters, then placed above immune cells located in the bottom compartment to mimic their presence in the lamina propria. After a defined period of coculture, the epithelial cell monolayer is mounted in Ussing chambers to study changes in physiology.

T_{84} and HT29 cells (both human colonic epithelial cells) cocultured with activated peripheral blood mononuclear cells (PBMC) consisting of monocytes and lymphocytes, display altered functions. PBMC in which the T cells were activated by anti-CD3 (antibody to a component of the TcR) induced ion transport changes (decreased short-circuit current $[I_{sc}]$ responses to carbachol and forskolin) and permeability changes (decreased resistance) in T_{84} monolayers.[6] Addition of conditioned media obtained from activated PBMC to HT29 cells resulted in a fourfold increase of the transepithelial flux of intact horseradish peroxidase (HRP, MW 44,000).[7] By contrast, monolayers that were cocultured with anti-CD3 activated T cells alone (in the absence of monocytes) did not show altered epithelial functions.[6] However, monocytes treated with conditioned media from such T cells evoked epithelial abnormalities, suggesting that mediators produced by T cells activated monocytes to affect the epithelial functions. Both IFN-γ and TNF-α were detected in supernatants obtained from activated PBMC.[6,7] Pretreating T cell–conditioned media with anti-IFN-γ or incubating the PBMC/enterocyte coculture system with anti-TNF-α totally abolished epithelial pathophysiology,[6] indicating that these cytokines play important roles.

A monocyte/enterocyte coculture system demonstrated that activating the monocytes with lipopolysaccharide (LPS) and/or the bacterial tripeptide formyl-methionyl-leucyl-phenylalanine (FMLP) resulted in a significant increase in T_{84} baseline I_{sc} (due to Cl^- secretion) and increased epithelial permeability to ^{51}Cr-EDTA.[8] Relatively few cells (< 1:20, monocytes to enterocytes) were needed to induce these effects. Anti-TNF-α antibodies inhibited the effect only when added into the coculture system, but not in the monocyte conditioned media, suggesting that TNF-α was acting via an autocrine mechanism. This study demonstrates that in the absence of T cells monocytes alone can evoke epithelial dysfunction.

A cell coculture study of superantigen-activated immune cells further demonstrates the role of PBMC in mediating epithelial functions. T_{84} cells cocultured with PBMC activated by *Staphylococcus aureus* enterotoxin B (SEB) showed decreased ion secretion in response to carbachol and forskolin, decreased resistance and increased permeability to small probes, mannitol (MW 182) and ^{51}Cr-EDTA (MW 362).[9] In further studies, it was demonstrated that TGF-β and IL-10 were able to inhibit the altered epithelial ion transport and barrier function induced by SEB activation of PBMC.[9,10] Studies of PBMC isolated from cow's milk–allergic infants have shown that conditioned media obtained from PBMC stimulated with cow's milk protein, added to HT29 cells, induced decreased resistance and increased flux of the intact macromolecule HRP and small probes (mannitol and ^{51}Cr-EDTA).[11] The drop in resistance was reversed by preincubating the conditioned media with anti-TNF-α, suggesting a direct role for TNF-α on epithelial function in certain cell lines. The direct effect of TNF-α on HT29 cells was found to be highly potentiated by IFN-γ.[11] Finally, a model of T_{84} cells cocultured with human mucosa-derived lymphocytes (MDL), where the immune cells localized within the intercellular spaces of the intestinal epithelial monolayer, provided a system that mimics the interaction of IEL with enterocytes.[12] The presence of MDL altered the physiology of the epithelial monolayer as shown by the decreased resistance and I_{sc} responses to secretogogues.

Effect of Immune Mediators

A wide array of cytokines and immune cell mediators have been reported to alter epithelial transport (summarized in TABLE 1). The summary is based on studies of cell coculture systems, epithelial cell lines, and tissue segments.

TABLE 1. Cytokines and immune mediators that alter epithelial physiology

Cell Source	Function Affected	
	Ion secretion	Macromolecular transport
Mast cell	IL-1, IL-3, IL-4, TNF-α, histamine, serotonin, PG, LT, adenosine, ROM, PAF	IL-4, RMCP II
Monocyte/macrophage	IL-1, TNF-α, ROM, PG, LT	IFN-γ, TNF-α
T cell	IL-1, IL-3, IL-4, IFN-γ	IL-4, IFN-γ, TNF-α
Granulocyte	IL-1, TNF-α, IFN-γ, PG, LT, adenosine, ROM, PAF	IFN-γ, TNF-α

ABBREVIATIONS: IL, interleukin; TNF-α, tumor necrosis factor–α; IFN-γ, interferon–γ; PG, prostaglandin; LT, leukotriene; ROM, reactive oxygen metabolites; PAF, platelet activating factor; RMCP-II, rat mast cell protease–II.

Cytokines. The direct action of cytokines on epithelial cells to alter ion transport has been examined using intestinal epithelial cell lines. Whereas neither IFN-γ nor IL-4 affected baseline ion transport in T_{84} cell monolayers, both cytokines reduced Cl$^-$ secretion induced by carbachol or forskolin.[13,14] Treatment with IFN-γ resulted in changes in T_{84} cell membrane protein composition,[13] decreased Cl$^-$ channel activity,[15] and decreased expression of the cystic fibrosis transmembrane regulator (CFTR).[16] CFTR is a cAMP-regulated Cl$^-$ channel located in the apical membrane of epithelial cells; it also regulates other Cl$^-$ channels.[17] More recently, it was found that IL-4 also resulted in a decreased ($\sim 65\%$) expression of CFTR.[14] Furthermore, IFN-γ induced a focal condensation of actin filaments in the perijunctional actomyosin ring.[18] Studies from Madara and coworkers[19,20] demonstrated that pharmacological stabilization of F-actin prevents cAMP-elicited Cl$^-$ secretion in T_{84} cells. Therefore, the reduction of forskolin-induced Cl$^-$ secretion by IFN-γ is likely mediated through the rearrangement of actin filaments.

A number of cytokines have been shown to directly affect the permeability of intestinal epithelial monolayers. Confluent T_{84} cells exposed to IL-4 for 48–72 h exhibited a marked attenuation in transepithelial barrier function as assessed by decreased resistance; this effect was inhibitable by anti-IL-4 and anti-IL-4 receptor antibodies.[21] Increased macromolecular transport in response to the addition of IL-4 has been demonstrated by Berin *et al.*[22] Treatment of T_{84} cells with IL-4 for 24 h upregulated transepithelial transport of HRP by both transcellular and paracellular pathways. Epithelial macromolecular uptake, assessed by measuring the area of HRP-containing endosomes within enterocytes, was severalfold higher in IL-4–treated epithelium compared with control monolayers. In addition, HRP was present in 70% of paracellular spaces in IL-4–treated monolayers, compared to 0% of nontreated monolayers. IL-13, which shares similar biological activities with IL-4, reduced epithelial resistance and increased the paracellular transport of a labeled dextran tracer probe.[14] HT29 cells incubated with TNF-α for 48 h in the presence of a low concentration of IFN-γ displayed reduced transepithelial resistance[11,23,24] and increased flux of intact HRP.[11,23] The synergistic effect of TNF-α and IFN-γ on barrier function was suggested to be through the upregulation of TNF-α receptor on intestinal epithelial cells by IFN-γ.[25] TNF-α has also been reported to disrupt actin filaments,[26] which may account for its effects on tight junctions to alter transepithelial resistance. One study demonstrated that treatment with a relatively high concentration of TNF-α (100 ng/mL) alone for 48 h induced a 30% reduction of the resistance of HT29 monolayers, whereas no significant effect was seen when using 0.1 to 10 ng/mL.[11]

Two cytokines have demonstrated an ability to prevent or correct barrier defects. IL-10 prevented the increased permeability to small probes induced by IFN-γ (pretreating T_{84} cells with IL-10 for three days prior to addition of IFN-γ and then concomitant treatment for four days).[27] In coculture, IL-10 acted to inhibit immune cell activation,[10] thus indirectly improving epithelial cell function. IFN-γ–induced and SEB-induced increases in T_{84} permeability have also been corrected by cotreatment of cells with TGF-β.[10,28]

Other mediators. Mast cells, macrophages, and other granulocytes secrete a large range of mediators that alter epithelial transport of ions and macromolecules. Activated mast cells release preformed mediators—for example, biogenic amines and TNF-α—and newly synthesized mediators—for example, eicosanoids and a wide array of cytokines: IL-1 to -6, IL-10, IFN-γ, TNF-α, TGF-β, and GM-CSF.[2] Histamine evokes chloride secretion in epithelial cell monolayers and intact tissues.[29,30] The effect of histamine was shown to be through an H1 receptor on intestinal epithelial cells, causing an elevation in cytosolic calcium level, protein kinase C activation, and protein phosphorylation.[31]

Eicosanoids are arachidonic acid (AA)–derived lipid mediators produced by many immune/inflammatory cells via cyclooxygenase and lipoxygenase pathways, including prostaglandins and leukotrienes, respectively. In studies of T_{84} cells, prostaglandins caused cAMP-dependent Cl^- secretion.[32] cAMP is a well-known secondary messenger in the induction of Cl^- secretion from epithelial cells.[33] Prostaglandins have been shown to increase the secretory response of intestinal epithelial monolayers directly or to serve as an intermediate step in the I_{sc} changes induced by several cytokines and mediators, such as TNF-α,[34] serotonin,[35] leukotrienes,[36] reactive oxygen metabolites,[37] and proteases.[38] The effect of these mediators on intestinal epithelial function was examined in tissue segments and will be discussed in a later section.

Adenosine causes a rapid rise in I_{sc} when administered to either the serosal or the mucosal side of T_{84} cells.[39] Increased cell content of cAMP was found after addition of adenosine,[39] and in a further study adenosine was identified to act via stimulation of the basolaterally located Na^+-K^+-ATPase pump, the Na^+-K^+-$2Cl^-$ cotransporter, and K^+ channels.[40]

Studies Using Tissue Segments

Cytokines. A number of cytokines, such as IL-1 and IL-3[41] and TNF-α,[21] have been reported to evoke epithelial ion secretion in intestinal mucosal tissues *in vitro*. Serosal addition of IL-1 and IL-3 to rabbit and chicken ileal segments caused transitory increases in I_{sc}.[41] These events were inhibited by cyclooxygenase blockers— for example, indomethacin and peroxicam—suggesting that a prostaglandin, presumably from a subepithelial source (myofibroblasts), was the terminal mediator.[41]

Short-term addition of TNF-α to the serosal side of colonic mucosal tissues induced a dose-dependent increase in I_{sc}.[21] However, no change in I_{sc} was detected in HT29 cell lines incubated with TNF-α alone for up to 100 min, suggesting that an intermediate step is involved.[34] The change in I_{sc} in tissue preparations induced by short-term incubation with TNF-α was abolished by pretreating the tissue with indomethacin.

A role for IFN-γ in increasing macromolecular transport in the small intestine was demonstrated in a suckling rat model.[42] Rat pups that were injected intraperitoneally with IFN-γ showed increased transport of intact HRP across jejunal segments. However, increased HRP flux was found only in segments containing Peyer's patches, whereas no effect was seen in patch-free segments.

Other mediators. Serotonin causes a transitory rise in I_{sc} in rat small intestine via calcium-dependent electrogenic Cl⁻ secretion; the effect was shown to be mediated by the release of prostaglandin E2.[35] The addition of leukotrienes (LTD$_4$ and LTE$_4$) to rabbit and rat ileum evoked a significant increase in Cl⁻ secretion, events that are also indomethacin sensitive.[36] Reactive oxygen metabolites (ROM), such as H_2O_2, resulted in a biphasic increase of I_{sc} in rat and rabbit colonic tissues, mediated by the release of PGI$_2$ and PGE$_2$.[37,43] Adenosine increased the I_{sc} when it was applied to the serosal side of rabbit ileal and colonic segments.[44] The serosal addition of adenosine analogues increased the tissue level of cAMP with time in rabbit ileum. Moreover, platelet activating factor (PAF) caused a biphasic increase in I_{sc} in rabbit colon and rat jejunum,[37,45] also mediated by the release of prostaglandin.[37] It has recently been shown that stimulation of proteinase-activated receptor–2 (PAR-2), which is a receptor for mast cell tryptase and pancreatic trypsin, induced an increase in I_{sc} in rat jejunum.[46] In addition, rat mast cell protease (RMCP)–II, used as an indicator of mast cell activation, has been suggested to increase macromolecular transport via the paracellular pathway.[47]

Activated Epithelium

Intestinal epithelial cells themselves are able to secrete a large range of cytokines and mediators (TABLE 2).[48,49] Receptors on intestinal epithelial cells include IL-2,[50] IL-4,[14,21] IL-13,[14,51] TNF-α,[11,52] IFN-γ,[18,52] and TGF-β,[53,54] implying an autocrine effect of at least some cytokines to alter epithelial function either directly and/or via stimulation of subepithelial cells to secrete factors that further modulate physiology. Thus epithelial cells may play a role in the exacerbation or maintenance of the inflammatory response.

Animal Models

Animal models of inflammation and hypersensitivity have confirmed the effects of immunological reactions on intestinal epithelial function. Here, we will comment on only two types of models: parasitic infection and food antigen sensitization. Nematode parasites, *Nippostrongylus brasiliensis* (Nb) and *Trichinella spiralis* (Ts) have been employed in studies of acute inflammation and anaphylaxis.

Parasite infection models. Studies of rats infected with Nb have shown that by day 7 postinfection (p.i.) alterations of intestinal ion transport occurred, shown by an increased basal I_{sc} and the reversal from absorption to secretion for Na⁺ and Cl⁻ ions.[55] The intestine of Nb-infected rats also demonstrated structural abnormalities, includ-

TABLE 2. Inflammatory mediators derived from epithelial cells

Epithelial cell mediators	
Cytokines	IL-1α, IL-1β, IL-1 receptor antagonist, IL-6, IL-8, TNF-α, TGFα, TGF-β, MCP-1, IGF-II, IGF-binding protein 1-4, GM-CSF, NGF, eotaxin
Lipid metabolites	arachadonic acid, PGE-2, PGF-2α, LTB$_4$, PAF
ROM	nitric oxide

ABBREVIATIONS: TGF, transforming growth factor; MCP, monocyte chemoattractant peptide; IGF, insulin growth factor; GM-CSF, granulocyte/monocyte colony-stimulating factor; NGF, nerve growth factor.

ing enterocyte detachment, villus atrophy, and crypt hyperplasia associated with biochemical abnormalities, such as decreased brush border enzymatic activity and increased thymidine kinase activity in crypt cells, indicating heightened crypt proliferation.[56–58] Moreover, mast cell activation, indicated by elevated serum and reduced tissue levels of RMCP-II as well as the decreased mast cell granule staining, was seen accompanying the morphological changes.[56,57] Similar functional and morphological changes were demonstrated when worm antigen was administered to the intestine of infected rats on day 35 p.i., after parasite expulsion was completed and mast cell hyperplasia had developed.[55,59,60] This study supports the hypothesis that mast cells play a role in altering ion transport following worm antigen challenge in the intestine of postinfection rats. Moreover, rats immunized against Ts also showed antigen-induced biphasic peaks in I_{sc} and increased Cl$^-$ secretion.[61] The biphasic I_{sc} peaks were mediated by histamine and serotonin (peak I) and prostaglandin E$_2$ (peak II), as assessed by using specific mediators to mimic the response as well as employing specific antagonists/inhibitors to block the effects.[61,62]

Infection of rats with Nb has been demonstrated to alter the barrier function of the intestine during acute inflammation. Increased permeability was indicated by enhanced recovery of the luminally administered probe ^{51}Cr-EDTA (MW 360), as well as a bystander macromolecule, ovalbumin (OVA, MW 45,000), in the blood.[58,63] The defective barrier function recovered by day 35, when worm expulsion was completed and healing occurred.[63] However, a secondary challenge with worm antigen induced anaphylactic responses. Increased permeability of the intestine to both ^{51}Cr-EDTA and OVA developed and peaked at 1 hour.[63] The size of OVA in the blood coincided with that of native OVA, suggesting that the intact macromolecules were getting access across the epithelium.[63]

The increased uptake of macromolecules in nematode-infected rats was postulated to be due to villus damage, which was demonstrated during acute inflammation. However, the situation was different on day 35, when morphology had returned to normal. Mast cell activation, demonstrated by high serum levels of RMCP-II, was seen following worm antigen challenge.[56,57] The release of RMCP-II from mast cells was associated with the increased paracellular permeability to macromolecules in jejunal epithelium.[47] Perfusing RMCP-II into the mesenteric artery in naive rats resulted in an increase in the accumulation of a macromolecule, human serum albumin (HSA), in the gut lumen.[47] Therefore, RMCP-II appears to play a role in enhanced macromolecular transport across the epithelium during anaphylaxis.

Food allergy models. Early studies of intraluminal antigen challenge in the small intestine of OVA-sensitized rats demonstrated a dramatic reduction of the absorption of water, Na^+, Cl^-, and K^+.[64] The response was antigen specific and IgE mediated.[64] Reduced net ion absorption results from either a decrease in mucosal-to-serosal flux or an increase in serosal-to-mucosal flux, or a combination of both. The use of radiolabled Na^+ and Cl^- ions showed that antigen challenge in sensitized intestine induced decreased mucosal-to-serosal flux and increased serosal-to-mucosal flux of Cl^-, while fluxes of Na^+ did not change.[65] The I_{sc} also increased after antigen challenge and was inhibited by Cl^- channel blockers and Cl^- free buffer.[65-67]

The transport change after antigen challenge was associated with reduced mucosal content of histamine and numbers of stained mucosal mast cells, suggesting mast cell activation.[64] The involvement of mast cells in altered ion transport in intestinal anaphylaxis was further confirmed by using mast cell–deficient animals[68] and mast cell stabilizing agents.[65,66,69] The antigen-induced change in I_{sc} in the jejunum of OVA-sensitized mast cell–deficient mice was ~70% less than that seen in +/+ mice, whereas repopulating mast cells by injection of bone marrow precursor cells from +/+ mice completely restored the I_{sc} response to antigen challenge.[68] Similar inhibition of the secretory response was found by treating rats with dozantrazole, an agent that inhibits histamine release from mucosal mast cells.[69] Moreover, histamine-1 antagonists, serotonin antagonists, and cyclooxygenase inhibitors also reduced the I_{sc} response.[66] These results suggest that active mediators released from mast cells are responsible for antigen-induced intestinal ion secretion.

In OVA-sensitized rats, no significant changes in the baseline I_{sc} and baseline net fluxes of ions were seen prior to antigen challenge, indicating that intestinal ion transport was not altered by sensitization per se. Antigen challenge to the serosal side of the intestine in Ussing chambers induced an immediate increase (~30 s) in I_{sc}, whereas the addition of antigen to the mucosal side of the tissue resulted in a lag phase of ~3 min before the rise in I_{sc}.[66] The luminal response was abolished when antigen was added to the serosal side first, suggesting that the mechanism of the I_{sc} change resulting from both luminal and serosal antigen challenge acts through a common mechanism or terminal target, likely to be mast cells.[70] The lag phase seen in response to luminal antigen challenge has been suggested to reflect the time for antigen to be transported across the intestinal epithelium to activate mast cells in the lamina propria.[66]

The mechanism of this rapid rise of I_{sc} to luminal antigen challenge has long been puzzling, inasmuch as transcellular transport of macromolecules normally requires ~20 min,[71,72] whereas paracellular transport is limited to small molecules.[1] In rats sensitized to OVA, blood recovery of luminally administered ^{51}Cr-EDTA was 60–80% higher than in control animals, suggesting that sensitization per se could result in increased intestinal permeability to small probes.[73] After luminal challenge with OVA (radius of 36 Å), its uptake was elevated 14-fold, indicating increased macromolecular transport in intestine of antigen-challenged sensitized animals.[74] Increased intestinal permeability in sensitized animals may account for the rapid rise of I_{sc} (~3 min) after luminal antigen challenge.

Evidence from guinea pigs orally sensitized to cow's milk further supports the concept of immunomodulation of intestinal transport of macromolecules. Antigen challenge in cow's milk–sensitized guinea pigs has shown that serosal challenge with β-lactoglobulin (βLG) induces an increase of the influx of an intact bystander protein, HRP, whereas no significant difference was seen in the level of degraded HRP.[75] Luminal challenge with specific antigen in OVA-sensitized rats showed an increased amount of bystander protein (HSA) in the serum.[76] A similar increased uptake of nonspecific protein has also been reported in immunized mice.[77] This may be due to release of certain cytokines—for instance, IL-4—during sensitization, which induces increased transport of nonspecific macromolecules.[22] These results suggest that antigen challenge in the intestine of sensitized animals may result in enhanced access of immunologically intact bystander macromolecules (e.g., food antigens and bacterial products) across the barrier.

The rapid change of I_{sc} (within 3 min) in response to luminal antigen challenge was suggested to be due to increased intestinal permeability following sensitization. The question of whether this enhanced permeability occurred via transcellular or paracellular pathways was addressed in a recent study.[78] Similarly to OVA-sensitized animals, the rapid rise in I_{sc} was also demonstrated in rats sensitized to HRP.[78] The use of HRP allows the researcher to monitor and visualize transepithelial transport of antigen (using kinetic enzymatic assays and electron microscopy). Electron microscopic studies showed enhanced endocytic uptake of specific antigen in enterocytes of HRP-sensitized rats at 2 min postchallenge.[78] Both the area of HRP-containing endosomes within the enterocytes and their rate of transepithelial transport were higher in HRP-sensitized animals compared with controls.[78] At this time, HRP-containing endosomes were distributed throughout the enterocytes in HRP-sensitized rats, while in OVA- and saline-treated animals HRP was restricted to only the apical region. Therefore, the epithelial uptake of antigen appeared to be specific within 3 min postchallenge (*phase I*).[78] At 30 min postchallenge, HRP was also visualized within the tight junctions and paracellular regions, suggesting that subsequent to mast cell activation, antigen transport is further enhanced by crossing through the paracellular pathway. This period (after mast cell activation) has been denoted as *phase II*. In addition, a severalfold increase in the luminal-to-serosal flux of HRP was detected in sensitized intestine in this phase, associated with increased tissue conductance.[78]

The mechanism of this enhanced transepithelial macromolecular transport was studied using mast cell–deficient rats in which phase I was retained but phase II was absent, confirming the role of mast cells in altering epithelial tight junctional permeability.[79] In phase I, the uptake of antigen was specific, indicating that antibody recognition may be involved. Antigen binding to antibody receptors may prevent the degradation of protein within endosomes and thus explain the large area of HRP-containing endosomes seen within enterocytes at 2 min postchallenge.

Role of Nerves in Immune Regulation of Epithelial Function

Immunoepithelial interactions can be modulated by nerves. The use of atropine (an antagonist to muscarinic cholinergic receptors) and the neurotoxin tetrodotoxin (TTX), a blocker to fast sodium channels, has been shown to inhibit the ion secretory response to antigen challenge,[61,68] histamine, and serotonin,[61] and to reduce the increased permeability to macromolecules.[73] These results suggest that nerves may be involved in epithelial ion secretion and permeability in response to the antigen challenge and mast cell activation. However, the relationship between neural and immunological regulation of intestinal physiology will not be described in detail in this paper.

CONCLUSIONS

The identification of a wide spectrum of immune cells and mediators that regulate intestinal epithelial functions—namely, transport of ions and macromolecules—stresses the importance of GALT in protecting the body from the constant exposure to potentially harmful materials in the gut lumen, such as antigenic food proteins and pathogenic microorganisms. Conversely, defective immunoregulation of intestinal function may lead to pathophysiology or pathology in the gut. Intestinal diseases, such as inflammatory bowel disease, celiac disease, and food allergy, may originate from abnormal control mechanisms, or may be exacerbated due to immune-mediated alterations in intestinal function. Increased ion secretion leads to the passive flux of water that can result in diarrhea; and altered macromolecular transport of luminal antigens enhances the potential for the occurrence of chronic inflammation. The study of immunophysiology may result in the development of novel therapeutic interventions for various intestinal diseases.

REFERENCES

1. MADARA, J.L. 1989. Loosing tight junctions: lessons from the intestine. J. Clin. Invest. **83:** 1089–1094.
2. PERDUE, M.H. & D.M. MCKAY. 1993. Immunomodulation of the gastrointestinal epithelium. *In* Immunopharmacology of the Gastrointestinal System. J.L. Wallace, Ed.: 15–39. Academic Press. New York.
3. FERGUSON, A. 1976. Models of intestinal hypersensitivity. Clinics Gastroenterol. **5:** 271–288.
4. MANSON-SMITH, D.F., R.G. BRUCE & D.M.V. PARROTT. 1979. Villous atrophy and expulsion of intestinal *Trichinella spiralis* are mediated by T cells. Cell. Immunol. **47:** 285–292.
5. MACDONALD, T.T. & J. SPENCER. 1987. Evidence that activated mucosal T cells play a role in the pathogenesis of enteropathy in human small intestine. Lancet **167:** 1341–1349.
6. MCKAY, D.M., K. CROITORU & M.H. PERDUE. 1996. T cell–monocyte interactions regulate epithelial physiology in a coculture model of inflammation. Am. J. Physiol. **270:** C418–C428.
7. HIRRIBAREN, A., M. HEYMAN, A. L'HELGOUACH & J.F. DESJEUX. 1993. Effect of cytokines on the epithelial function of the human colon carcinoma cell line HT29 cl 19A. Gut **34:** 616–620.
8. ZAREIE, M., D.M. MCKAY, G.G. KOVARIK & M.H. PERDUE. 1998. Monocyte/macrophages evoke epithelial dysfunction: indirect role of tumor necrosis factor-α. Am. J. Physiol. **275:** C932–C939.
9. MCKAY, D.M. & P.K. SINGH. 1997. Superantigen activation of immune cells evokes epithelial (T_{84}) transport and barrier abnormalities via IFN gamma and TNF alpha. J. Immunol. **159:** 2382–2390.
10. LU, J., D.J. PHILPOTT, P.R. SAUNDERS *et al.* 1998. Epithelial ion transport and barrier abnormalities evoked by superantigen-activated immune cells are inhibited by interleukin-10 but not interleukin-4. J. Pharmacol. Exp. Ther. **287:** 128–136.
11. HEYMAN, M., N. DARMON, C. DUPOT *et al.* 1994. Mononuclear cells from infants allergic to cow's milk secrete tumor necrosis factor alpha, altering intestinal function. Gastroenterology **106:** 1514–1523.
12. KAOUTZANI, P., S.P. COLGAN, K.L. CEPEK *et al.* 1994. Reconstitution of cultured intestinal epithelial monolayers with a mucosal-derived T lymphocyte cell line. J. Clin. Invest. **94:** 788–796.
13. HOMGREN, J., J. FRYKLUND & H. LARSSON. 1989. Gamma-interferon–mediated down-regulation of electrolyte secretion by intestinal epithelial cells: a local immune mechanism? Scand. J. Immunol. **30:** 499–503.

14. ZÜND, G., J.L. MADARA, A.L. DZUS et al. 1996. Interleukin-4 and interleukin-13 differentially regulate epithelial chloride secretion. J. Biol. Chem. **271:** 7460–7464.

15. COLGAN, S.P., C.A. PARKOS, J.B. MATTHEWS et al. 1994. Interferon-gamma induces a cell surface phenotype switch on T_{84} intestinal epithelial cells. Am. J. Physiol. **267:** C402–C410.

16. BESANÇON, F., G. PRZEWLOCKI, I. BARO et al. 1994. Interferon-gamma downregulated CFTR gene expression in epithelial cells. Am. J. Physiol. **267:** C1398–C1404.

17. BEAR, C.E., C. LI, N. KARTNER et al. 1992. Purification and functional reconstitution of the cystic fibrosis transmembrane regulator (CFTR). Cell **68:** 809–818.

18. MADARA, J.L. & J. STAFFORD. 1989. Interferon-gamma directly affects barrier function of cultured intestinal epithelial monolayers. J. Clin. Invest. **83:** 724–727.

19. SHAPIRO, M., J. MATTHEWS, G. HECHT et al. 1991. Stabilization of F-actin prevents cAMP-elicited Cl^- secretion in T_{84} cells. J. Clin. Invest. **87:** 1903–1909.

20. MATTHEW, J.B., J.A. SMITH & B.J. HRNJEZ. 1997. Effects of F-actin stabilization or disassembly on epithelial Cl^- secretion and Na-K-2Cl transport. Am. J. Physiol. **272:** C254–C262.

21. COLGAN, S.P., M.B. RESNICK, C.A. Parkos et al. 1994. IL-4 directly modulates function of a model human intestinal epithelium. J. Immunol. **153:** 2122–2129.

22. BERIN, M.C., P.C. YANG, L. CIOK et al. 1998. Role of IL-4 in macromolecular transport across human intestinal epithelium. Gastroenterology **114:** A930.

23. MAHRAOUI, L., M. HEYMAN, O. PLIQUE et al. 1997. Apical effect of diosmectite on damage to the intestinal barrier induced by basal tumor necrosis factor-alpha. Gut **40:** 339–343.

24. MULLIN, J.M. & K.V. SNOCK. 1990. Effect of tumor necrosis factor on epithelial tight junctions and transepithelial permeability. Cancer Res. **50:** 2172–2176.

25. TSUJIMOTO, M., Y.K. YIP & J. VILCEK. 1986. Interferon-gamma enhances expression of cell receptors for tumor necrosis factor. J. Immunol. **136:** 2441–2444.

26. GOLDBLUM, S.E., X. DING & J. CAMPBELL-WASHINGTON. 1993. TNF-α induces endothelial cell F-actin depolymerization, new actin synthesis, and barrier dysfunction. Am. J. Physiol. **264:** C894–C905.

27. MADSEN, K.L., S.A. LEWIS, M.M. TAVERNINI et al. 1997. Interleukin-10 prevents cytokine-induced disruption of T_{84} monolayer barrier integrity and limits chloride secretion. Gastroenterology **113:** 151–159.

28. PLANCHON, S.M., C.A.P. MARTINS, R.L. GUERRANT & J.K. ROCHE. 1994. Regulation of intestinal epithelial barrier function by TGF-β1. J. Immunol. **153:** 5730–5739.

29. HARDCASTLE, J. & P.T. HARDCASTLE. 1987. The secretory actions of histamine in rat small intestine. J. Physiol. **388:** 521–532.

30. WASSERMAN, S.I., K.E. BARRETT, P.A. HUOTT et al. 1988. Immune-related intestinal Cl^- secretion. I. Effect of histamine on the T_{84} cell line. Am. J. Physiol. **254:** C53–C62.

31. COHN, J.A., N.C. DOUGHERTY & W.F. KING. 1989. Histamine stimulates calcium-mediated protein phosphorylation in a colonic epithelial cell line. Biochem. Biophys. Res. Commun. **165:** 810–816.

32. HALM, D.R., G.R. RECHKEMMER & R.A. SCHOUMACHER. 1988. Apical membrane chloride channels in a colonic cell line activated by secretory agonists. Am. J. Physiol. **254:** C505–C511.

33. BARRETT, K.E. & K. DHARMSATHAPHORN. 1990. Mechanisms of chloride secretion in a colonic epithelial cell line. In Textbook of Secretory Diarrhea. E. Lebenthal & M. Duffey, Eds.: 59–66. Raven Press. New York.

34. SCHMITZ, H., M. FROMM, H. BODE et al. 1996. Tumor necrosis factor-α induces Cl^- and K^+ secretion in human distal colon driven by prostaglandin E_2. Am. J. Physiol. **271:** G669–G674.

35. BEUBLER, E., K. BUKHAVE & J. RASK-MADSEN. 1986. Significance of calcium for the prostaglandin E_2–mediated secretory response to 5-hydroxytryptamine in the small intestine of the rat in vivo. Gastroenterology **90:** 1972–1977.

36. SMITH, J.A., D.P. MONTZKA, G.P. MCCAFFERTY et al. 1988. Effect of sulfidopeptide leukotrienes D_4 ad E_4 on ileal ion transport in the rat and rabbit. Am. J. Physiol. **255:** G175–G183.

37. BERN, M.J., C.W. STURBAUM, S.S. KARAYALCIN et al. 1989. Immune system control of rat and rabbit colonic electrolyte transport. J. Clin. Invest. **83:** 1810–1820.

38. KONG, W., K. MCCONALOGUE, L.M. KHITIN *et al.* 1997. Luminal trypsin may regulate enterocytes through proteinase-activated receptor 2. Proc. Natl. Acad. Sci. USA **94:** 8884–8889.
39. BARRETT, K.E., P.A. HUOTT, S.S. SHAH *et al.* 1989. Differing effects of apical and basolateral adenosine on colonic epithelial cell line T_{84}. Am. J. Physiol. **256:** C197–C203.
40. BARRETT, K.E., J.A. COHN, P.A. HUOTT *et al.* 1990. Immune-related intestinal chloride secretion. II. Effect of adenosine of T_{84} cell line. Am. J. Physiol. **258:** C902–C912.
41. CHANG, E.B., M.W. MUSCH & L. MAYER. 1990. Interleukin 1 and 3 stimulate anion secretion in chicken intestine. Gastroenterology **98:** 1518–1524.
42. SÜTAS, Y., S. AUTIO, I. RANTALA & E. ISOLAURI. 1997. IFN-gamma enhances macromolecular transport across Peyer's patches in suckling rats: implication for natural immune responses to dietary antigens early in life. J. Pediatr. Gastroenterol. Nutr. **24:** 162–169.
43. KARAYALCIN, S.S., C.W. STURBAUM, J.T. WACHSMAN *et al.* 1990. Hydrogen peroxide stimulates rat colonic prostaglandin production and alters electrolyte transport. J. Clin. Invest. **86:** 60–68.
44. DOBBINS, J.W., J.P. LAURENSON & J.N. FORREST. 1984. Adenosine and adenosine analogues stimulate adenosine cyclic 3′, 5′-monophosphate–dependent chloride secretion in the mammalian ileum. J. Clin. Invest. **74:** 929–935.
45. HANGLOW, A.C., J. BIENENSTOCK & M.H. PERDUE. 1989. Effects of platelet-activating factor on ion transport in isolated rat jejunum. Am. J. Physiol. **257:** G845–G850.
46. VERGNOLLE, N., W.K. MACNAUGHTON, B. AL-ANI *et al.* 1998. Proteinase-activated receptor 2 (PAR_2)–activating peptides: identification of a receptor distinct from PAR_2 that regulates intestinal transport. Proc. Natl. Acad. Sci. USA **95:** 7766–7771.
47. SCUDAMORE, C.L., E.M. THORNTON, L. MCMILLAN *et al.* 1995. Release of the mucosal mast cell granule chymase, rat mast cell protease–II, during anaphylaxis is associated with rapid development of paracellular permeability to macromolecules in rat jejunum. J. Exp. Med. **182:** 1871–1881.
48. PERDUE, M.H. & D.M. MCKAY. 1994. Integrative immunophysiology in the intestinal mucosa. Am. J. Physiol. **267:** G151–G165.
49. PERDUE, M.H. & D.M. MCKAY. 1998. Epithelium. *In* Allergy and Allergic Diseases: the Mechanisms and Therapeutics. J.A. Denberg, Ed.: 281–303. Humana Press. Totowa, NJ.
50. CIACCI, C., Y.R. MAHIDA, A. DIGNASS *et al.* 1993. Functional interleukin-2 receptors on intestinal epithelial cells. J. Clin. Invest. **92:** 527–532.
51. SANDERS, S.E., J.L. MADARA, D.K. MCGUIRK *et al.* 1995. Assessment of inflammatory events in epithelial permeability: a rapid screening method using fluorescein dextrans. Epithelial Cell Biol. **4:** 25–34.
52. TAYLOR, C.T., A.L. DZUS & S.P. COLGAN. 1998. Autocrine regulation of epithelial permeability by hypoxia: role for polarized release of tumor necrosis factor alpha. Gastroenterology **114:** 657–668.
53. KUROKAWA, M., K. LYNCH & D.K. PODOLSKY. 1987. Effects of growth factors on an intestinal epithelial cell line: transforming growth factor β inhibits proliferation and stimulates differentiation. Biochem. Biophys. Res. Commun. **142:** 775–782.
54. BARNARD, J.A., R.D. BEAUCHAMP, R.J. COFFEY & H.L. MOSES. 1989. Regulation of intestinal epithelial cell growth by transforming growth factor type β. Proc. Natl. Acad. Sci. USA **86:** 1578–1582.
55. PERDUE, M.H., J. MARSHALL & S. MASSON. 1990. Ion transport abnormalities in inflamed rat jejunum: involvement of mast cells and nerves. Gastroenterology **98:** 561–567.
56. PERDUE, M.H., J.K. RAMAGE, D. BURGET *et al.* 1989. Intestinal mucosal injury is associated with mast cell activation and leukotriene generation during *Nippostrongylus*-induced inflammation in the rat. Dig. Dis. Sci. **34:** 724–731.
57. D'INCA, R., J.K. RAMAGE, R.H. HUNT & M.H. PERDUE. 1990. Antigen-induced mucosal damage and restitution in the small intestine of the immunized rat. Int. Arch. Allergy Appl. Immunol. **91:** 270–277.
58. RAMAGE, J.K., R.H. HUNT & M.H. PERDUE. 1988. Changes in intestinal permeability and epithelial differentiation during inflammation in the rat. Gut **29:** 57–61.

59. MILLER, H.R.P., R.G. WOODBURY, J.F. HUNTLEY & G. NEWLANDS. 1983. Systemic release of mucosal mast-cell protease in primed rats challenged with *Nippostrongylus brasiliensis*. Immunology **49:** 471–479.
60. KING, S.J. & H.R.P. MILLER. 1984. Anaphylactic release of mucosal mast cell protease and its relationship to gut permeability in *Nippostrongylus*-primed rats. Immunology **51:** 653–659.
61. CASTRO, G.A., Y. HARARI & D. RUSSEL. 1987. Mediators of anaphylaxis-induced ion transport changes in small intestine. Am. J. Physiol. **253:** G540–G548.
62. ZHANG, S., S. MYERS & G.A. CASTRO. 1991. Inhibition of anaphylaxis-evoked intestinal fluid secretion by the dual application of an H1 antagonist and cyclooxygenase inhibitor. Gastroenterology **100:** 922–928.
63. RAMAGE, J.K., A. STANISZ, R. SCICCHITANO et al. 1988. Effect of immunologic reactions on rat intestinal epithelium: correlation of increased permeability to chromium 51–labeled ethylenediamine tetraacetic acid and ovalbumin during acute inflammation and anaphylaxis. Gastroenterology **94:** 1368–1375.
64. PERDUE, M.H., M. CHUNG & D.G. GALL. 1984. Effect of intestinal anaphylaxis on gut function in the rat. Gastroenterology **86:** 391–397.
65. PERDUE, M.H. & D.G. GALL. 1986. Intestinal anaphylaxis in the rat: jejunal response to *in vitro* antigen exposure. Am. J. Physiol. **250:** G427–G431.
66. CROWE, S.E., P. SESTINI & M.H. PERDUE. 1990. Allergic reactions of rat jejunal mucosa: ion transport responses to luminal antigen and inflammatory mediators. Gastroenterology **99:** 74–82.
67. CUTHBERT, A.W., P. MCLAUGHLAN & R.A.A. COOMBS. 1983. Immediate hypersensitivity reaction to β-lactoglobulin in the epithelium lining the colon of guinea pigs fed cow's milk. Int. Arch. Allergy Appl. Immunol. **72:** 34–40.
68. PERDUE, M.H., S. MASSON, B.K. WERSHIL & S.J. GALLI. 1991. Role of mast cells in ion transport abnormalities associated with intestinal anaphylaxis. J. Clin. Invest. **87:** 687–693.
69. PERDUE, M.H. & D.G. GALL. 1983. Transport abnormalities during intestinal anaphylaxis in the rat: effect of antiallergic agent. J. Allergy Clin. Invest. **76:** 498–503.
70. BAIRD, A.W., R.A.A. COOMBS, P. MCLAUGHLAN & A.W. CUTHBERT. 1984. Immediate hypersensitivity reactions to cow milk proteins in isolated epithelium from ileum of milk-drinking guinea-pigs: comparisons with colonic epithelia. Int. Arch. Allergy Appl. Immunol. **75:** 255–263.
71. KELJO, D. & J.R. HAMILTON. 1983. Quantitative determination of macromolecular transport rate across intestinal Peyer's patches. Am. J. Physiol. **244:** G637–G644.
72. BOSMEL, M., K. PRYDZ, R.G. PARTON et al. 1989. Endocytosis in filter-grown Madin-Darby Canine Kidney cells. J. Cell Biol. **109:** 3243–3258.
73. CROWE, S.E., K. SODA, A.M. STANISZ & M.H. PERDUE. 1993. Intestinal permeability in allergic rats: nerve involvement in antigen-induced changes. Am. J. Physiol. **264:** G617–G623.
74. KITAGAWA, S., S. ZHANG, Y. HARARI & G.A. CASTRO. 1995. Relative allergenicity of cow's milk and cow's milk–based formulas in an animal model. Am. J. Med. Sci. **310:** 183–187.
75. HEYMAN, M., M. ANDRIANTSOA, A.M. CRAIN-DENOYELLE & J.F. DESJEUX. 1990. Effect of oral or parenteral sensitization to cow's milk on mucosal permeability in guinea pigs. Int. Arch. Allergy Appl. Immunol. **92:** 242–246.
76. BLOCH, K.J. & W.A. WALKER. 1981. Effect of locally induced intestinal anaphylaxis on the uptake of a bystander antigen. J. Allergy Clin. Immunol. **67:** 312–316.
77. KLEINMAN, R.E., P.R. HARMATZ, R.A. HATZ et al. 1989. Divalent hapten-induced intestinal anaphylaxis in the mouse: uptake and characterization of a bystander protein. Immunology **68:** 464–468.
78. BERIN, M.C., A.J. KILIAAN, P.C. YANG et al. 1997. Rapid transepithelial antigen transport in rat jejunum: impact of sensitization and the hypersensitivity reaction. Gastroenterology **113:** 1–9.
79. BERIN, M.C., A.J. KILIAAN, P.C. YANG et al. 1998. The influence of mast cells on pathways of transepithelial antigen transport in rat intestine. J. Immunol. **161:** 2561–2566.

Carbachol-Induced Ca^{2+} Entry into Rat Colonic Epithelium

G. SCHULTHEISS,[a] M. FRINGS, AND M. DIENER

Institut für Veterinär-Physiologie, Justus-Liebig-Universität Giessen, 35392 Giessen, Germany

INTRODUCTION

Intracellular Ca^{2+} is an important regulator of intestinal ion transport.[1] In the case of excitable cells, Ca^{2+} entry is mediated through voltage-operated Ca^{2+} channels. Epithelial cells do not express voltage-sensitive Ca^{2+} channels; instead in nonexcitable tissue a store-operated Ca^{2+} entry is thought to be responsible for the influx of extracellular Ca^{2+}. The signal for activation of this pathway is an emptying of the intracellular stores, which then in turn induces the Ca^{2+} entry.[2] No information is available about this pathway in the native colonic epithelium.

RESULTS AND DISCUSSION

Increasing the Ca^{2+} buffering capacity of the pipette solution in whole-cell patch-clamp experiments at isolated rat colonic crypts induced a depolarization of about 20–30 mV compared to control. Administration of La^{3+} (10^{-5} mol $\cdot$ 1^{-1}), a blocker of nonselective cation channels, induced a strong hyperpolarization of 22.1 ± 5.1 mV (FIG. 1a; $n = 17$), restoring the potential to normal values. Ion substitution experiments revealed that among Ca^{2+} sodium ions can pass through the conductance, demonstrating an even six-times higher permeability for Na^{+} than for Ca^{2+}. The reversal potential of the lanthane-sensitive current amounted to -8.7 ± 4.8 mV, which is not significantly different from 0 mV, as can be expected for a nonselective cation conductance with equal cation concentrations on both sides of the membrane.[3] The hyperpolarization of the crypt cells was accompanied by a decrease in the membrane conductance by 1.6 ± 0.6 nS (FIG. 1B, $n = 17$).

Indirect evidence suggests that the nonselective cation conductance is activated after stimulation of muscarinic receptors. The administration of carbachol ($5 \cdot 10^{-5}$ mol $\cdot$ 1^{-1}) resulted in a transient hyperpolarization and an increase in whole-cell current due to the opening of Ca^{2+}-sensitive K^{+} channels. Both, that is, the membrane potential and the membrane current, decreased within about 2 min to 50% of the peak values. Pretreatment with La^{3+} (10^{-5} mol $\cdot$ 1^{-1}) diminished the half-time of the decay by about 50% (FIG. 1c; $n = 6$). Measurement of changes in the intracellular Ca^{2+} concentration with the fura-2 technique revealed that carbachol induced a

[a]Address for correspondence: G. Schultheiss, Institut für Veterinär-Physiologie, Justus-Liebig-Universität Giessen, Frankfurter Str. 100, 35392 Giessen, Germany. Voice: +49-641-99-38160; fax: +49-641-99-38159.

gerhard.schultheiss@vetmed.uni-giessen.de

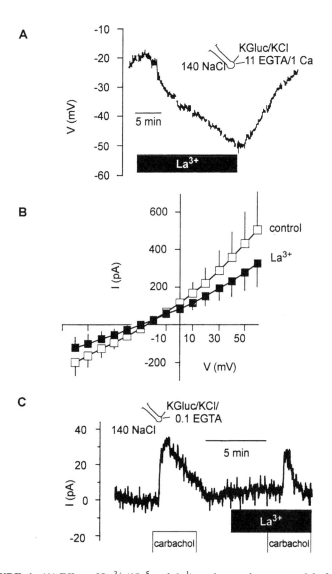

FIGURE 1. (**A**) Effect of La^{3+} (10^{-5} mol·l^{-1}) on the membrane potential of a crypt cell located in the fundus region. The interruptions in the voltage tracing are caused by measurements of IV-relations in the voltage-clamp mode. The pipette solution possessed a high Ca^{2+}-buffering capacity (11 mmol·l^{-1} EGTA/1 mmol·l^{-1} Ca^{2+}). (**B**) Membrane currents pooled from 6 cells from the fundus region in the absence (control: *open squares*) and presence of La^{3+} (*filled squares*). Values are means (symbols) ± SEM, $n = 6$. (**C**) Original tracing of the carbachol ($5·10^{-5}$ mol·l^{-1})-induced effect on membrane current measured at a holding potential of -35 mV in the absence and presence of La^{3+} (10^{-5} mol·l^{-1}). Crypt cell located in the fundus region studied in the whole-cell mode using a nystatin-perforated patch and a pipette solution with a low Ca^{2+}-buffering capacity (0.1 mmol·l^{-1} EGTA, nominally 0 mmol·l^{-1} Ca^{2+}). Note that in the presence of La^{3+} the decay of the carbachol-induced responses is accelerated.

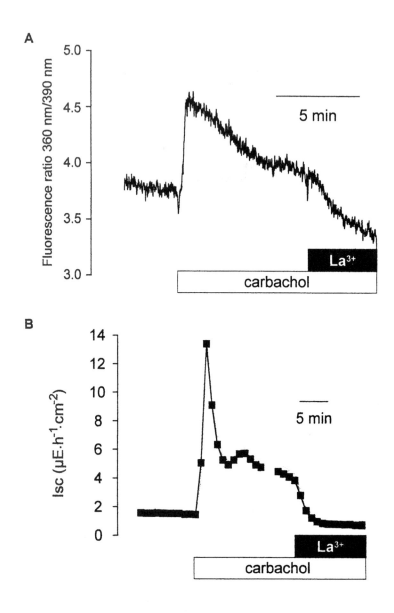

FIGURE 2. (A) La^{3+} (10^{-5} mol·l^{-1}) inhibits the increase in the ratio of the fluores-cence signal of fura-2 (excited at 360 and 390 nm) induced by carbachol ($5·10^{-5}$ mol·l^{-1}). Typical response measured in the fundus region of an isolated crypt. Data were sampled with 2 Hz and digitally low-pass filtered with a cut-off frequency of 0.125 Hz. **(B)** Inhibition of the increase in short-circuit current (I_{sc}) induced by carbachol ($5·10^{-5}$ mol·l^{-1} at the serosal side) by a high concentration of La^{3+} (10^{-3} mol·l^{-1} at the serosal side). La^{3+} was adminis-tered during the plateau phase of the carbachol-stimulated I_{sc}. Values are means (*symbols*) ± SEM (*shaded area*), $n = 6$.

biphasic increase of the fluorescence ratio to a peak followed by a long-lasting plateau phase. Administration of La^{3+} (10^{-5} mol·1^{-1}, FIG. 2A; $n = 8$) during the plateau phase strongly accelerated the decay, yielding a fluorescence ratio after 90 s that was not significantly different from the baseline before the administration of carbachol. The peak of fluorescence signal induced by carbachol represents the release of Ca^{2+} from the intracellular stores, whereas the plateau is based on the entry of extracellular Ca^{2+}.[4] Therefore, La^{3+} exerts an inhibitory influence on the Ca^{2+}-entry pathway.

Ussing chamber experiments demonstrated that the indirect activation of Cl^- secretion by carbachol[5] could be suppressed by the lanthanide. Addition of La^{3+} (10^{-3} mol·1^{-1}, on the serosal side) during the plateau phase suppressed the carbachol-induced changes in the short-circuit current (I_{sc}, FIG. 2B, $n = 6$). In addition, the carbachol-induced peak in I_{sc} was also significantly lowered by pretreatment with La^{3+}, demonstrating a partial inhibition of Ca^{2+} refilling or an inhibitory influence of La^{3+} on the carbachol-induced release of intracellular Ca^{2+}. Taken together these results demonstrate that the activation of a nonselective cation conductance by store depletion is involved in the regulation of electrolyte transport by agonists of the Ca^{2+} signaling pathway.

REFERENCES

1. BINDER, H.J. & G.J. SANDLE. 1994. Electrolyte transport in the mammalian colon. *In* Physiology of the Gastrointestinal Tract. L.R. Johnson, Ed.: 2133–2171. Raven Press. New York.
2. PAREKH, A.B. & R. PENNER. 1997. Store depletion and calcium influx. Physiol. Rev. **77**: 901–930.
3. HILLE, B. 1992. Ionic Channels in Excitable Membranes. 2nd edit.: 337–361. Sunderland. Sinauer.
4. Lindqvist, S.M., P. Sharp, I.T. Johnson *et al.* 1998. Acetylcholine-induced calcium signaling along the rat colonic crypt axis. Gastroenterology **115**: 1131–1143.
5. BÖHME, M., M. DIENER & W. RUMMEL. 1991. Calcium- and cyclic-AMP-mediated secretory responses in isolated colonic crypts. Pflügers Arch. Eur. J. Physiol. **419**: 144–151.

Effects of Endotoxin on Human Large Intestine

SABINE BÜHNER,[a,b] BASTIAN MAYR,[a] HAGEN BODE.[c] HEINZ SCHMITZ,[c]
JÖRG-DIETER SCHULZKE,[c] MICHAEL FROMM,[d] AND HERBERT LOCHS[a]

[a]Medizinische Klinik mit Schwerpunkt Gastroenterologie, Hepatologie, Endokrinologie,
Universitätsklinikum Charité, Humboldt Universität zu Berlin, 10117 Berlin, Germany

[c]Medizinische Klinik I Gastroenterologie und Infektiologie, and [d]Institut für Klinische
Physiologie, Universitätsklinikum Benjamin Franklin, Freie Universität Berlin,
12200 Berlin, Germany

INTRODUCTION

Several gastrointestinal diseases, including inflammatory bowel diseases (IBD), are associated with increased endotoxin levels (lipopolysaccharides, LPS). In a recent study of 64 patients with IBD, 88% of patients with ulcerative colitis and 94% with Crohn's disease showed systemic endotoxemia during clinical relapse.[1] This correlated positively with the spatial distribution and clinical activity of the disease. Similar results were found by others.[2,3] LPS have a strongly immunomodulating effect. Therefore, secondary mediators, such as cytokines, were often considered to be the cause of many pathophysiological effects following endotoxemia. It is known that LPS increases the production of proinflammatory cytokines like TNF-α, IL-6, IL-8, and IL-1β in the gut wall.[4] Because LPS are present in high concentrations in the lumen of the gut and may be absorbed through the impaired mucosal barrier in IBD or HIV infection, it is of interest to know whether LPS have similar effects. The aim of this study was to analyze the effects of LPS (*E. coli* 055:B5) on epithelial transport and barrier function of intact and injured human distal colon *in vitro*.

METHODS

The study was performed on macroscopically normal human distal colon (sigma/rectum). Tissue samples were obtained from patients with colorectal carcinoma or adenoma who underwent abdominal surgery. Specimens were partially stripped and mounted into Ussing chambers. Details are given in References 5 and 6. Briefly, experiments were performed using a computer-controlled voltage clamp device (CVC6, Fiebig, Berlin). Short circuit current (I_{sc}) and transepithelial resistance (R^t) were recorded for at least 300 minutes. Mucosal injury was induced by HCl (35 mM) in the mucosal bathing solution (5 mL) for 10 min (pH = 2.0).

In addition, Ussing experiments were performed using the colorectal cell line HT-29/B6, which grows as highly differentiated polarized monolayers.[7] The bathing

[b]Address for correspondence: Dr. Sabine Bühner, Universitätsklinikum Charité, Medizinische Klinik SP Gastroenterologie, Hepatologie, Endokrinologie, Schumannstr. 20-21, 10117 Berlin, Germany. Voice: +49-30-2802-3192; fax: +49-30-2802-8978.
sabine.buehner@charite.de

solution was RPMI 1640 (Biochrom), 5% normal human serum (Sigma), and human serum albumin (1 mg/mL; Sigma). The experiments were carried out at days 8 and 9; the transepithelial resistance of monolayers was $562 \pm 9 \; \Omega \cdot cm^2$.

RESULTS

Simultaneous addition of LPS (10 µg/mL; *E. coli* 055:B5; Sigma) to both mucosal and serosal bathing solutions resulted in a maximum I_{sc} increase of 2.5 ± 0.3 $\mu mol \cdot h^{-1} \cdot cm^{-2}$ ($n = 5$; m $\pm$ SEM, $p < 0.05$; paired t-test), which was reached within 93 ± 6 minutes. I_{sc} remained elevated during the course of recording. Although exclusive serosal addition of LPS resulted in a similar response, there was no significant reaction to a mucosal addition alone. R^t was $90 \pm 5 \; \Omega \cdot cm^2$ and was not significantly affected by LPS during the 300-min measuring period.

Preincubation (15 min) with tetrododoxin (TTX, 1 µM) or preincubation (360 min) with antiinflammatory cytokines IL-4 (1000 U/mL) and IL-10 (200 U/mL; Genzyme) had no significant effects on the LPS-mediated I_{sc} response.

LPS (10 µg/mL, both sides) did not significantly affect I_{sc} in HT-29/B6 cells, suggesting that LPS does not act directly at the enterocyte level. Theophylline (10 mM), a potent stimulator of active Cl^- secretion, caused I_{sc} to increase, indicating intact secretory responsiveness.

HCl (35 mM), added to the mucosal bathing solution, induced a transient reduction of transepithelial resistance. Thirty minutes after HCl exposure, resistance was reduced to 40% of the initial value. It partly recovered to 78% of initial value 180 min after HCl. Only after HCl-induced mucosal injury, LPS (10 µg/mL), added to mucosal bathing solution 30 min after the end of HCl exposure, resulted in a significant increase of I_{sc} ($0.40 \pm 0.20 \; \mu mol \cdot h^{-1} \cdot cm^{-2}$ ($n = 5$) vs. $1.04 \pm 0.3 \; \mu mol \cdot h^{-1} \cdot cm^{-2}$ ($n = 4$); $p < 0.05$, compared to intact mucosa).

CONCLUSIONS

LPS, after contacting the serosal tissue side, strongly stimulates electrogenic secretion in the human distal colon. The mechanism is independent of the enteric nervous system. Whether or not cytokines are involved in the mediation of the LPS response remains open. At least antiinflammatory cytokines like IL-10 and IL-4 had no inhibitory effect.

In the presence of an intact epithelial barrier, luminal LPS has no significant effect. By contrast, when the epithelial barrier is impaired, as, for example, in inflammatory bowel diseases, luminal LPS can also stimulate secretion. Therefore, LPS could be an important pathophysiological factor in diarrheal diseases.

REFERENCES

1. GARDINER, K.R., M.I. HALLIDAY, G.R. BARCLAY *et al.* 1995. Significance of systemic endotoxaemia in inflammatory bowel disease. Gut **36:** 897–901.
2. AOKI, K. 1978. A study of endotoxaemia in ulcerative colitis and Crohn's disease. I. clinical study. Acta Med. Okayama **32:** 147–158.

3. WELLMANN, W., P.C. FINK, F. BENNER & F.W. SCHMIDT. 1986. Endotoxaemia in active Crohn's disease. Treatment with whole gut irrigation and 5-aminosalicylic acid. Gut **27:** 814–820.
4. OGLE, C.K., X. GUO, P.O. HASSELGREN et al. 1997. The gut as a source of inflammatory cytokines after stimulation with endotoxin. Eur. J. Surg. **163:** 45–51.
5. BODE, H., H. SCHMITZ, M. FROMM et al. 1998. IL-1β, TNF-α, but not IFN-α, IL-6 or IL-8, are secretory mediators in human distal colon. Cytokine **10:** 457–465.
6. SCHMITZ, H., M. FROMM, H. BODE et al. 1996. Tumor necrosis factor alpha induces Cl$^-$ and K$^+$ secretion in human distal colon driven by prostaglandin E$_2$. Am. J. Physiol. **271:** G669–G674.
7. KREUSEL, K.M., M. FROMM, J.D. SCHULZKE & U. HEGEL. 1991. Cl$^-$ secretion in epithelial monolayers of mucus-forming human colon cells (HT-29/B6). Am. J. Physiol. **261:** C574–C582.

Increased Paracellular Macromolecular Transport and Subnormal Glucose Uptake in Duodenal Biopsies of Patients with Microvillus Inclusion Disease

Comparisons to Other Chronic Diarrhea Patients and to Nondiarrhea Patients

P. B. BIJLSMA,[a,b] A. J. KILIAAN,[a] G. SCHOLTEN,[a] A. VAN DER WAL,[c]
M. HEYMAN,[a] H. S. A. HEYMANS,[a] J. A. GROOT,[a] AND J. A. J. M. TAMINIAU[a]

[a]Department of Pediatric Gastroenterology and Nutrition, Academic Medical
Center/Institute of Neurobiology, University of Amsterdam, the Netherlands
[c]Faculté Necker-Enfants Malades, INSERM CJF 9710, 75730 Paris, France

INTRODUCTION

Microvillus inclusion disease (MVID) is a severe small intestinal disorder characterized by endosomal inclusion of microvilli beneath the luminal surface of the enterocytes. The most prominent clinical symptom is chronic diarrhea, and patients require parenteral nutrition. It is generally assumed that the diarrhea in MVID is caused by impaired nutrient uptake due to a lack of small intestinal absorptive surface area. Other etiologic factors cannot be excluded: for example, an increased macromolecular permeability. We studied duodenal macromolecular permeability and Na-glucose carrier activity in 3 MVID patients, in 7 other children with chronic diarrhea fed by combined enteral and parenteral nutrition (disease controls), and in 3 nondiarrhea patients with transient reflux symptoms (controls).

METHODS

Duodenal endoscopical forcipal biopsies were oriented and mounted in special Ussing chambers, with an exposed serosal surface area of 1.7 mm^2 or 1.1 mm^2. Transepithelial potential difference and resistance (R) were recorded, equivalent short circuit current (I_{sc}) was calculated, and I_{sc} responses to bilateral addition of 20 mM D-glucose were analyzed. Permeability to horseradish peroxidase (HRP, 40 kDa, 10^{-5} M added mucosally, and serosal appearance measured enzymatically) was studied for 2 hours, and tissues were fixed for electron-microscopical HRP staining.

[b]Address for correspondence: P. B. Bijlsma, Institute of Neurobiology, University of Amsterdam, Kruislaan 320, 1098 SM Amsterdam, the Netherlands. Fax: +31-20-6659125.
bijlsma@bio.uva.nl

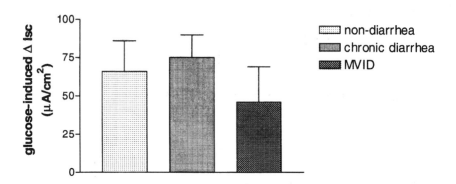

FIGURE 1. Changes in short circuit current after bilateral addition of 20 mM glucose in 3 nondiarrhea patients, 7 chronic diarrhea patients, and 3 MVID patients. No significant differences were observed between these patient groups.

RESULTS

Glucose-induced I_{sc} responses (mean ± SEM) in MVID patients tended to be lower, but were not significantly different from those in controls or disease controls (respectively: 46 ± 23; 66 ± 20; 75 ± 15 µA/cm^2) as shown in FIGURE 1. Baseline I_{sc} values (resp.: 47 ± 16; 57 ± 16; 44 ± 4 µA/ cm^2) and R values (resp.: 18 ± 5; 19 ± 1; 21 ± 3 Ohm-cm^2) in the three groups were comparable. HRP fluxes (60–120 min) in the MVID patients were significantly larger than those in controls and in disease controls—resp.: 24.2 ± 2.5; 8.6 ± 0.8; 14.9 ± 2.9 pmol/cm^2-h ($p < 0.05$, Welch's t test) as shown in FIGURE 2. Electron microscopy revealed obvious paracellular leak of HRP through numerous pores in the tight junctions of the enterocytes in the

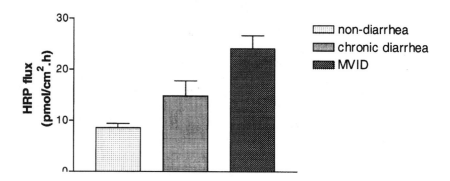

FIGURE 2. Mucosal to serosal fluxes of HRP from 60 to 120 minutes in 3 nondiarrhea patients, 7 chronic diarrhea patients, and 3 MVID patients. Values in MVID patients were significantly larger compared to both other patient groups ($p < 0.05$).

MVID patients. This paracellular leak of HRP was never observed in the controls and only observed in 3 of the 7 disease controls, indicating that paracellular leak of macromolecules is not necessarily a side effect of chronic diarrhea. One MVID patient was tested twice and the permeability defect appeared to be consistent.

DISCUSSION

We conclude that the diarrhea observed in MVID patients may not be primarily caused by a lack of Na-glucose carriers resident in apical membranes of the enterocytes or by the inability of the enterocytes to transport glucose, in contrast to suggestions by Schmitz *et al.*[1] and Rhoads *et al.*[2] from single case studies. This brings us to the finding of an increased macromolecular permeability in MVID patients as a factor that may contribute to the diarrhea in these patients. An increased HRP flux was observed in all 3 patients examined, and this result appeared to be reproducible in 1 patient tested twice. Moreover, the increased HRP flux showed only a small variation between the 2 to 3 biopsies tested per patient. This variation was, in general, larger in other patients with increased macromolecular permeability, in which also a more patchy pattern of paracellular HRP staining was observed by electron microscopic examination compared to a more even distribution in the tissues of the MVID patients. Interestingly, the formation of inclusions of microvilli has been suggested to be a result of abnormalities in the enterocyte cytoskeleton. Carruthers *et al.*[3] have shown a decreased amount of myosin and vinculin, and Cutz *et al.*[4] observed decreased staining of actin in MVID patients. These cytoskeletal proteins are also intimately associated with the tight junctional complex and are associated with the regulation of paracellular permeability.[5] Thus, the increased paracellular leak of HRP in MVID patients is likely due to the same abnormalities in the cytoskeleton as the assembly of inclusions of microvilli.

The observed paracellular macromolecular leak in MVID patients may lead to an increased antigenic load to the mucosal immune system. This may induce an immunological hyperresponsiveness, which may be a major stimulating factor for water and electrolyte secretion and thus contribute to the diarrhea observed in these patients.

REFERENCES

1. SCHMITZ, J., J.L. GINIES *et al.* 1982. Congenital microvillus atrophy, a rare cause of neonatal intractable diarrhea [abstract]. Pediatr. Res. **16:** 1041.
2. RHOADS, J.M., R.C. VOGLER *et al.* 1991. Microvillus inclusion disease: *in vitro* jejunal electrolyte transport. Gastroenterology **100:** 811–817.
3. CARRUTHERS, L., A.D. PHILLIPS *et al.* 1985. Biochemical abnormality in brush border membrane protein of a patient with congenital microvillus atrophy. J. Pediatr. Gastroenterol. Nutr. **4:** 902–907.
4. CUTZ, E., J.M. RHOADS *et al.* 1989. Microvillus inclusion disease: an inherited defect of brush-border assembly and differentiation. N. Engl. J. Med. **320:** 646–651.
5. ANDERSON, J.M. & C.M. VAN ITALLIE. 1995. Tight junctions and the molecular basis for regulation of paracellular permeability. Am. J. Physiol. **269:** G467–G475.

Apoptosis and Intestinal Barrier Function

CHRISTIAN BOJARSKI,[a] KERSTIN BENDFELDT,[b] ALFRED H. GITTER,[b] JOACHIM MANKERTZ,[a] MICHAEL FROMM,[b] SIEGFRIED WAGNER,[c] ERNST-OTTO RIECKEN,[a] AND JÖRG-DIETER SCHULZKE[a,d]

Departments of [a]Gastroenterology and [b]Clinical Physiology, Universitätsklinikum Benjamin Franklin, Freie Universität Berlin, Hindenburgdamm 30, 12200 Berlin, Germany

[c]*Department of Gastroenterology and Hepatology, Medizinische Hochschule Hannover, 30623 Hannover, Germany*

BACKGROUND AND AIM

In the intestine, the epithelium is exposed to several toxins and inflammatory cytokines that can affect barrier function by both necrosis and apoptosis. Substances as TNF-α or IF-γ have been described as affecting tight junctions and the apoptotic rate.[1] However, the role of apoptosis for epithelial barrier function is not clear so far. Therefore, we aimed to characterize the functional effect of induced apoptosis on barrier function in a large intestinal model epithelium HT-29/B6. For induction of apoptosis we used the topoisomerase-I inhibitor camptothecin.

MATERIAL AND METHODS

Cell Culture

Experiments were performed on HT-29/B6 cells subcloned from the human colon carcinoma cell line HT-29, which grows as highly differientiated polarized monolayers.[2] HT-29/B6 cells were routinely cultured in 25-cm^2 culture flasks in RPMI 1640 (Biochrom, Berlin) containing 2% stabilized L-glutamine and supplemented with 10% FCS at 37°C in an atmosphere of 95% O_2 and 5% CO_2. For electrophysiological measurements cells were seeded on Millicell filters (effective membrane area 0.6 cm^2, Millipore PCF). Confluence of the monolayers was reached after 7 days, and experiments were performed on days 9 or 10.

Measurement of Electrical Parameters

Transepithelial resistance (R^t) and permeability were measured in Ussing chambers by means of a computer-controlled voltage clamp device. Flux measurements

[d]Address for correspondence: Prof. Dr. med. Jörg-Dieter Schulzke, Medizinische Klinik I, Gastroenterologie und Infektiologie, Universitätsklinikum Benjamin Franklin, Freie Universität-Berlin, Hindenburgdamm 30, 12200 Berlin, Germany. Voice: +49-30-8445-2347; fax: +49-30-8445-4481.
Schulzke@medizin.fu-berlin.de

from mucosa to serosa (m→s) were performed under short circuit conditions with the marker of paracellular permeability [^{3}H]mannitol.

Induction and Detection of Apoptosis

Apoptosis was induced by means of the topoisomerase-I inhibitor camptothecin. Apoptosis was assessed by two independent methods: DNA-specific fluorochrome staining with 4', 6'-diamidino-2'-phenylindola-dihydrochloride (DAPI) and enzyme-linked immunosorbent assay (ELISA) with detection of cytosolic oligonucleosome-bound DNA. LDH assay was performed to investigate cytotoxic effects of camptothecin.

Conductance Scanning

The local current density, generated by transepithelial current (AC, 0.3 mA·cm^{-2}, 24 Hz) in the solution on the mucosal side of the epithelium, was detected with a microelectrode sensor that was positioned under microscopical control. In nonapoptotic areas the distribution of transepithelial current was evenly distributed, but above an area of apoptosis (surrounded by clusters suggestive of a rosette) the current density was elevated. From integration of the current density exceeding that of nonapoptotic areas, the conductance associated with the single apoptosis was determined.[3]

Statistical Analysis

Results are given as mean ± SEM or as percentage of initial R^t value ± SEM. Significance was tested by means of the two-tailed Student's t test; $p < 0.05$ was considered significant.

RESULTS

Detection of Apoptosis in HT-29/B6 Cells Treated with Camptothecin

After incubation for 48 hours with camptothecin, HT-29/B6 cells showed typical morphological changes (FIG. 1). The amount of apoptotic cells as a percent of all cells in the monolayer was determined by counting apoptotic nuclei stained with the DNA-specific fluorochrome DAPI. In control monolayers 4.82 ± 0.5% of the cells were apoptotic, whereas after treatment with 2 μg/mL camptothecin, the frequency of apoptoses was increased to 11.2 ± 0.1% ($p < 0.05$, $n = 2$).

Camptothecin treatment altered time and dose dependently the absorbance values (O.D. 405 nm) detected with ELISA. A minimum effective concentration of camptothecin was 2 μg/mL, and the most effective one was 20 μg/mL. In untreated cells initial absorbance values (O.D. 405 nm) increased within 72 hours about threefold, from 0.05 ± 0.03 to 0.15 ± 0.01 ($p = 0.03$, $n = 3$). After incubaction with 20 μg/mL camptothecin initial absorbance values increased 60-fold, from 0.04 ± 0.01 to 2.3 ± 0.08 ($p < 0.001$, $n = 3$). Taken together, treatment with camptothecin induced an increase in absorbance values ($p = 0.001$).

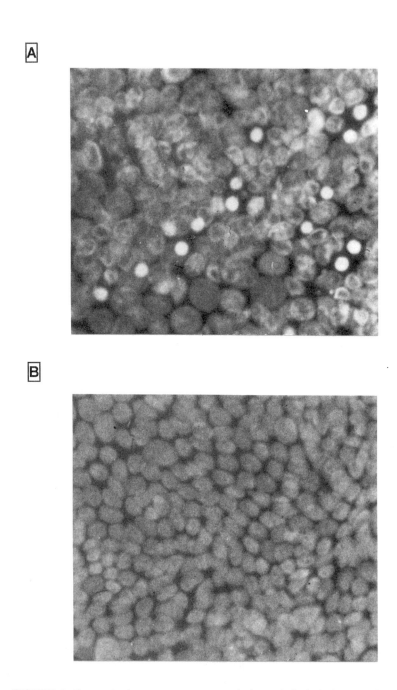

FIGURE 1. Camptothecin treatment causes typical morphological changes. **(A)** Condensed nuclei after treatment with camptothecin. **(B)** Control cells show no relevant changes.

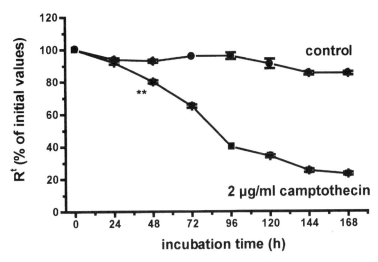

FIGURE 2. Time course of transepithelial resistance (R^t) induced by 2 μg/mL camptothecin. Changes in R^t are expressed as percentage of inital R^t (**$p = 0.01$, $n = 3$).

Effect of Camptothecin on Transepithelial Resistance

Camptothecin dose dependently altered transepithelial resistance with a minimum effect at 0.2 μg/mL and a maximum effect at 20 μg/mL. After an incubation period of 168 hours with 2 μg/mL camptothecin R^t was reduced from 445 ± 5 to 101 ± 7 Ω·cm^2 (23% ± 2 of initial R^t, $p < 0.001$, $n = 3$). At this time R^t of controls was reduced in this interval from 457 ± 4 to 389 ± 1 Ω·cm^2, corresponding to 85 ± 1% of initial resistance values ($p = 0.05$, $n = 3$, FIG. 2). Camptothecin at a dose of 2 μg/mL led to a reduction of transepithelial resistance compared to control ($p = 0.01$).

Effect of Camptothecin on Unidirectional Fluxes of [³H]Mannitol and LDH Assay

Mucosa-to-serosa fluxes of [³H]mannitol increased from 63 ± 17 in controls to 362 ± 176 nmol·h^{-1}·cm^{-2} ($p = 0.05$, $n = 3$) in cells treated with 20 μg/mL camptothecin.

In untreated controls LDH release was 1.2 ± 1.5%. This value was not significantly different after 20 μg/mL (1.4 ± 3.2%) or after 2 μg/mL camptothecin (1.2 ± 2.2%) compared to control (both not significant). This indicates that camptothecin did not induce necrosis. The postexperimental LDH content of the cells after treatment with 4% Triton X 100 was also not different in both groups (4495 ± 235 U/L and 4990 ± 90 U/L, respectively, $p = 0.19$).

Conductance Scanning

After treatment with 20 μg/mL camptothecin the conductance of single apoptoses was dramatically increased compared to that of untreated cells (750 ± 230 nS vs. 48

$\pm$ 19 nS, $p = 0.01$, $n = 11$). The density of apoptotic rosettes increased from 3700 to 5400 cm^{-2}.

SUMMARY AND CONCLUSION

The signal transduction pathways of the induction of apoptosis in the gastrointes-tinal tract have in part been discovered.[4,5] However, almost nothing is known about the functional influence of apoptotic signals on intestinal barrier function. In this study the effect of camptothecin-induced apoptosis in HT-29/B6 monolayers and the influence of apoptosis on epithelial barrier function were characterized. We demon-stated that camptothecin causes a decrease of transepithelial resistance and an in-crease in fluxes of the paracellular marker [^{3}H]mannitol. Camptothecin increased the apoptotic rate and the conductance of single-cell apoptosis as measured by the conductance scanning technique. We conclude that in our model of HT-29/B6 cells camptothecin is a potent inductor of apoptosis that causes significant barrier defects measured by the Ussing chamber technique and the conductance scanning tech-nique. Based on these results we are able to investigate the effect of other cytok-ines—TGF-β, for instance, and its role in apoptotic conditions.

REFERENCES

1. SCHMITZ, H. et al. 1999. Tumor necrosis factor–α (TNFα) regulates the epithelial bar-rier in human intestinal cell line HT-29/B6. J. Cell Sci. **112:** 137–146.
2. KREUSEL, K.M. et al. 1991. Cl secretion in epithelial monolayers of mucus-forming human colon cells (HT-29/B6). Am. J. Physiol. **261**(Cell Physiol. **30**): C574– C582.
3. GITTER, A.H. et al. 1997. Measurement of paracellular epithelial conductivity by con-ductance scanning. Pflügers Arch. **434:** 830–840.
4. MULLIN, J.M. et al. 1997. Different size limitations for increased transepithelial para-cellular solute flux across phorbol ester and tumor necrosis factor–treated epithelial cell sheets. J. Cell Physiol. **171:** 226–233.
5. STRÄTER, J. et al. 1997. CD95 (APO-1/Fas)–mediated apoptosis in colon epithelial cells: a possible role in ulcerative colitis. Gastroenterology **113:** 160–167.

p53-Independent Apoptosis Induced by Menadione in the Human Colon Carcinoma Cell Line Caco-2

J. M. KARCZEWSKI,[a,b] J. A. M. VET,[c] D. HESSELS,[c] AND J. NOORDHOEK[d]

[a]Institute for Neurobiology, University of Amsterdam, Amsterdam, the Netherlands

[c]Department of Toxicology, University of Nijmegen, Nijmegen, the Netherlands

[d]Department of Urology, University Hospital Nijmegen, Nijmegen, the Netherlands

INTRODUCTION

Apoptosis in the intestine has been identified as a disposal mechanism for redundant cells.[1] With a genotoxic insult like irradiation and exposure to alkylating agents, the rate of apoptosis was dramatically increased.[2] Such DNA damage causes p53 levels to rise, which induces growth arrest and allows the cell to repair its damaged DNA or to undergo apoptosis. The efficiency of anticancer drugs is believed to be closely related to the p53 status of the tumor. As a complication, p53 mutations are found in 75% of human colorectal cancers.[3]

Little is known about the involvement of oxidative stress in the apoptotic mechanism in the intestine and in particular of cells derived from the intestinal tract. Menadione (MEN) is toxic to cells by endogenous generation of reactive oxygen species and arylation to cellular thiols.[4] Hence, we used the colon-derived colon carcinoma cell line Caco-2 as a model to study the effects of exposure to MEN on the intestinal epithelium. In this report, we describe the p53-independent apoptotic pathway that is activated by oxidative stress and inhibited by antioxidants.

MATERIAL AND METHODS

Determination of Apoptosis

Caco-2 cells were cultured and treated as described previously.[5] Cells were incubated with 20 µM propidium iodide (PI) and 20 µM Hoechst 33258 (bisbenzimide) for 30 min at 37°C. Morphology of about 500 nuclei was evaluated with fluorescence microscopy (Axioskop with Plan-Neofluar objectives, Zeiss, Germany). Apoptosis was expressed as the percentage of cells with apoptotic nuclei compared to the total number of cells. Penetration of PI into the nucleus was considered as a marker for necrosis.

[b]Address for correspondence: J. M. Karczewski, Institute for Neurobiology, University of Amsterdam, Kruislaan 320, NL-1098 SM, Amsterdam, the Netherlands. Voice: +31-20-5257650; fax: +31-20-5257709.

karczewski@bio.uva.nl

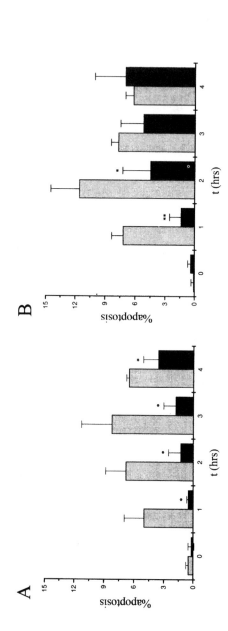

FIGURE 1. Effect of antioxidants on MEN-induced apoptosis: 10 mM *N*-acetylcysteine (**A**) or 10 mM trolox (**B**) were applied to Caco-2 cells at 1 hour prior to 25 µM MEN exposure. After 1 hour, antioxidant containing buffer was removed and cells were washed three times. Control cells were pretreated with KH-buffer alone. Depicted are control (*light boxes*) and antioxidant pretreated (*dark boxes*) Caco-2 cells. Shown are the means ± SD of three independent experiments (ANOVA: *$p \leq 0.05$; **$p \leq 0.01$).

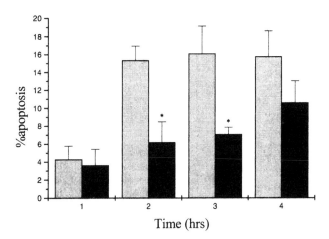

FIGURE 2. Effect of cycloheximide on MEN-induced apoptosis. Apoptosis was determined as described in the text. Caco-2 cells were incubated with 25 μM MEN in the absence (*light boxes*) or presence (*dark boxes*) of the protein synthesis inhibitor cycloheximide (50 μg/mL). Shown are the means ± SD of three independent experiments (ANOVA: *$p \leq 0.05$).

PCR-SSCP and Sequence Analysis

PCR-SSCP and sequence analysis were performed to investigate p53 mutations in exons 5–9.[6]

RESULTS AND DISCUSSION

MEN is toxic to cells by its ability to generate intracellular free oxygen radicals. FIGURE 1 shows that about 15% of the cell culture is apoptotic after 1 hour of exposure to 25 μM MEN and remained so in the further course of the experiments. We were interested if pretreatment of cells with antioxidants could prevent apoptosis. For this purpose, we used the water-soluble vitamin E analogue, trolox, and the thiol and GSH precursor, *N*-acetylcysteine (NAC). FIGURE 1 shows that the rate of apoptosis is diminished by both antioxidants, although the effect of NAC is stronger.

Apoptosis is generally considered to be an active process and, in many cases, protein synthesis is a precondition for this type of cell death. In order to investigate the involvement of protein synthesis in MEN-induced apoptosis in Caco-2 cells, the protein synthesis inhibitor, cycloheximide (CHX), was applied together with MEN. CHX was able to reduce the rate of apoptosis induced by MEN (FIG. 2). This effect was significant for up to 4 hours of exposure. The results show that protein synthesis is involved in MEN-induced apoptosis.

It has been shown that 98% of base substitution mutations in p53 found in tumors occur in a 600-base-pair region of the gene product (codons 110 to 307). This sequence encompasses exons 5 through 8, where most of the evolutionarily conserved

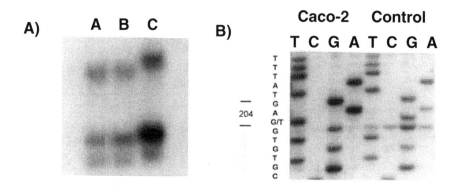

FIGURE 3. Genetic analysis of the p53 of the Caco-2 colon carcinoma cell line. (**A**) PCR-SSCP analysis of exon 6 of p53. Lanes A and B are wild-type p53; lane C is from the Caco-2 cell line. (**B**) Sequence analysis of exon 6. Note the G→T transition at codon 204.

amino acids are concentrated. Considering the fact that the Caco-2 cell line is derived from a colon carcinoma, we wanted to know if the p53 was still intact and involved in the induction of apoptosis. PCR-SSCP analysis of exons 5 through 9 was performed on DNA isolated from Caco-2 cells. FIGURE 3A shows in lane C, next to the wild-type fragments (A and B) corresponding to exon 6, a clear mobility shift. Further sequence analysis revealed a G→T transition at codon 204, resulting in a change of a glutamine to a stop codon (FIG. 3B). The corresponding protein in the Caco-2 cell line may therefore be truncated in the central domain that contains the sequence-specific DNA binding site. Furthermore, the protein lacks the C-terminal domain that is responsible for p53 oligomerization and binding to both single-stranded DNA and RNA. Considering the gravity of the mutation, it is unlikely that p53 is of biological importance in the Caco-2 cell line

   In conclusion, Caco-2 cells are able to undergo p53-independent apoptosis after exposure to MEN. The mechanisms of this apoptotic process involve protein synthesis and are inhibitable by the antioxidants, trolox and NAC.

REFERENCES

1. HALL, P.A. *et al.* 1994. Regulation of cell number in the mammalian gastrointestinal tract: the importance of apoptosis. J. Cell Sci. **107:** 3569–3577.
2. POTTEN, C.S. 1992. The significance of spontaneous and induced apoptosis in the gastrointestinal tract of mice. Cancer Metastasis Rev. **11:** 179–195.
3. HOLLSTEIN, M. *et al.* 1991. p53 mutations in human cancers. Science **253:** 49–53.
4. O'BRIEN, P. 1991. Molecular mechanisms of quinone cytotoxicity. Chem. Biol. Interactions **80:** 1–41.
5. KARCZEWSKI, J.M. *et al.* 1999. Prevention of oxidant-induced cell death in Caco-2 colon carcinoma cells after inhibition of poly(ADP-ribose) polymerase and Ca^{2+} chelation: involvement of a common mechanism. Biochem. Pharmacol. **57:** 19–26.
6. ORITA, M. *et al.* 1989. Rapid and sensitive detection of point mutations and DNA polymorphisms using the polymerase chain reaction. Genomics **5:** 874–879.

Epidermal Growth Factor, Polyamines, and Epithelial Remodeling in Caco-2 Cells

VLADAN MILOVIC, RUTH BAUSKE, LYUDMILA TURCHANOWA,
AND JÜRGEN STEIN[a]

2nd Department of Medicine, Johann Wolfgang Goethe University, Frankfurt, Germany

Epidermal growth factor (EGF), a 53–amino acid peptide, is a potent mitogen in the gastrointestinal tract. In addition to its effects related to cellular proliferation, EGF exerts maturative actions by increasing intestinal lactase activity and upregulating intestinal glucose and electrolyte transport.[1] The interrelation between EGF and polyamines has been well established: EGF stimulates polyamine biosynthesis in the gut,[2] as well as polyamine uptake in enterocyte-like Caco-2 cells.[3] In turn, polyamine deprivation dysregulates distribution of EGF receptor, depletes the cells of their cytoskeleton, and thickens the actin cortex.[4] All this implies that EGF and polyamines, in their stimulatory roles in cell proliferation and maturation in the gut, may act synergistically and be interrelated on several levels of action. Since exogenous polyamines are highly available in the intestinal lumen (originating from food, pancreaticobiliary secretions, and sloughed epithelial cells), interactions between EGF and exogenous/luminal intestinal polyamines may be of a greater importance than in other cell types and systems of the body.

In a preliminary study,[5] we showed that treatment of Caco-2 cells with 100 ng/ mL EGF for 12 h stimulated ornithine decarboxylase activity twofold, and S-adenosyl-methionine activity by 60%. Polyamine content in these cells was increased three- to fourfold, and putrescine uptake by more than fourfold. Therefore, it appears that, at least in Caco-2 cells, polyamine uptake might contribute to EGF-induced polyamine accumulation more markedly than synthesis itself.

Our earlier study[3] showed that EGF evoked a marked increase in polyamine uptake in proliferating Caco-2 cells. In differentiated Caco-2 cells the effect was less prominent but still significant. EGF-stimulated polyamine uptake was blocked neither by cycloheximide (indicating that it was not due to newly synthesized carrier protein) nor by brefeldin A (indicating no translocation of the carrier protein from the Golgi apparatus). However, the effect was completely abolished by both anti–EGF receptor neutralizing antibody and genistein, indicating that it occurs as a consequence of tyrosine phosphorylation of the EGF receptor. Both capacities and affinities of the putrescine and spermidine transporters were increased, indicating that exposure to EGF led to its structural modification.[3]

To investigate the mechanisms of EGF-stimulated polyamine uptake further, Caco-2 cells were stimulated with EGF under the same conditions known to have

[a]Address for correspondence: P.D. Dr. Dr. J. Stein, Med. Klinik II, J. W. Goethe-Universität, Theodor Stern Kai 7, 60590 Frankfurt, Germany. Voice: +49-69-63015197; fax: +49-69-63016246.
j.stein@em.uni-frankfurt.de

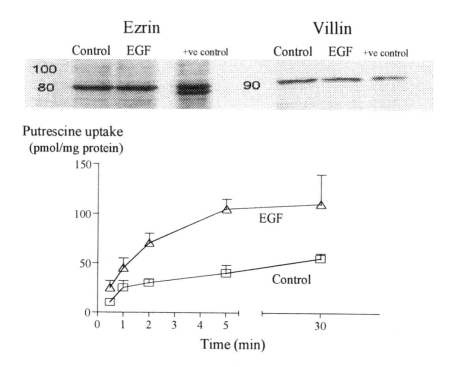

FIGURE 1. Effect of EGF (100 ng/mL for 12 h) on ezrin and villin expression and putrescine uptake into brush border vesicles isolated from EGF-treated Caco-2 cells. **Upper panel:** Western blots for ezrin (*left*) and villin (*right*). **Lower panel:** Putrescine uptake in brush border membrane vesicles isolated from Caco-2 cells pretreated with 100 ng/ml EGF for 12 h. Mean ± SEM, $n = 3$.

maximal effect on polyamine uptake.[3] After incubation with EGF, the cells were harvested and analyzed for ezrin (a product of tyrosine phosphorylation, incorporated into the apical membrane cytoskeleton) and villin (a normal constituent of the apical membrane cytoskeleton). In another series of experiments, brush border membrane vesicles were isolated from differentiated Caco-2 cells under control conditions and after treatment with 100 ng/mL EGF for 12 h. As shown on FIGURE 1, there was no enrichment in ezrin and villin after 12-h treatment with EGF. Still, there was a significant increase in putrescine incorporation into brush border vesicles isolated from Caco-2 cells treated with EGF.

In conclusion, over a period sufficient to induce maximal increase in polyamine uptake, EGF did not induce apparent remodeling of the brush border membrane cytoskeleton. It is therefore likely that EGF upregulates polyamine uptake in the enterocyte by enhancing the incorporation of polyamine transporter from the protein recruited from surrounding cytosol.

REFERENCES

1. OPLETA-MADSEN, K., J. HARDIN & D.G. GALL. 1991. Epidermal growth factor upregulates intestinal electrolyte and nutrient transport. Am. J. Physiol. **260:** G807–G814.
2. FITZPATRICK, L.R., P. WANG & L.R. JOHNSON. 1987. Effect of epidermal growth factor on polyamine synthesizing enzymes in rat enterocytes. Am. J. Physiol. **252:** G209–G214.
3. MILOVIC, V., C. DEUBNER, S. ZEUZEM *et al.* 1995. EGF stimulates polyamine uptake in Caco-2 cells. Biochem. Biophys. Res. Commun. **206:** 962–968.
4. MCCORMACK, S.A., P.M. BLANNER, B.J. ZIMMERMANN *et al.* 1998. Polyamine deficiency alters EGF receptor distribution and signalling effectiveness in IEC-6 cells. Am. J. Physiol. **274:** C192–C205.
5. MILOVIC, V., L. TURHANOWA, F.A. FARES *et al.* 1998. S-adenosylmethionine decarboxylase activity and utilization of exogenous putrescine are enhanced in colon cancer cells stimulated to grow by EGF. Z. Gastroenterol. **36:** 947–954.

Effect of TNFα and IFNγ on Epithelial Barrier Function in Rat Rectum *in Vitro*

INGO GROTJOHANN,[a] HEINZ SCHMITZ,[b] MICHAEL FROMM,[a] AND JÖRG-DIETER SCHULZKE[b,c]

[a]*Institut für Klinische Physiologie, Universitätsklinikum Benjamin Franklin, Freie Universität Berlin, 12200 Berlin, Germany*

[b]*Medizinische Klinik I, Gastroenterologie und Infektiologie, Universitätsklinikum Benjamin Franklin, Freie Universität Berlin, 12200 Berlin, Germany*

INTRODUCTION

In inflammatory bowel disease (IBD), multiple changes in transport and barrier features of the intestinal epithelium have been observed. For instance, in ulcerative colitis, an altered tight junction structure was found to impair the barrier function of human colonic epithelium.[1] A major role in this regulatory process has been ascribed to cytokines, for example, tumor necrosis factor alpha (TNFα), which has been shown to downregulate the barrier properties in different cell monolayer models, including the human colonic epithelial cell line HT-29/B6.[2] Furthermore, using the renal cell line LLC-PK1, TNFα was shown to induce morphological changes that were interpreted as evidence for this cytokine to be involved in tissue remodeling.[3] Another cytokine that plays a major role in this regard is interferon gamma (IFNγ), which was shown to induce a phenotype change in T84 intestinal epithelial cells.[4] The present study aimed to investigate the role of these cytokines not only in cell models, but also in mammalian large intestine *in vitro*.

METHODS

Late distal colon was obtained from male albino Wistar rats (280–320 g). The animals were killed by 10-min inhalation of carbon dioxide. Then, the colon was removed, rinsed with Ringer's solution, and "totally" stripped of serosa and muscular layers. The tissue was mounted to standard Ussing chambers, kept in Ringer's solution (in mM: Na^+ 140, Cl^- 123.8, K^+ 5.4, Ca^{2+} 1.2, Mg^{2+} 1.2, HPO_4^{2-} 2.4, $H_2PO_4^-$ 0.6, HCO_3^- 21, D-glucose 10, D-mannose 10, glutamine 2.5, β-OH-butyrate 0.5; 50 mg/L azlocillin; pH 7.4), supplemented with 10% fetal calf serum (FCS) at 37°C, and oxygenated with 95% O_2 and 5% CO_2. Thirty minutes after starting the experiment, either TNFα (100 ng/mL) or IFNγ (1000 units/mL) alone or in combination was added to the basolateral side of the tissue in comparison to a control

[c]Address for correspondence: Jörg-Dieter Schulzke, Medizinische Klinik I, Gastroenterologie und Infektiologie, Universitätsklinikum Benjamin Franklin, FU Berlin, 12200 Berlin, Germany. Voice: +49-30-8445-2347; fax: +49-30-8445-4481.
schulzke@medizin.fu-berlin.de

group (all groups, $n = 8$). During [^{3}H]mannitol fluxes, the bathing medium additionally contained 10 mM mannitol. After the end of the experiment, the tissue was fix-
ated, and cross-sections of the epithelium were produced.

RESULTS AND INTERPRETATION

Regarding electrophysiological measurements, application of TNFα or IFNγ
alone did not lead to significant changes in the electrical conductance (G) compared
to the control group at the end of a 20-h experiment [$G = 5.3 \pm 0.4$ mS/cm^2 (control),
5.4 ± 0.4 mS/cm^2 (TNFα), and 6.5 ± 0.8 mS/cm^2 (IFNγ)]. By contrast, TNFα and
IFNγ together showed a marked increase in G to 11.3 ± 1.6 mS/cm^2 starting at about
8–10 h after the addition of the cytokines. As the latter group was the only one showing significant conductance changes, further experiments were restricted to this
group in comparison to control. FIGURE 1 shows the development of the conductance
in the TNFα + IFNγ treated group compared to control. During the first 10 hours,
both groups showed similar conductance values. The peak at 5 hours of experimental
time is due to an initial peak in the short-circuit current (I_{sc}) after mounting the epithelium and addition of FCS. Whereas the control tissue reached a baseline value

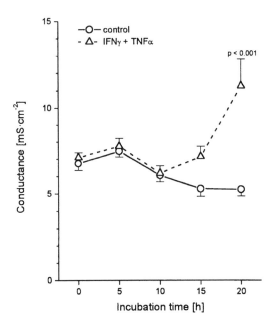

FIGURE 1. Cytokine effect on epithelial conductance. During the first 10 h of incubation, rat rectal epithelia exposed to IFNγ and TNFα exhibited similar conductances as the
control tissues. The peak at $t = 5$ h corresponds to the short-circuit current (I_{sc}) maximum
(I_{sc} data not shown). After 10 h, the conductance of control tissues stabilized at a plateau
level, whereas that of cytokine-treated tissues continuously increased, reaching a conductance twice as high after 20 h at the end of the experiment.

after about 15 hours, conductance increased in the cytokine group, and this change did not reach a plateau level within the experimental period of 20 hours.

Changes in conductance can be due to different reasons. As TNFα is known to induce a secretory response in mammalian large intestine,[5] an increased conductance may in part reflect a higher I_{sc}. However, at the time of the 20-h experiment, the cytokine-treated group showed an even smaller I_{sc} of 11 ± 3 μmol h^{-1} cm^{-2} compared to the control group (19 ± 2 μmol h^{-1} cm^{-2}). Thus, changes in I_{sc} are an unlikely reason for the observed change in conductance. The second possible explanation, a change in paracellular conductance, can be further investigated by [^{3}H]mannitol flux measurements. These experiments revealed a pronounced increase in [^{3}H]mannitol flux rate from 0.07 ± 0.01 in control to 0.25 ± 0.08 μmol h^{-1} cm^{-2} in TNFα + IFNγ treated tissues, indicating a significant change in paracellular conductance. This means that the combination of TNFα and IFNγ was able to impair the barrier function of the colonic epithelium.

At the end of the experiment, the tissues were removed from the Ussing chamber. Those tissues incubated with both cytokines showed a much thinner appearance than control tissues. Thus, histological cross sections of these epithelia were investigated in order to check for morphological changes (FIG. 2). Whereas control epithelia after 20 h of incubation showed a normal appearance with well-formed crypts, those incubated with TNFα and IFNγ showed marked changes. As with the [^{3}H]mannitol fluxes, the histological appearance also showed a high degree of variation in this group exposed to both cytokines. However, the mean thickness of the mucosal layer in this TNFα + IFNγ treated group was reduced to about 60% of that of control tissues. Crypt depth was dramatically reduced and, in some tissues, crypts even seemed to disappear completely (FIG. 2b), which led to a mucosa with a flat appearance.

Inasmuch as mucosal architecture reflects the equilibrium of cell loss and proliferation, this pronounced change towards an almost flat mucosa in the final stage of this process may predominantly reflect the pronounced loss of epithelial cells due to induction of apoptosis by these cytokines, whereas acceleration of epithelial cell proliferation is insufficient to come up with loss of epithelial cells. In line with this interpretation, TNFα is known to induce morphological changes in renal cell lines induced by apoptosis,[3] and an increased loss of cells from the epithelial layer is also indicated in the histological cross-sections in the present study in rat large intestine (FIG. 2; the cells seen above the epithelial cell layer). This is also supported by lactate dehydrogenase (LDH) release assays, where LDH is a marker of cell death. The bath solution on the mucosal side of the tissue contained an increased amount of LDH in the TNFα + IFNγ group [125 ± 3 vs. 96 ± 3 units/mL (control)]. The fact that the amount of LDH detected in the serosal bath solution did not differ significantly (40 ± 1 vs. 44 ± 1 units/mL) may indicate that the TNFα- and IFNγ-induced increase in the rate of cell death is restricted to the epithelial cell population, whereas subepithelial cells (e.g., fibroblasts) are less severely concerned. The normal rate of cell turnover within the epithelium is reflected by the LDH content, which is higher in the mucosal than in the serosal bath solution. However, some cytokine-induced changes of the fibroblast population can be observed too. TNFα and IFNγ induced an increase in the number of nuclei in the fibroblast layer combined with smaller cell bodies. This may be due to a proliferative signal imposed on the fibroblasts, although further investigation is needed regarding this point.

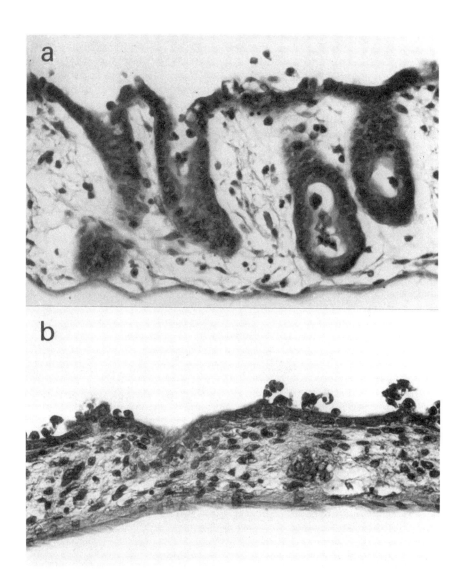

FIGURE 2. Histology of rat rectal epithelia after 20 h of incubation with IFNγ and TNFα. **(a)** Control tissues: Crypts showed normal appearance; the average diameter of the epithelium above the muscular layer, that is, the epithelial cell layer and the lamina propria, was 30 ± 1 μm. **(b)** Tissues exposed to TNFα (100 ng/mL) + IFNγ (1000 units/mL): The figure shows an advanced state of epithelial remodeling after 20 h. The average diameter of the epithelium was reduced to 18 ± 1 μm, and crypt length was dramatically decreased. The pit at the left of the middle may mark actual crypt depth, although it cannot be ruled out that it is only a cut through the edge. Note the change of the appearance of the fibroblastic nuclei. Magnification: ×2000.

CONCLUSIONS

The rise in conductance in tissues incubated with TNFα and IFNγ is the result of several effects acting in parallel. On the one hand, TNFα may alter tight junction permeability, as observed, for example, in the HT-29/B6 cell model.[2] On the other hand, TNFα and IFNγ induce cell death in colonic epithelial cells. Although apoptotic spots of high conductance may significantly contribute to the barrier disturbance, this could be partially compensated for by the reduction of the mucosal surface area. Later on, however, the disturbed equilibrium of apoptotic cell loss and insufficient cell renewal may overrule this compensatory morphological change, and erosions arise that may additionally contribute to the severe barrier defect.

The similarity of the phenomena observed here and detected for the impaired epithelial barrier function in ulcerative colitis with crypt rarefaction, apoptosis, and (at least in severely inflamed tissues) erosions, as well as tight junction alterations, stresses the importance of these cytokines for the barrier dysfunction in inflammatory bowel disease.

TNFα and IFNγ are able to impair the barrier function of the colonic mucosa by a rise in paracellular permeability in association with a marked tissue remodeling as a consequence of increased cell death partially compensated for by the reduction of mucosal area.

REFERENCES

1. SCHMITZ, H., C. BARMEYER, M. FROMM et al. 1999. Altered tight junction structure contributes to the impaired epithelial barrier function in ulcerative colitis. Gastroenterology **116:** 301–309.
2. SCHMITZ, H., M. FROMM, C.J. BENTZEL et al. 1999. Tumor necrosis factor-alpha (TNF-alpha) regulates the epithelial barrier in the human intestinal cell line HT-29/B6. J. Cell Sci. **112:** 137–146.
3. PERALTA-SOLER, A., J.M. MULLIN, K.A. KNUDSEN & C.W. MARANO. 1996. Tissue remodeling during tumor necrosis factor–induced apoptosis in LLC-PK1 renal epithelial cells. Am. J. Physiol. **270:** F869–F879.
4. COLGAN, S.P., C.A. PARKOS, J.B. MATTHEWS et al. 1994. Interferon-gamma induces a cell surface phenotype switch on T84 intestinal epithelial cells. Am. J. Physiol. **267:** C402–C410.
5. SCHMITZ, H., M. FROMM, H. BODE et al. 1996. Tumor necrosis factor-alpha induces Cl⁻ and K⁺ secretion in human distal colon driven by prostaglandin E2. Am. J. Physiol. **271:** G669–G674.

Colectomy and Ileal Pouch

Transport and Barrier in Pouchitis

A. J. KROESEN,[a,b] M. STOCKMANN,[c] J. D. SCHULZKE,[c] M. FROMM,[d]
AND H. J. BUHR[a]

Departments of [a]Surgery, [c]Gastroenterology, and [d]Clinical Physiology,
Universitätsklinikum Benjamin Franklin, Freie Universität Berlin, Berlin, Germany

INTRODUCTION

The causes of pouchitis after ulcerative colitis are still unknown. The main risk factors for the development of pouchitis are the preoperative presence of a back-wash-ileitis[1] and the presence of a primary sclerosing cholangitis.[2] As is the case with ulcerative colitis, smokers have a lower risk of pouchitis.[3]

There is some evidence that there is a common pathogenesis with ulcerative colitis. The common pathogenesis is based on a similar clinical appearance of pouchitis, similar histological alterations, and simultaneous immunological changes. The changes in permeability are still unknown. Schulzke et al.[4] have shown decreased barrier function after pouch formation in the rat, but a detailed analysis of human specimens is still lacking. There exists one clinical study, by Merret et al.[5] ^{51}Cr-EDTA was administered into the pouch; urinary recovery over 24 hours was taken as an indicator of permeability (barrier function). Histologic analysis of pouch biopsy specimens was undertaken. Pouchitis was associated with increased permeability (5.9% of administered dose absorbed) compared to that of a healthy pouch.

MATERIAL AND METHODS

Twenty-four patients were included in this study. All suffered from ulcerative colitis. The mean age was 35.2 ± 12.5. The sex distribution was 15 females, 9 males. Mucosal biopsies were taken with a 3.4-mm endoscopic biopsy forceps. The different groups and the biopsy sites are shown in TABLE 1.

Alternating Current Impedance Analysis

As described in former publications, alternating current (AC) impedance analysis makes it possible to distinguish between the epithelial (R^e) and subepithelial (R^{sub}) portions of the total wall resistance (R^t).[6–10] The voltage responses after transepithelial application of a 35-μA/cm^2 sine wave AC of 48 discrete frequencies ranging from 1 to 65 kHz were detected by phase-sensitive amplifiers (Model 1250 frequen-

[b]Address for correspondence: Anton J. Kroesen, M.D., Universitätsklinikum Benjamin Franklin, Department of Surgery I, Hindenburgdamm 30, 12200 Berlin, Germany. Voice: +49-30-8445-2543; fax: +49-30-8445-2740.
kroesen@medizin.fu-berlin.de

TABLE 1. Site and status of the specimen taken

Group	n	Biopsy Site	Status
1	5	terminal ileum	control
2	5	terminal ileum	ileum colitis
3	8	pouch corpus	deviation
4	5	pouch corpus	intact pouch
5	6	pouch corpus	pouchitis

cy response analyzer and Model 1286 electrochemical interface; Solartron Schlumberger, Farnborough, Hampshire, England). The impedance values were calculated and corrected for the resistance of the bathing solution and the frequency behavior of the measuring setup for each frequency. For every specimen, the impedance locus was then plotted in a Nyquist diagram and a circle segment was fitted by least-square analysis. From this circle segment, three variables of an electric equivalent circuit were obtained, a resistor and a capacitor in parallel representing the epithelium and a resistor in series with this unit representing the subepithelium. Because of the frequency-dependent electrical characteristics of the capacitor, R^t is obtained at low frequencies, whereas R^{sub} is obtained at high frequencies. R^e was calculated as $R^e = R^t - R^{sub}$.

Mannitol Fluxes

Mucosal barrier function of the different pouch status and controls was determined by measurements of unidirectional mucosal-to-serosal flux of ^{3}H-mannitol. This setup was also performed under short circuit conditions as previously described.[11]

Electrogenic Chloride Secretion

One aspect of mucosal transport function as part of the barrier was examined by electrogenic Cl$^-$ secretion. The tissues were stimulated by prostaglandin E$_2$ (10^{-6} M) and theophylline (10^{-2} M) dissolved in the bathing media on both sides for theophylline and on the serosal side for prostaglandin-E$_2$. The increase in I_{sc} (ΔI_{sc}) was measured thereafter. After reaching a steady state, the effect of theophylline and prostaglandin-E$_2$ was antagonized by bumetanide (10^{-5} M) and the decrease in I_{sc} measured. ΔI_{sc}^{max} (maximum velocity [V_{max}]) and Michaelis constant (K_m) were determined from Eadie-Hofstee plots for each tissue specimen separately and ΔI_{sc}^{max}, and K_m was calculated for each experimental group.

Na-Glucose Cotransport

The other aspect of mucosal transport function is measured by glucose-dependent sodium absorption. 3-o-Methyl-glucose is added for the detection of glucose-dependent sodium absorption. 3-o-Methyl-glucose is a hexose that is transported by the glucose carrier but not by the intestine, as described earlier.[12] Aliquots of standard medium supplement with 3-*o*-methyl-glucose were added at about 10-min intervals, resulting in concentrations of 4, 8, 16, 32, and 48 mM. ΔI_{sc}^{max} and K_m were determined from the reciprocal plot (Lineweaver & Burk) for each specimen separately. For each experimental group the $\pm$ SEM of ΔI_{sc}^{max} and K_m were calculated. Finally, ΔI_{sc} values for each glucose concentration and ΔI_{sc}^{max} values were corrected for subepithelial resistance contributions by multiplication with the respective correction factor—which, owing to error propagation, also accounts of the magnitude of the SEM of these variables.

Statistical Analysis

Results are given as means $\pm$ SEM. Student-Newman-Keul's test was used for the multiple comparisons when the null hypothesis was rejected by Friedman's test. $p <$ 0.05 was considered significant.

RESULTS

Alternating Current Impedance Analysis

Numeric results and statistical evaluation of the impedance analysis are shown in TABLE 2. Total resistance was raised in all states of the ileoanal pouch (intact pouch 41.3 $\pm$ 1.1 Ω·cm^2; pouchitis 40.3 $\pm$ 2.6 Ω·cm^2) a part of the pouch under deviation (28.6 $\pm$ 5.1 Ω·cm^2) compared to controls (24.6 $\pm$ 1.8 Ω·cm^2). R^e remained more or less unchanged in the five compared groups. R^{sub} was significantly ($p < 0.05$) increased in the ileum of colitis (24.4 $\pm$ 8.4 Ω·cm^2), intact pouch (21.2 $\pm$ 2.7 Ω·cm^2), and pouchitis (22.4 $\pm$ 2.0 Ω·cm^2) compared to deviation (13.3 $\pm$ 1.1 Ω·cm^2) and controls (10.8 $\pm$ 0.6 Ω·cm^2).

Mannitol Fluxes

The numeric results and statistical evaluation are also shown in TABLE 2. No statistical differences were found between the five groups. Also, in the deviation group there was no significant change in porosity.

Chloride Secretion

Chloride secretion was determined from the increase in I_{sc} after administration of theophylline and prostaglandin E$_2$ as V_{max} (TABLE 2). Reduction of transport rates ($p < 0.05$) was found for the pouchitis (38.6 $\pm$ 16.0 μA·cm^{-2}) and the ileum colitis group (36.6 $\pm$ 16.3 μA·cm^{-2}) as compared to controls (113.9 $\pm$ 24.75 μA·cm^{-2}), deviation (90.1 $\pm$ 18.5 μA·cm^{-2}), and intact pouch (120.9 $\pm$ 34.8 μA·cm^{-2}).

Na-Glucose Cotransport

In the different states of pouch ileum the 3-*o*-methyl-glucose–dependent increase in I_{sc} showed saturation. The data and statistical analysis are given in TABLE 2. The data in the reciprocal plot fit a straight line, indicating Michaelis-Menten kinetics.

TABLE 2. Transmural electrical resistance and mannitol fluxes

	n	R^t [Ω cm^2]	R^e [Ω cm^2]	R^s [Ω cm^2]	Mannitol Flux J (nmol h^{-1}cm^{-2})	Cl Secretion ΔI_{sc} (μA cm^{-2})	Na-Glucose Cotransport V_{max} (μA cm^{-2})
Ileum colitis	5	39.7 ± 7.8	15.2 ± 4.4	24.4 ± 8.4	382.0 ± 113.9	36.6 ± 16.3	55.1 ± 29.2
Deviation	8	28.6 ± 5.1	15.7 ± 4.4	13.3 ± 1.1	404.0 ± 53.5	90.1 ± 18.5	69.6 ± 29.8
Intact pouch	5	41.3 ± 1.1	18.1 ± 0.6	21.2 ± 2.7[a]	245.5 ± 35.3	120.9 ± 34.8	138.4 ± 25.0
Pouchitis	6	40.3 ± 2.6	17.9 ± 1.5	22.4 ± 2.0[a]	249.4 ± 45.9	38.6 ± 16.0[b]	41.8 ± 14.37[c]
Control	5	24.6 ± 1.8	13.8 ± 1.7	10.73 ± 0.6	293.0 ± 26.4	113.9 ± 24.7	264.4 ± 45.7

$p < 0.05$: [a]versus deviation, intact pouch, and control; [b]versus deviation and control; [c]versus intact pouch and control.

DISCUSSION

Our results did not show any significant differences between pouchitis and the other states of the terminal ileum simultaneous to U.C. or as transformed to a neorectum. This is confirmed by ion permeability as well as by porosity. These findings are in contrast to the findings of Merret *et al.*,[5] who reported increased porosity for ^{51}Cr-EDTA in pouchitis compared to healthy pouches. In addition, they found a negative correlation to colonic metaplasia and mucin type. Merret *et al.* also demonstrated increased permeability for the deviation group, which, at least for the mannitol flux, is concordant with our results. We explain the difference concerning the barrier function by the more reliable method of mannitol flux and the alternating current impedance analysis.[6–10] Apart from that, Merret examines only the aspect of porosity (MW of ^{51}Cr-EDTA 340; mannitol MW 182).

In the present study we observed increased subepithelial resistance in the ileum of ulcerative colitis, the intact pouch, and in pouchitis. We interpret this as adaptational changes due to the chronic inflammation of the terminal ileum. Functional analysis in pouchitis and terminal ileum prior to ileoanal pouch anastomosis showed significant reduction of electrogenic chloride secretion and Na-glucose cotransport. This leads to the conclusion that pouchitis affects only the mucosal transport function, whereas the mucosal barrier function remains unchanged.

REFERENCES

1. SCHMIDT, C.M., A.J. LAZENBY, R.J. HENDRICKSON & J.V. SITZMANN. 1998. Preoperative terminal ileal and colonic resection histopathology predicts risk of pouchitis in patients after ileoanal pull-through procedure. Ann. Surg. **227:** 654–662; discussion 663–665.
2. PENNA, C., R. DOZOIS, W. TREMAINE *et al.* 1996. Pouchitis after ileal pouch-anal anastomosis for ulcerative colitis occurs with increased frequency in patients with associated primary sclerosing cholangitis. Gut **38:** 234–239.
3. MERRET, M.N. & M. KETTLEWELL. 1996. Smoking may prevent pouchitis in patients with restorative proctocolectomy for ulcerative colitis. Gut **38:** 362–364.
4. SCHULZKE, J.D., M. FROMM, W. GOGARTEN *et al.* 1993. Epithelial ion transport in the ileal J-Pouch after proctocolectomy in the rat. Scand. J. Gastroenterol. **28:** 533–539.
5. MERRET, M.N., N. SOPER, N. MORTENSEN & D.O. JEWELL. 1996. Intestinal permeability in the ileal pouch. Gut **39:** 226–230.
6. FROMM, M., J.D. SCHULZKE & U. HEGEL. 1985. Epithelial and subepithelial contributions to transmural electrical resistance of intact rat jejunum, *in vitro*. Pflügers Arch. **405:** 400–402.
7. GITTER, A.H., J.D. SCHULZKE, D. SORGENFREI & M. FROMM. 1997. Ussing chamber for high-frequency transmural impedance analysis of epithelial tissues. J. Biochem. Biophys. Methods **35:** 81–88.
8. SCHULZKE, J.D, M. FROMM & U. HEGEL. 1986. Epithelial and subepithelial resistance of rat large intestine: segmental differences, effect of stripping, time course, and action of aldosterone. Pflügers Arch. **407:** 632–637.
9. SCHULZKE, J.D., M. FROMM, H. MENGE & E.O. RIECKEN. 1987. Impaired intestinal sodium and chloride transport in the blind loop syndrome of the rat. Gastroenterology **92:** 693–698.
10. SCHULZKE, J.D., M. FROMM, C.J. BENTZEL *et al.* 1992. Ion transport in the experimental short bowel syndrome of the rat. Gastroenterology **102:** 497–504.

11. SCHMITZ, H., C. BARMEYER, M. FROMM, *et al.* 1999. Altered tight junction structure contributes to the impaired epithelial barrier function in ulcerative colitis. Gastroenterology **116:** 301–309.
12. SCHULZKE, J.D., M. FROMM, H. MENGE & E.O. RIECKEN. 1987. Impaired intestinal sodium and chloride transport in the blind loop syndrome of the rat. Gastroenterology **92:** 693–698.

Mechanisms of Epithelial Barrier Impairment in HIV Infection

MARTIN STOCKMANN,[a] HEINZ SCHMITZ,[a] MICHAEL FROMM,[b]
WOLFGANG SCHMIDT,[a] GEORG PAULI,[c] PETER SCHOLZ,[d]
ERNST-OTTO RIECKEN,[a] AND JÖRG-DIETER SCHULZKE[a,e]

*Departments of [a]Gastroenterology and [b]Clinical Physiology,
Universitätsklinikum Benjamin Franklin, Freie Universität Berlin,
12200 Berlin, Germany*

[c]Robert Koch Institut, Berlin, Germany

[d]Schering AG, Berlin, Germany

ABSTRACT: Diarrhea and malabsorption due to intestinal dysfunction are common symptoms in HIV infection. The pathophysiologic mechanisms of these alterations are often not known, and the role of HIV per se is still controversially discussed. We measured the epithelial transport and barrier function by means of a miniaturized Ussing chamber system in the duodenum of HIV-infected patients in different disease stages, determined by the CD4 cell count in the serum as well as symptoms in patients with and without diarrhea. We could show that diarrhea induced by HIV per se is caused by a leak flux mechanism due to impaired epithelial barrier function. Antisecretory therapy does not seem to be useful in these patients, because we did not find increased active ion secretion. Along the course of the HIV infection, the epithelial transport and barrier function varies with HIV disease stage (expressed by CD4 cell status). In addition, an *in vitro* model was studied to characterize the effect of HIV-infected human immune cells on the epithelial barrier function using the human colonic epithelial cell line HT-29/B6. HIV infection of human immune cells induced an increase in cytokine release—for example, TNF-α, IL-1β, IFN-α, and IFN-γ—downregulating the epithelial barrier function of the human colonic epithelial cell line HT-29/B6. Taken together we postulate a specific stage-dependent cytokine pattern released from HIV-infected immune cells in the mucosa, which, corresponding to the HIV disease stage, is responsible for the variation in epithelial function.

INTRODUCTION

The gastrointestinal tract is important as a reservoir and entrance for the human immunodeficiency virus (HIV). During the course of HIV infection, the majority of patients suffer from gastrointestinal disease.[1,2] Common clinical features are diarrhea, malabsorption, and weight loss, which are associated with low CD4 cell counts and occur with or without detected enteropathogens. Even if an enteropathogen is

[e]Address for correspondence: Prof. Dr. Jörg-Dieter Schulzke, Gastroenterologie u. Infektiologie, Universitätsklinikum Benjamin Franklin, Freie Universität Berlin, 12200 Berlin, Germany. Voice: +49-30-8445-2347; fax: +49-30-8445-4481.
schulzke@medizin.fu-berlin.de

detected, the meaning can be uncertain, as enteropathogens are found also in asymptomatic HIV patients.[1] Because HIV has been detected in lamina propria mononuclear cells as well as in enterochromaffin cells or lymphocytes, changes in the intestinal immune system and the release of inflammatory mediators could be induced.[3–5] Released cytokines could be important for the development of gastrointestinal symptoms. In the intestinal mucosa of HIV-infected patients, indeed, tumor necrosis factor–α (TNF-α), interleukin-1β (IL-1β), and interferon-γ (IFN-γ) have been found to be increased.[6–8] It has already been shown in animal studies that these cytokines can induce active Cl$^-$ secretion (TNF-α and IL-1β) or can directly affect the epithelial barrier function (TNF-α and IFN-γ).[9–12] Whether these effects are relevant, and, if so, at what stage along the course of HIV infection alterations of the epithelial transport and barrier function occur, is not known. Also, the pathophysiologic role of HIV per se for these alterations is not known.

In the first part of this paper we show measurements of the epithelial transport and barrier function in the duodenum of HIV-infected patients in different disease stages determined by the CD4 cell count in the serum as well as by symptoms in patients with and without diarrhea. In the second part we used an *in vitro* model of HIV-infected human immune cells to measure the cytokine release induced by HIV infection and the effect of the respective supernatants on the epithelial barrier function of the human colonic epithelial cell line HT-29/B6 as a model for analysis of the effect of HIV per se.

MATERIAL AND METHODS

Study Population

Twenty-seven HIV-seropositive patients (confirmed by ELISA and Western blot) from the University Hospital Benjamin Franklin in Berlin, who underwent endoscopy because of diarrhea or other symptoms, were included. Patients who underwent upper endoscopy because of follow-up after *Helicobacter pylori* eradication or cancer search, but had neither macroscopical nor histological abnormalities, served as controls. The study was approved by the local ethics committee, and all patients gave written informed consent for taking extra biopsies for the study.

Miniaturized Biopsy Container

A miniaturized container insert with an opening diameter of 2.5 mm (exposed area 0.05 cm^2) was used, which was described recently.[13] The container can be placed between the two halves of Ussing chambers designed for conventional short circuit current, flux measurements, and impedance analysis.[14,15] The chamber preserves the advantages of conventional low-volume, gas lift–stirred flux chambers.[16] Current electrodes were silver rings located far from the epithelium to form a homogeneous electrical field across the epithelium. The voltage electrodes were positioned axially within the electrical field. Voltage electrodes were covered by a driven shield and consisted of commercial microelectrode holders (MEH3SF, WP-Instruments, New Haven, CT, USA) connected to glass micropipettes (not pulled; inner/

outer diameter 1.0/1.4 mm; length 25 mm). The pipettes were filled with 3 g/dL Agar-Agar and 0.5 M KCl.

Experimental Procedures with Human Duodenal Biopsy Samples

Biopsy specimens were obtained by 3.4-mm biopsy forceps from the distal duodenum. Under a dissection microscope, biopsy specimens were spread out and a support disc (stamped out from tables of Astralon N™, Hüls Troisdorf, Troisdorf, Germany), equipped with a centered 2.5-mm hole, was glued on the serosal side by Histoacryl™ tissue glue (B. Braun, Melsungen, Germany). This disc was inserted in the biopsy sample container and mounted between the two halves of an Ussing-type chamber (for details see Stockmann *et al.*[13]). The time between taking the biopsy and mounting it into the Ussing chamber was about half an hour; meanwhile it was kept in oxygenated standard medium at 4 °C. Short circuit current (I_{sc}) and transepithelial resistance (R^t) were recorded continuously. The standard medium contained (mM): Na^+ 140, Cl^- 123.8, K^+ 5.4, HPO_4^{2-} 2.4, $H_2PO_4^-$ 0.6, Ca^{2+} 1.2, Mg^{2+} 1.2, HCO_3^- 21, and as substrates D(+)-glucose 10.0, β-OH-butyrate 0.5, glutamine 2.5, and D(+)-mannose 10.0 (gassed with 95% O_2 and 5% CO_2; pH 7.4 at 37° C). For determination of epithelial transport function after reaching steady state values for I_{sc}, $5 \cdot 10^{-4}$ M phlorizin (Sigma, St. Louis, MO, USA) was added to the mucosal side, and subsequently 10^{-5} M bumetanide (Sigma, St. Louis, MO, USA) was given serosally. Flux measurements were performed as previously described.[15] Fluxes (*J*) were calculated by standard formula, as described by Schultz and Zalusky.[16] All I_{sc} values were corrected for the bath resistance, as described by Tai and Tai.[17]

Alternating Current Impedance Analysis

Alternating current impedance analysis was applied to discriminate between epithelial (R^e) and subepithelial resistance (R^s) without mechanical preparation.[18,19] Measurements were performed as described previously.[14] Briefly, sine wave alternating currents (35 µA/cm² eff.) in the range of 1 Hz to 65 kHz were applied and the voltage responses detected. Each recording consisted of 48 frequencies and took about 1 minute. The complex impedance was calculated on-line with a personal computer. Correction for the impedance of the experimental setup (including bath resistance) was performed in each experiment by measuring the impedance of the fluid-filled chamber and subtracting it for each frequency from subsequently measured impedance data. The impedance locus was plotted in a Nyquist diagram. For further explanation refer to FIGURE 1.

Cell Culture of Human Immune Cells

Blood samples of six healthy volunteers were sampled separately, and peripheral blood monocytes were cultured and differentiated to macrophages as described by Rokos and Pauli.[20] Briefly, erythrocytes were separated from the other blood cells by Ficoll-Paque gradient centrifugation (Pharmacia Biotech, Uppsala, Sweden). Then, the leukocyte-containing layer was sucked by a glass pipette. Thereafter, the cells were rinsed in PBS, centrifuged, and decanted three times. Finally the cells were seeded in 12-well plates with RPMI 1640 medium and 20% fetal calf serum and cultured for 7 days. This procedure has been proved to yield pure macrophage cultures.[20] After maturation the monocyte-derived macrophages were infected with HIV strain HTLV-IIIb, which shows lymphocyte tropism.[21] After seven days of infection the macrophages were cocultured with homologues peripheral blood leukocytes (PBL), which were obtained again by Ficoll-Paque gradient centrifugation for another 24-hour period. Then, the supernatants were collected. Success of infection was determined by measuring the p24 content of the supernatants with a sandwich ELISA (R & D Systems, Abingdon, UK).

Determination of Cytokine Content

By means of commercial ELISA kits the content of TNF-α, IL-1 (both R & D Systems Abingdon, UK), IFN-α, and IFN-γ (both BioSource, USA) was determined in the supernatants of the infected macrophage/PBL cocultures and the respective uninfected controls.

Cell Culture of Intestinal HT-29/B6 Cells

HT-29/B6 cells were routinely cultured in 25-cm^2 culture flasks (Nunc).[22] The culture medium (RPMI 1640, Biochrom KG, Berlin, Germany) contained 2% stabilized L-glutamine and was enriched with 10% fetal calf serum (FCS). Culture was performed at 37°C in a 95% air, 5% CO_2 atmosphere. Cells were seeded on Millicell filters (Millipore, effective membrane area 0.6 cm^2), and three inserts were placed together into one conventional culture dish (OD 60 mm). Confluence of the polarized monolayers was reached after 7 days. Experiments were performed on day 11 or 12, giving transepithelial resistances (R^t) of 250–400 Ω·cm^2. The apical compartment was routinely filled with 500 μL culture medium; the basolateral compartment contained 10 milliliters.

Monitoring of Transepithelial Resistance

Transepithelial resistance of the monolayers was measured by a modification of the method described by Kreusel et al.[22] Briefly, electrical measurements were performed in the culture dishes by two fixed pairs of electrodes (STX-2, World Precision Instruments, USA) connected with an impedance meter (D. Sorgenfrei, Inst. Klinische Physiologie). R^t was calculated from the voltage deflections caused by an external ±10 μA, 21 Hz rectangular current. The temperature was maintained at 37°C during the measurements by a temperature-controlled warming plate. Resistance values were corrected for the resistance of the empty filter and of the bathing solution.

Cytotoxicity Assay

As a measure of cell deterioration, the lactate dehydrogenase (LDH) release from the cells was used.[10,23] Briefly, the postexperimental LDH content in the supernatant of controls and of TNF-α–treated cells was determined. After detergent extraction with 2% Triton X-100 for 20 minutes, the total LDH content of the residual cells was measured. Then, the percentage of LDH released into the supernatant was calculated.

TABLE 1. Epithelial transport function measured on human duodenal biopsy specimens of controls and HIV-infected patients[a]

	Patients Classified by CD4 Count				Patients Classified by Symptoms	
	Control	HIV E	HIV I	HIV L	HIV asymp.	HIV diarrhea
ΔI_{sc} Phlorizin	31 ± 3	38 ± 5	$57 \pm 13^*$	36 ± 4	34 ± 4	37 ± 6
Basal I_{sc}	72 ± 7	64 ± 7	$103 \pm 14^{*\#}$	80 ± 9	65 ± 6	72 ± 10
ΔI_{sc} Bumetanide	25 ± 4	26 ± 3	$44 \pm 5^{*\#}$	28 ± 5	24 ± 2	22 ± 5
Diarrhea (%)	0	6	20	70	0	100

[a]HIV E, early-stage HIV patients with CD4 cells above 250/μL; HIV I, intermediate-stage HIV patients with CD4 cells between 100 and 250/μL; HIV L, late-stage HIV patients with CD4 cells below 100/μL; HIV asymp., asymptomatic HIV-infected patients; HIV diarrhea, HIV-infected patients with diarrhea. All values are means ± SEM.
** $p < 0.01$ versus control; [#] $p < 0.05$ versus HIV E.

INTESTINAL TRANSPORT AND BARRIER FUNCTION IN HIV-INFECTED PATIENTS

Epithelial Transport Function in Human Duodenal Biopsies

Phlorizin is a blocker of the Na^+-glucose symporter in the apical enterocyte membrane. Thus, the decrease in I_{sc} after addition of phlorizin is a measure of Na^+-glucose cotransport. The baseline I_{sc} remaining after addition of phlorizin is a measure of active electrogenic transport, including electrogenic chloride and bicarbonate secretion. Bumetanide is a blocker of the basolateral $Na^+2Cl^-K^+$ symporter of the enterocytes, which is part of the active chloride secretory system.

When the patients were classified by the symptom *diarrhea*, we found no significant differences in spontaneous glucose absorption (ΔI_{sc} phlorizin) and spontaneous chloride/bicarbonate secretion (basal I_{sc} and ΔI_{sc} bumetanide) in HIV-infected patients with diarrhea or in asymptomatic HIV-infected patients, as already shown in a recent paper[15] (TABLE 1). Classification by CD4 cell count in the serum also revealed no alteration of ΔI_{sc} phlorizin, basal I_{sc}, or ΔI_{sc} bumetanide in early-stage HIV patients but an increased ΔI_{sc} phlorizin, basal I_{sc}, and ΔI_{sc} bumetanide in intermediate-stage HIV patients. The increase of these parameters was lacking again in late-stage HIV patients, although in this group diarrhea was more common (TABLE 1).

Taken together we could not find impaired glucose absorption at the enterocyte level, at least in HIV-infected patients with diarrhea or in late-stage HIV patients. First of all, this is in contrast to oral *in vivo* tests for glucose malabsorption,[24–26] but here other factors might be taking place due to the *in vivo* condition. These studies measured the urinary recovery rates of 3-*O*-methyl-D-glucose or D-xylose but did not test the Na^+-glucose cotransport system directly on the cellular level. The increased ΔI_{sc} phlorizin in intermediate-stage HIV patients might be due to an unspecific activation together with other active transport mechanisms—for example, by a cAMP-dependent pathway induced by immune modulators (e.g., TNF-α–induced Cl^-

secretion[9]). Active chloride and bicarbonate secretion was not altered in the early phase of the HIV infection (classified by symptoms or CD4 cells) and was not activated in HIV-infected patients with diarrhea.[15] This result points strongly against a secretory type of diarrhea in HIV-infected patients by HIV itself but cannot rule out that in some of the patients (e.g., with detected enteropathogens) a secretory type of diarrhea is present. An important clinical consequence of this result is that an antisecretory drug therapy (e.g., with octreotide) may not be useful in the majority of HIV patients with diarrhea without a detected enteropathogen (often described as HIV enteropathy). This is underlined by a multicenter trial of octreotide therapy in AIDS patients with diarrhea, which did not find octreotide therapy to be effective in reducing diarrhea in the majority of patients.[27] Interestingly, we found in intermediate-stage HIV patients increased active ion secretion in the duodenum, which, again, was lacking in late-stage HIV infection. An explanation for the observation that this increase of active ion secretion in the duodenum of intermediate-stage HIV patients was not strongly associated with diarrhea might be that the absorptive function of the colon is still intact in this group.

Epithelial Barrier Function in Human Duodenal Biopsies

Pure epithelial resistance (determined by impedance analysis) and the tracer flux of lactulose (J_{lac}) represent the epithelial permeability for small ions and macromol-

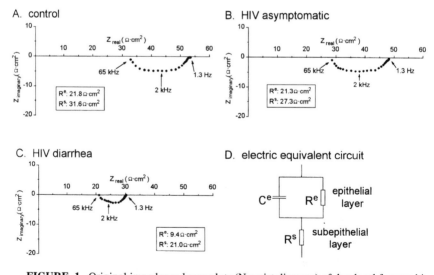

FIGURE 1. Original impedance locus plots (Nyquist diagram) of duodenal forceps biopsy specimens from a control (**A**), an asymptomatic HIV patient (**B**), and an HIV patient with diarrhea (**C**). Z_{real} gives the ohmic component and $Z_{imaginary}$ the reactive component of the complex impedance. **D** shows a lumped electrical equivalent circuit for the intestine. Extrapolation of the high- and low-frequency ends of the impedance curve to the abscissa yielded subepithelial resistance (R^s) and total resistance (R^t), respectively. Thus, the impedance technique discriminates between the epithelial ($R^e = R^t - R^s$) and the subepithelial (R^s) contribution to the total tissue resistance (R^t).

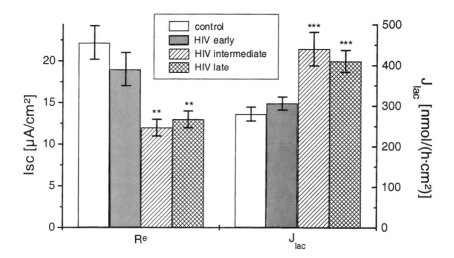

FIGURE 2. Epithelial resistance (R^e) and lactulose flux (J) of duodenal biopsy specimens from control and HIV-infected patients in three different stages of the HIV infection (HIV early = early-stage HIV patients with CD4 cells above 250 μL^{-1}; HIV intermediate = intermediate-stage HIV patients with CD4 cells between 100 and 250 μL^{-1}; late-stage HIV patients with CD4 cells below 100 μL^{-1}). All values are means ± SEM. ** $p < 0.01$ versus control; *** $p < 0.005$ versus control.

ecules, respectively.[15,28] Classification of patients by symptoms showed no alteration in asymptomatic HIV-infected patients compared to controls. However, a noticeably impaired epithelial barrier (decreased epithelial resistance and increased lactulose flux) was found in HIV-infected patients with diarrhea, as shown in a previous paper.[15] FIGURE 1 shows original impedance locus plots of these three groups and also a brief description of the alternating current impedance analysis. Together with the above-mentioned transport data, these results indicate that the diarrheal mechanism in HIV infection by HIV per se is not a secretory one but is caused by leak flux.[29]

Classification of patients by HIV disease stage (determined by CD4 cells) revealed no alteration of R^e and J_{lac} in early-stage HIV patients but a decrease of R^e and an increase of J_{lac} in intermediate- and late-stage HIV patients (FIG. 2). This alteration of the epithelial barrier function shows a tendency to a more pronounced alteration in patients with a lower count of CD4 cells: when R^e is directly correlated to the CD4 cell, we found decreasing epithelial resistance with decreasing number of CD4 cells, with a weak but significant correlation (data not shown). Taken together with the data of the transport function, these results showed that classification of patients according to their CD4-cell status revealed a variation of the epithelial transport and barrier function with HIV disease stage. In early stage we found no change; in intermediate stage we detected an increased active ion secretion and impaired epithelial barrier function; and in late stage only an impaired epithelial barrier function

TABLE 2. Cytokine concentrations in the cell culture supernatant of HTLV-IIIb–infected immune cells and uninfected controls (pg/mL)

	Control	HTLV-IIIb
TNF-α	35 ± 11	5592 ± 1355
IL-1β	25 ± 10	28769 ± 8168
IFN-α	0.2 ± 0.2	503 ± 151
IFN-γ	5 ± 1	638 ± 338

was still observed. Because diarrhea was more common in late-stage HIV patients, mechanisms like alteration of colonic fluid absorption must, in addition, take place in these patients. The small intestinal villus surface area could also be reduced, which would lead to a reduction in absorption and increase in resistance, the latter of which may be superposed by the concomitant epithelial barrier disturbance, leaving this parameter unchanged in the further progress of the HIV infection. Preliminary data indeed seem to support a reduced villus surface area in our late-stage HIV patients.

As already noted in the introduction, distinct cytokine profiles could be responsible for alterations of the epithelial transport and barrier function. Reka *et al.* can show in a study that the expression of cytokine mRNA in rectal mucosa varies with HIV disease stage.[30] However, classification of patients was not done in the CD4 cell range we investigated. Indeed, these authors found TNF-α and IL-1β to be increased in late-stage HIV patients, which can influence epithelial transport and barrier function. We postulate a specific stage-dependent cytokine pattern that could be responsible for the variation of epithelial function with HIV disease stage. For induction of active Cl⁻ secretion, TNF-α, and IL-1β, and for alteration of the epithelial barrier function, TNF-α and IFN-γ could be important. In the second part of our paper, we show evidence that HIV infection of immune cells can cause release of cytokines, which is capable of impairing epithelial barrier function.

HIV INFECTION OF HUMAN IMMUNE CELLS *IN VITRO*

All immune cell cultures were successfully infected with HTLV-IIIb, as indicated by p24 (8–157 pg/mL). Measurements of cytokine concentrations in the cell culture supernatant showed an elevation of TNF-α, IL-1β, IFN-α, and IFN-γ, although with high interindividual variability (TABLE 2).

The HTLV-conditioned supernatants added to HT29/B6 cell monolayers decreased dose dependently the transepithelial resistance (R^t, FIG. 3). The effect appeared only if the supernatant was added basolaterally; serosal addition resulted in no significant change of R^t. Twelve hours after incubation of the HT29/B6 cells with the supernatant in a dilution of 1:1, R^t decreased to 42 ± 3% ($p < 0.001$, $n = 4$) of the initial resistance. After 24 hours R^t was decreased to 20% ($p < 0.001$, FIG. 4). Supernatants of uninfected immune cells showed no change.

LDH activity in the supernatant of the HT29/B6 cells was 78 ± 7 U/L in the control group and 91 ± 8 U/L (n.s.) in the supernatant of the HT29/B6 cells with addition

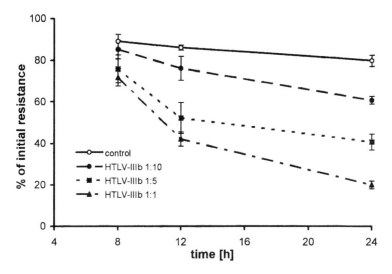

FIGURE 3. Effect of HTLV-IIIb–conditioned supernatants of human immune cells on transepithelial resistance (R^t) of HT-29/B6 cells in the dilutions of 1:10, 1:5, and 1:1. All values are means ± SEM, $n = 4$.

of the HTLV-conditioned supernatant. Total LDH activity of all cells on the filter was the same in both groups. This indicates no significant difference in cell turnover in both groups. Therefore the decrease in R^t in the HTLV group was not based on increased cell death (cytotoxicity) but seems to be due to increased tight junction permeability.

In order to test whether liberation of TNF-α is crucial for the R^t decrease of HIV-induced cytokines, TNF-α was inhibited by addition of the soluble TNF receptor sp55 (10 μg/mL) to the basolateral compartment in advance of the HTLV conditioned supernatant. Presence of sp55 partially inhibited the R^t effect of the HTLV-IIIb–conditioned supernatant (FIG. 4). Thus, TNF-α may be especially important for the epithelial barrier impairment in HIV infection. Indeed, in recent studies, we can demonstrate that TNF-α impairs the epithelial barrier by an alteration of the tight junction meshwork and also by increasing the occurrence and conductance of epithelial apoptoses.[12,31] As already mentioned above, an impaired epithelial barrier in HIV infection could lead to a leak flux–induced diarrhea by a passive leak flux of ions, substrates, and water towards the intestinal lumen and could increase bacterial invasion and as a result mucosal inflammation.

CONCLUSION

Diarrhea by HIV per se is caused by a leak-flux mechanism due to impaired epithelial barrier function. Inasmuch as there was no increased active ion secretion, attempts for antisecretory therapy were not useful in these patients. Along the course

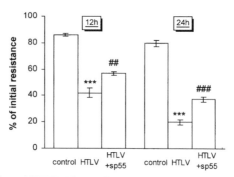

FIGURE 4. Effect of HTLV-IIIb–conditioned supernatants alone and with addition of the soluble TNF receptor sp55 (10 mg/mL) on transepithelial resistance (R^t) of HT-29/B6 after 12 and 24 hours. All values are means ± SEM, $n = 4$.

of the HIV infection, the epithelial transport and barrier function varies with HIV disease stage (expressed by CD4 cell status). HIV infection of human immune cells induces a cytokine release that can impair the epithelial barrier function of the human colonic epithelial cell line HT-29/B6. Taken together, these data lead us to postulate a specific stage-dependent cytokine pattern in the HIV-infected mucosa, which could be responsible for the variation of the epithelial function with HIV disease stage.

REFERENCES

1. MAYER, H.B. & C.A. WANKE. 1994. Diagnostic strategies in HIV-infected patients with diarrhea. AIDS **8:** 1639–1648.
2. RENÉ, E., C. MARCHE, B. REGNIER *et al.* 1989. Intestinal infections in patients with acquired immunodeficiency syndrome—a prospective study in 132 patients. Dig. Dis. Sci. **34:** 773–780.
3. BIGORNIA, E., D. SIMON, L.M. WEISS *et al.* 1992. Detection of HIV-1 protein and nucleic acid in enterochromaffin cells of HIV-1–seropositive patients. Am. J. Gastroenterol. **87:** 1624–1628.
4. EHRENPREIS, E.D., B.K. PATTERSON, J.A. BRAINER *et al.* 1992. Histopathologic findings of duodenal biopsy specimens in HIV-infected patients with and without diarrhea and malabsorption. Am. J. Clin. Pathol. **97:** 21–28.
5. ULLRICH, R., M. ZEITZ, W. HEISE *et al.* 1989. Small intestinal structure and function in patients infected with human immunodeficiency virus (HIV): evidence for HIV-induced enteropathy. Ann. Int. Med. **111:** 15–21.
6. KOTLER, D.P., S. REKA & F. CLAYTON. 1993. Intestinal mucosal inflammation associated with human immunodeficiency virus infection. Dig. Dis. Sci. **38:** 1119–1127.
7. McGOWAN, I., G. RADFORD-SMITH & D.P. JEWELL. 1994. Cytokine gene expression in HIV-infected intestinal mucosa. AIDS **8:** 1569–1575.
8. VYAKARNAM, A., P. MATEAR, G. KELLY *et al.* 1991. Altered production of tumour necrosis factors alpha and beta and interferon gamma by HIV-infected individuals. Clin. Exp. Immunol. **84:** 109–115.
9. SCHMITZ, H., M. FROMM, H. BODE *et al.* 1996. Tumor necrosis factor-α induces Cl⁻ and K⁺ secretion in human distal colon driven by prostaglandin E2. Am. J. Physiol. **271:** G669–G674.
10. MADARA, J.L. & J. STAFFORD. 1989. Interferon-γ directly affects barrier function of cultured intestinal epithelial monolayers. J. Clin. Invest. **83:** 724–727.

11. BODE, H., H. SCHMITZ, M. FROMM *et al.* 1998. IL1β and TNFα, but not IFNα, IFNγ, IL6 or IL8, are secretory mediators in human distal colon. Cytokine **10:** 457–465.
12. SCHMITZ, H., M. FROMM, C.J. BENTZEL *et al.* 1999. Tumor necrosis factor–alpha (TNFα) regulates the epithelial barrier in the human intestinal cell line HT-29/B6. J. Cell Sci. **112:** 137–146.
13. STOCKMANN, M., A.H. GITTER, D. SORGENFREI *et al.* 1999. Low edge damage container insert that adjusts intestinal forceps biopsies into Ussing chamber systems. Pflügers Arch. **438:** 107–112.
14. GITTER, A.H., J.D. SCHULZKE, D. SORGENFREI & M. FROMM. 1997. Ussing chamber for high-frequency transmural impedance analysis of epithelial tissues. J. Biochem. Biophys. Meth. **35:** 81–88.
15. STOCKMANN, M., M. FROMM, H. SCHMITZ *et al.* 1998. Duodenal biopsies of HIV-infected patients with diarrhea exhibit epithelial barrier defects but no active secretion. AIDS **12:** 43–51.
16. SCHULTZ, S.G. & R. ZALUSKY. 1964. Ion transport in isolated rabbit ileum. I. Short-circuit current and Na fluxes. J. Gen. Physiol. **47:** 567–584.
17. TAI, Y.H. & C.Y. TAI. 1981. The conventional short-circuiting technique under-short-circuits most epithelia. J. Membr. Biol. **59:** 173–177.
18. FROMM, M., J.D. SCHULZKE, U. HEGEL. 1985. Epithelial and subepithelial contributions to transmural electrical resistance of intact rat jejunum, *in vitro*. Pflügers Arch. **405:** 400–402.
19. GITTER, A.H., M. FROMM & J.D. SCHULZKE. 1998. Impedance analysis for determination of epithelial and subepithelial resistance in intestinal tissues. J. Biochem. Biophys. Meth. **37:** 35–46.
20. ROKOS, K. & G. PAULI. 1991. Interferon induction by cell-cell interaction of HIV-infected monocytes/macrophages with lymphocytes. Res. Virol. **142:** 221–225.
21. COLLMAN, R., N.F. HASSAN, R. WALKER *et al.* 1989. Infection of monocyte-derived macrophages with human immunodeficiency virus type 1 (HIV-1). Monocyte-tropic and lymphocyte-tropic strains of HIV-1 show distinctive patterns of replication in a panel of cell types. J. Exp. Med. **170:** 1149–1163.
22. KREUSEL, K.M., M. FROMM, J.D. SCHULZKE *et al.* 1991. Cl⁻ secretion in epithelial monolayers of mucus-forming human colon cells (HT-29/B6). Am. J. Physiol. **261:** C574–C582.
23. HEYMAN, M., N. DARMON, C. DUPONT *et al.* 1994. Mononuclear cells from infants allergic to cow's milk secrete tumor necrosis factor α, altering intestinal function. Gastroenterology **106:** 1514–1523.
24. KEATING, J., I. BJARNASON, S. SOMASUNGERAM *et al.* 1995. Intestinal absorptive capacity, intestinal permeability and jejunal histology in HIV and their relation to diarrhea. Gut **37:** 623–629.
25. BJARNASON, I., D.R. SHARPSTONE, N. FRANCIS *et al.* 1996. Intestinal inflammation, ileal structure and function in HIV. AIDS **10:** 1385–1391.
26. LIM, S.G., S. MENZIES, C.A. LEE *et al.* 1993. Intestinal permeability and function in patients infected with human immunodeficiency virus. Scand. J. Gastroenterol. **28:** 573–580.
27. SIMON, D.M., J.P. CELLO, J. VALENZUELA *et al.* 1995. Multicenter trial of octreotide in patients with refractory acquired immunodeficiency syndrome-associated diarrhea. Gastroenterology **108:** 1753–1760.
28. BJARNASON, I., A. MACPHERSON & D. HOLLANDER. 1995. Intestinal permeability: an overview. Gastroenterology **108:** 1566–1581.
29. STOCKMANN, M., M. FROMM, E.O. RIECKEN *et al.* 1998. Non-malabsorptive mechanisms of diarrhea in HIV infection. Pathobiology **66:** 165–169.
30. REKA, S., M.L. GARRO & D.P. KOTLER. 1994. Variation in the expression of human immunodeficiency virus RNA and cytokine mRNA in rectal mucosa during the progression of infection. Lymphokine Cytokine Res. **13:** 391–398.
31. GITTER, A.H., K. BENDFELDT, J.D. SCHULZKE & M. FROMM. 2000. Leaks in the epithelial barrier caused by spontaneous and TNFα-induced single-cell apoptosis. FASEB J. **14:** 1749–1753.

Cytokine-Induced Alteration of the Epithelial Barrier to Food Antigens in Disease

M. HEYMAN[a,b] AND J. F. DESJEUX[c]

[b]INSERM CJF97-10, Faculté Necker, 75730 Paris Cedex 15, France

[c]CNAM, 2 rue Conté, Paris, France

ABSTRACT: The alteration of the intestinal epithelial barrier is often a consequence of various intestinal diseases but may also be the starting point of these
diseases. Undigested food antigens are transported across the intestinal epithelium by a transcytotic mechanism, including a processing within the enterocytes, and leading to the passage of intact proteins, peptides, and amino acids
to the underlying mucosa. Inflammation and infection lead to the upregulation
of the transport and processing of food proteins; for example, IFNγ increases
the rate of transcytosis and alters, like TNFα, the tight junction permeability.
Infection of gastric digestive epithelia with *Helicobacter pylori* also increases
the antigenic load transmitted to the underlying immune system by inhibiting
the enterocytic lysosomal degradation of proteins. In allergic diseases, such as
cow's milk allergy, TNFα may be involved in the intestinal dysfunction and the
associated enteropathy.

INTRODUCTION

Food antigens are often involved in immune system–related intestinal diseases.
This is the case in cow's milk allergy, in celiac disease, and also perhaps in intestinal
bowel disease (IBD), where specific or unknown antigens induce mucosal alteration.
Despite the efficiency of the gut barrier, food proteins that are not totally degraded
by the gastric and pancreatic enzymes[1] can be taken up by the intestinal epithelium
and transported to the mucosal compartment to arouse the local immune system. As
a function of genetic and environmental factors, the activation of the immune system
can in certain cases lead to the release of mediators or cytokines acting directly or
indirectly on the intestinal layer, leading to barrier disruption and diarrhea. Here, we
present some examples of cytokine regulation of the epithelial barrier to food-type
antigens and the cross-talk between luminal antigens and mucosal immune cells
through the main line of defense of the intestinal tract, the intestinal epithelium.

THE INTESTINAL BARRIER TO FOOD ANTIGENS IS NOT ABSOLUTE

The transport across the intestinal epithelium is the first step before the interaction between food antigens and the mucosal immune system. Food antigens can be

[a]Address for correspondence: Dr. M. Heyman, CJF 97-10 INSERM, Faculté Necker, 156 rue de
Vaugirard, 75730 Paris Cedex 15, France.Voice: +33-01-40-61-56-33; fax: +33-01-40-61-56-38.
heyman@necker.fr

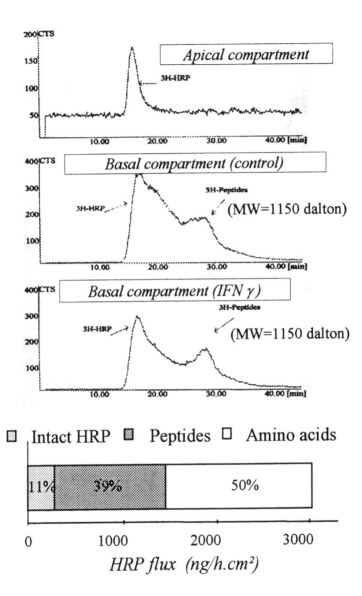

FIGURE 1. Transport and processing of food antigens across the enterocyte measured in Ussing chambers and HPLC chromatographic analysis of the compounds formed during transepithelial transport of ^{3}H-HRP (MW: 40,000) across the HT29-19A intestinal cell monolayers. In physiological conditions, the transport is transcellular via a transcytotic mechanism leading to the passage of small amounts of intact antigen (10%) and to the formation of peptides (40%) and amino acids (50%), released on the basal (internal) compartment. It is intriguing to note the appearance of small peptides with a molecular weight of 1150 daltons, compatible with an association with MHC molecules.

transported by transcytosis and processed by the enterocyte. This was shown initially in Ussing chambers using the reference protein marker horseradish peroxidase (HRP, MW 40,000).[2] The use of tritiated or ^{14}C-labeled proteins allowed us to discriminate between two main transport pathways, a direct pathway allowing the transport of small amounts of intact proteins (0.2 μg/h·cm^2) and a major degradative pathway constituting 90% of the total transport. The overall transport remains very small (2 μg/h·cm^2), confirming the efficiency of the epithelial barrier in physiological conditions.

To analyze the capacity of the enterocyte to process antigens into peptides potentially presentable to T cells, the nature of the metabolites formed during ^{3}H-HRP transport across the HT29-19A cell line was analyzed using steric exclusion HPLC chromatography. Of the total amount of protein transported across the cell monolayers, 10% was in an intact form, 40% was degraded into peptides, and 50% was completely degraded into amino acids. The detection of the radioactive material using HPLC showed that, although HRP was not degraded in the apical compartment of the Ussing chambers after 4 hours, remarkably, in the basal compartment after transport a wide range of tritiated degradation products with a continuous molecular mass gradient was eluted. Most interestingly, a group of peptides with a MW of 1150 daltons, comprising approximately eight amino acid residues, were also eluted. The peptide MW is compatible with the association to MHC class I (or class II?) molecules[3] (FIG. 1).

Thus, on the one hand the intestinal epithelium constitutes an efficient barrier to the penetration of massive amounts of food antigens, but on the other hand it may also regulate the antigenic information transmitted to the mucosal immune system.

INFLAMMATION AND/OR INFECTION LEAD TO THE UPREGULATION OF TRANSPORT AND PROCESSING OF FOOD ANTIGENS

In inflammatory conditions, the expression of molecules that are involved in antigen presentation, such as MHC class I or class II molecules, is upregulated on enterocytes. The antigen presenting capacity of the enterocyte could therefore be dependent on the inflammatory environment.[4] To present an antigen, antigen presenting cells process intact proteins into peptides, which are capable of association with MHC class I or class II molecules, before the complex is expressed at the cell surface.

It is therefore important to determine whether the stimulation of the epithelial barrier by inflammatory cytokines could play a role in the antigenic information transmitted to the immune system. The inflammatory cytokine IFN-γ has been shown to disrupt tight junctions and to increase the paracellular permeability of the intestinal epithelium.[5,6] We have tested the hypothesis that it may also modify the transcellular transport and processing of food antigens because of its capacity to upregulate expression of MHC class I and class II molecules in enterocytes, possibly acting on the antigen uptake and processing of food antigens during transepithelial transport (FIG. 2).

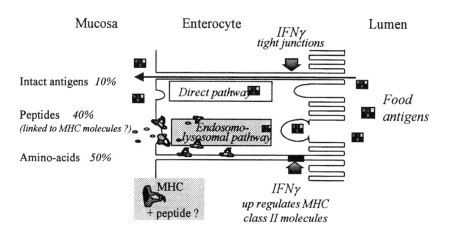

FIGURE 2. In inflammatory conditions, cytokines such as IFN-γ open the paracellular pathway by disrupting the tight junctions and may modify the intracellular processing by up-regulating MHC class I and class II molecules. The increase in the paracellular transport is attested by the rise in ionic conductance and mannitol/Na fluxes, and the rise in transcellular transport is attested by the rise in degraded HRP fluxes.

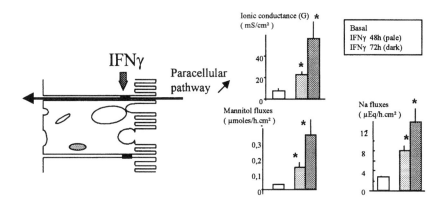

FIGURE 3. IFN-γ increases the paracellular transport pathway; this is shown by a progressive increase in ionic conductance and in mannitol and Na fluxes, as markers of the paracellular transport pathway.

In the HT29-19A intestinal cell line and as initially described in the T_{84} cell line,[5] IFN-γ was able to increase the paracellular permeability as indicated by a progressive increase with time in the ionic conductance and mannitol and Na fluxes, which are markers for the paracellular pathway (Fig. 3). At the same time, IFN-γ also modified the transcellular transport of HRP, as attested by a significant increase in total fluxes of tritiated HRP, which was related not only to the increase of the intact protein fluxes but also to the increase in the degraded HRP fluxes, indicating an in-

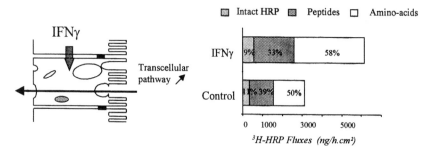

FIGURE 4. IFN-γ increases the transcellular transport pathway of antigens. ³H-HRP was placed in the apical compartment bathing HT29-19A cell monolayers mounted in Ussing chambers. The total fluxes as well as the degradation during transport were evaluated by HPLC chromatography: IFN-γ increased the total amount of HRP transported across the HT29-19A cell line, and the relative proportions of intact protein, peptides, and amino acids formed during transport did not differ strikingly, nor did the HPLC peptidic profile shown in FIGURE 1 (*lower panel*).

creased rate of transcytosis of the protein. By HPLC chromatography, it was shown that in control conditions and in the presence of IFN-γ, the relative proportion of intact protein, peptides, and amino acids were not changed; and the peptidic profile obtained in the presence of IFN-γ was not strikingly modified.[3] However, these results suggest that IFN-γ amplifies the antigenic load within the mucosa and probably the activation of the immune system, by allowing both the paracellular and transcellular antigen transport pathways to be upregulated (FIG. 4).

Another situation where the intestinal barrier capacity can be altered is the interaction of enterocytes with pathogenic bacteria. *Helicobacter pylori* is a bacteria attaching to the gastric epithelium and responsible for the development of gastric ulcers. We have recently shown, using the model of HT29-19A cells infected with the bacteria, that *H. pylori* modifies the antigen processing by the epithelial cell, by increasing the relative proportion of the intact protein transported without modifying the total transport. The bacteria did not modify the electrical resistance of the intestinal cell monolayers, indicating that the bacteria itself does not alter the tight junctional complex. In fact, *H. pylori* secretes large amounts of urease, leading to the local production of ammonia. Ammonium chloride is a lysosomotropic compound that increases the endosomal pH and therefore inhibits the acid proteolytic activity (cathepsins). This finally allows the intact proteins to be transmitted in larger amounts to the submucosa[7] (FIG. 5). This example shows that an increased antigenic load can be transmitted to the submucosa by a transcellular mechanism without alteration of the tight junctions.

FOOD ALLERGY: TNF-α AS A KEY FACTOR IN INTESTINAL DYSFUNCTION?

Food allergy and particularly cow's milk allergy (CMA) in children is another case in which cytokines are implicated in the alteration of the intestinal barrier ca-

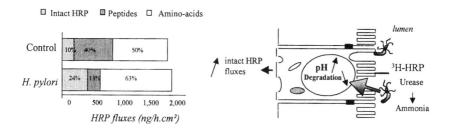

FIGURE 5. *Helicobacter pylori* modifies the intracellular processing of HRP across HT29-19A digestive cell line. When bacteria are placed for 24 h in the apical compartment of the transwell-grown HT29-19A cells, the total ^{3}H-HRP fluxes are not modified (neither is the electrical resistance of the monolayers), but the intracellular degradation by cathepsins (acid proteases) is inhibited, allowing higher amounts of intact protein to be transported. This is due to the rise of endosomal pH by ammonia, the end product of ureasic activity of the bacteria.

pacity. Cow's milk allergy occurs in about 3% of the pediatric population during the first years of life. In children in the active phase of CMA, two main dysfunctions of the intestinal epithelium are observed: first, an increase in antigen absorption and second, an activation of chloride secretion.[8] In Ussing chambers, intestinal biopsies of children with CMA show an increased permeability to the milk protein β-lacto-globulin; levels of this protein return to normal on a diet free from cow's milk.[9] Con-comitantly, the short circuit current is stimulated by the addition of β-lactoglobulin (but not by human α-lactalbumin). We know from animal models of intestinal ana-phylaxis that this phenomenon is compatible with the release of mast cell mediators. By contrast, the mechanism by which antigen absorption is increased is not clearly understood. Recently, the different cytokines that may be involved in the alteration of the intestinal function during CMA were investigated. Using isolated blood lym-phocytes (PBMC) cultured in the presence or absence of intact cow's milk proteins, the cytokines secreted in the culture medium were analyzed. Among IFN-γ, TNF-α, IL-4, and IL-6, only TNF-α presented a specific profile of secretion in allergic chil-dren compared to tolerant children.[10] TNF-α was further implicated at the intestinal level by the presence of TNF-α in the feces of allergic children after oral ingestion of milk.[11] In order to analyze the effect of the TNF-α–enriched supernatants on the intestinal barrier function, filter-grown HT29 cells were treated with such condi-tioned medium and the barrier function was assessed in a Ussing chamber. The weakening of the barrier function was demonstrated by the decrease in the electrical resistance and the increase in mannitol and Na fluxes. The involvement of TNF-α was further confirmed by preincubation of the conditioned medium with anti–TNF-α antibodies, which completely abrogated the deleterious effect observed. The im-portance of TNF-α in the alteration of the epithelial barrier was confirmed by plac-ing recombinant h-TNF-α on the basal side of transwell-grown HT29 cells. Surprisingly, increasing amounts of TNF-α did not alter the electrical resistance of the cell monolayers when TNF-α was added alone, but the addition of a very small amount of IFN-γ (5 U/ml), which by itself did not alter the electrical resistance, al-lowed TNF-α to decrease the epithelial resistance. This indicates that TNF-α and

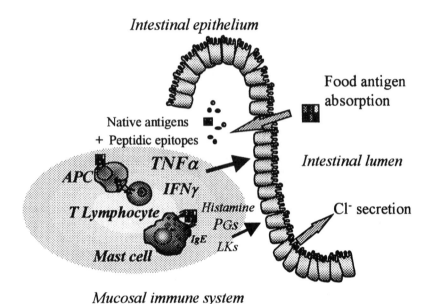

FIGURE 6. Intestinal diseases are often associated with an increased permeability to macromolecular food antigens. This allows more proteins and peptidic epitopes to reach the mucosa and to stimulate the underlying immune system. The release of cytokines and inflammatory mediators further enhances the degradation of the epithelial barrier, leading to a vicious circle where the enterocyte appears as a key factor in the cross-talk between luminal antigens and the mucosal immune system.

IFN-γ have a synergistic effect on the alteration of the epithelial barrier, IFN-γ being known to upregulate the TNF-α receptors. Finally, to analyze the implication of TNF-α in the alteration of tight junctions, a freeze-fracture electron microscopy study was done. In control monolayers there was an intricate network of strands and grooves that was significantly reduced in the presence of TNF-α, as shown by a significant decrease in the number of strands and in the junctional depth.[12]

In conclusion, intestinal diseases are often associated with an increased permeability to macromolecular food antigens (FIG. 6). This allows more proteins and peptidic epitopes to reach the mucosa and stimulate the underlying immune system. The release of cytokines and inflammatory mediators further enhances the degradation of the epithelial barrier leading to a vicious circle in which the enterocyte appears as a key factor in the cross-talk between luminal antigens and the mucosal immune system.

REFERENCES

1. MAHE, S., B. MESSING, F. THUILLIER & D. TOME. 1991. Digestion of bovine milk proteins in patients with a high jejunostomy. Am. J. Clin. Nutr. **54:** 534–538.
2. HEYMAN, M., R. DUCROC, J.F. DESJEUX & J.L. MORGAT. 1982. Horseradish peroxidase transport across adult rabbit jejunum *in vitro*. Am. J. Physiol. **242:** G558–G564.

3. TERPEND, K., M. BLATON, C. CANDALH, *et al.* 1998. Intestinal barrier function and cow's milk sensitization in guinea-pigs fed milk or fermented milk. J. Pediatr. Gastroenterol. Nutr. **28:** 191–198.
4. HERSHBERG, R.M., P.E. FRAMSON, D.H. CHO *et al.* 1997. Intestinal epithelial cells use two distinct pathways for HLA class II antigen processing. J. Clin. Invest. **100:** 204–215.
5. MADARA, J.L. & J. STAFFORD. 1989. Interferon-gamma directly affects barrier function of cultured intestinal epithelial monolayers. J. Clin. Invest. **83:** 724–727.
6. ADAMS, R.B., S.M. PLANCHON & J.K. ROCHE. 1993. IFN-gamma modulation of epithelial barrier function. Time course, reversibility, and site of cytokine binding. J. Immunol. **150:** 2356–2363.
7. MATYSIAK-BUDNIK, T., K. TERPEND, S. ALAIN *et al.* 1998. *Helicobacter pylori* alters exogenous antigen absorption and processing in a digestive tract epithelial cell line model. Infect. Immun. **66:** 5785–5791.
8. HEYMAN, M., E. GRASSET, R. DUCROC & J.F. DESJEUX. 1988. Antigen absorption by the jejunal epithelium of children with cow's milk allergy. Pediatr. Res. **24:** 197–202.
9. SAIDI, D., M. HEYMAN, O. KHEROUA *et al.* 1995. Jejunal response to beta-lactoglobulin in infants with cow's milk allergy. C. R. Acad. Sci. III. **318:** 683–689.
10. HEYMAN, M., N. DARMON, C. DUPONT *et al.* 1994. Mononuclear cells from infants allergic to cow's milk secrete tumor necrosis factor alpha, altering intestinal function. Gastroenterology **106:** 1514–1523.
11. MAJAMAA, H., A. MIETTINEN, S. LAINE & E. ISOLAURI. 1996. Intestinal inflammation in children with atopic eczema: faecal eosinophil cationic protein and tumour necrosis factor–alpha as non-invasive indicators of food allergy. Clin. Exp. Allergy **26:** 181–187.
12. RODRIGUEZ, P., M. HEYMAN, C. CANDALH *et al.* 1995. Tumour necrosis factor–alpha induces morphological and functional alterations of intestinal HT29 cl.19A cell monolayers. Cytokine **7:** 441–448.

Epithelial Barrier and Transport Function of the Colon in Ulcerative Colitis

H. SCHMITZ,[a] C. BARMEYER,[a] A. H. GITTER,[b] F. WULLSTEIN,[a] C. J. BENTZEL,[c]
M. FROMM,[b] E. O. RIECKEN,[a] AND J. D. SCHULZKE[a,d]

[a]Medizinische Klinik I, Gastroenterologie und Infektiologie, Universitätsklinikum
Benjamin Franklin, Freie Universität Berlin, 12200 Berlin, Germany

[b]Institut für Klinische Physiologie, Universitätsklinikum Benjamin Franklin,
Freie Universität Berlin, 12200 Berlin, Germany

[c]Department of Medicine, East Carolina University, Greenville, North Carolina

INTRODUCTION

The function of the colonic epithelium includes the regulation of water and ion transport as well as the formation of an effective barrier against the invasion of noxious agents or antigens. Both functions seem to be impaired in ulcerative colitis (UC), leading to intestinal inflammation and diarrhea in the course of the disease. The pathomechanisms of diarrhea in UC are not clarified in detail, though various investigations have been made concerning the contributory role of (1) the activation of ion secretion,[1,2] (2) the reduction of ion absorption,[3–5] and (3) a defect in epithelial barrier function.[6] However, only little information exists on the extent and precise mechanism of epithelial barrier dysfunction in UC.

In our study, we determined the extent of epithelial barrier dysfunction and correlated the findings with alterations in tight junction structure in the same specimen. Furthermore, the first measurements with the conductance scanning technique are presented, which allowed for the determination of the regional distribution of conductivity of an epithelium. In particular, the influence of an erosion for local conductivity (= leakiness) of the intestinal epithelium in UC is presented.

METHODS

Tissue Preparation, Solution, and Drugs

Measurements were performed on human distal colon from patients with (a) ulcerative colitis (UC) who underwent surgery for colectomy and (b) resection of sigmoidal or rectal cancer, adenoma, or diverticulosis. All tissues were "partially stripped" (i.e., the muscularis propria was removed) and mounted in Ussing-type chambers (area = 0.28 cm^2) as described earlier.[7,8] In the UC group, only tissues with macroscopically mild to moderate disease activity (without ulcers) were includ-

[d]Address for correspondence: Jörg-Dieter Schulzke, Medizinische Klinik I, Gastroenterologie und Infektiologie, Universitätsklinikum Benjamin Franklin, Hindenburgdamm 30, 12200 Berlin, Germany. Voice: +49-30-8445-2347; fax: +49-30-8445-4239.
schulzke@medizin.fu-berlin.de

ed, and subsequent histological grading demonstrated in 11 of 18 tissues a mild to moderate histological appearance with an intact epithelium.[9] These tissues were used for further functional characterization and EM analysis. The remaining 7 tissues with severe histological appearance, which showed (only) a few erosions, were only studied by impedance analysis and are presented separately in TABLE 1.

The bathing solution contained (in mM) Na^+ 140, Cl^- 123.8, K^+ 5.4, Ca^{2+} 1.2, Mg^{2+} 1.2, HPO_4^{2-} 2.4, $H_2PO_4^-$ 0.6, HCO_3^- 21, D(+)-glucose 10, β-OH-butyrate 0.5, glutamine 2.5, and D(+)-mannose 10 [23]. The solution was gassed with 95% O_2 and 5% CO_2; the temperature was maintained at 37°C and the pH was 7.4 in all experiments. Antibiotics (azlocillin, 50 mg/L; and tobramycin, 4 mg/L) served to prevent bacterial growth and had no effect on I_{sc} in the concentration used. All substances were obtained from Sigma (St. Louis, MO). If not stated otherwise, drugs were added to the serosal side.

Alternating Current Impedance Analysis

Whereas in conventional Ussing experiments only total wall resistance (R^t) of a tissue can be determined, the alternating current impedance analysis technique can be used for evaluation of the epithelial (R^e) and the subepithelial (R^{sub}) portion of R^t (see also Stockmann *et al.*, this volume).[10] Briefly, voltage responses after transepithelial application of 35 μA/cm² eff. sine-wave alternating current of 48 discrete frequencies in a range from 1 to 65 kHz were detected by phase-sensitive amplifiers (Model 1250 Frequency Response Analyzer and Model 1286 Electrochemical Interface; Solartron Schlumberger, Farnborough Hampshire, UK). Complex impedance values were calculated and corrected for the resistance of the bathing solution and the frequency behavior of the measuring setup for each frequency. Due to the capacitor characteristics of the epithelial layer after application of alternating current, the epithelial layer was electrically short-circuited at high frequencies (representing R^{sub}), whereas at low frequencies R^t was obtained. The epithelial resistance, R^e, could be calculated by $R^t - R^{sub}$.

Active Electrogenic Chloride Secretion

Active electrogenic chloride secretion was determined by conventional Ussing experiments using a computer-controlled voltage-clamp device (CVC 6, Fiebig, Berlin, Germany). Short-circuit current (I_{sc}), open-circuit transepithelial voltage, and transepithelial resistance were recorded to a hard disk. I_{sc} values were corrected for bath resistance. All experiments were performed in the presence of amiloride (10^{-4} M) in order to block electrogenic Na^+ absorption, which could have been elicited during anesthesia by elevated plasma aldosterone levels. The transport capacity of active electrogenic Cl^- secretion was characterized as described earlier.[11] The bathing solution was initially Cl^- and HCO_3^- free (in mM: Na^+ 125.4, SO_4^{2-} 65.1, K^+ 5.4, HPO_4^{2-} 2.4, $H_2PO_4^-$ 0.6, Ca^{2+} 1.2, Mg^{2+} 1.2, TRIS 4, HEPES 4, and mannitol 120; gassed with 100% O_2, pH 7.4, at 37°C). Then, the tissues were maximally stimulated by simultaneous addition of carbachol (10^{-4} M), PGE$_2$ (10^{-6} M), and theophylline (10^{-2} M) on both sides. Thereafter, the Cl^- concentration was stepwise increased on both sides to final Cl^- concentrations of 10, 20, and 40 mM by partly replacing the sulfate/mannitol medium with a medium containing 120 mM Cl^-, and the increase in short-circuit current was recorded. ΔI_{sc}^{max} (V_{max}) and K_m were determined from Eadie-Hofstee plots for each tissue specimen separately. Thereafter, ΔI_{sc}^{max} and K_m were calculated for each experimental group.

Freeze-Fracture Electron Microscopy

Freeze-fracture analysis was carried out as described earlier.[11] Briefly, 9 control and 9 UC tissues were fixed with phosphate-buffered 2% glutaraldehyde, frozen in Freon 22 and liquid nitrogen (−100°C), and fractured under a vacuum condition. These freeze fractures were shadowed with platinum and carbon in a Denton CV-502 apparatus and examined in a Phillips 200 electron microscope. Morphometric analysis was performed using coded prints of freeze-fracture electron micrographs (×60,000 magnification) on all tight junction regions in which both an apical and a contra-apical strand of the meshwork could be clearly demarcated. Vertical grid lines were drawn at 5-mm intervals (equivalent to 83 nm of the tight junction meshwork) perpendicular to the most apical strand. Intersections of strand and grid line served to determine the number of horizontally oriented strands in the main meshwork of the tight junction (= strand number). The distance between the most apical and the most contra-apical strand along each vertical grid line within the main compact meshwork was measured as tight junctional depth. For quantification of strand discontinuities, strand interruptions of more than 25 nm were counted within the main compact meshwork of the tight junction and expressed per 1000-nm tight junctional strand length. This restriction to 25-nm length was introduced to exclude small fixation artifacts from the analysis, although smaller strand discontinuities might also be of functional importance.

Conductance Scanning

The conductance scanning technique allows for the determination of the regional distribution of conductivities of surface epithelia[12] (see Gitter *et al.*, this volume). Briefly, after the stripping procedure, tissues were mounted horizontally (mucosal surface up, exposed area = 0.283 cm^2) in a special-type Ussing chamber that allowed observation of the specimen by an upright light microscope. Both halves of the chamber were continuously perfused by oxygenated bathing solution at 37°C (95% and 5% CO_2). In the upper half of the chamber, a glass microelectrode, which could be moved vertically by a piezoelectric micromanipulator, served to detect potential gradients in different heights above the mucosal surface. By means of a second micromanipulator for horizontal movements, tissue areas could be scanned for their distribution of local potential gradients. The conductivity was then calculated by the equation, $G = (\Delta V/\Delta x)/(p \cdot U)$, where p is the specific resistivity of the Ringer's solution and U is the transepithelial voltage measured by the voltage electrodes of the Ussing chamber. In order to detect gross lesions of conductivity, like erosions or ulcers, we used the low resolution method in our study, scanning the horizontally mounted tissues in a height of 100 μm and at distances of 250 to 500 μm. During the electrophysiological measurements, regions of high conductivity were ink-marked. Thereafter, all tissues were investigated histologically, especially in order to correlate regions of high conductivity to histological findings like erosions or ulcers.

Statistical Analysis

Values are given as the mean ± SEM (except conductance scanning data). The unpaired two-tailed t test was used to determine the significance of differences. $p < 0.05$ was considered significant.

RESULTS

Alternating Current Impedance Analysis and Tracer Flux Studies

FIGURE 1 shows typical impedance locus plots of control sigmoid colon (A) and colon from a patient with UC (B). Numerical results and the statistical evaluation of the impedance analysis are presented in TABLE 1. As indicated in TABLE 1, R^t was already decreased in UC to 51% compared to control. However, AC impedance analysis uncovered an overproportional decrease of R^e from $95 \pm 5 \ \Omega \cdot cm^2$ in control to $20 \pm 3 \ \Omega \cdot cm^2$ in UC. In parallel, R^{sub} was enhanced from $14 \pm 1 \ \Omega \cdot cm^2$ in control to $36 \pm 3 \ \Omega \cdot cm^2$ in mild to moderately inflamed tissues and even to $54 \pm 4 \ \Omega \cdot cm^2$ in severely inflamed UC. Cross-sections demonstrated an inflammatory reactive hypertrophy and inflammatory cell infiltration; as a consequence of this, the thickness of all subepithelial tissue layers was increased from $494 \pm 89 \ \mu m$ in control to 1276

A. control

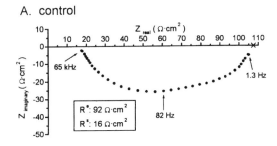

B. ulcerative colitis

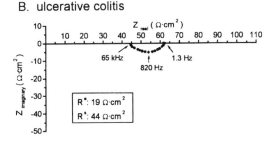

FIGURE 1. Original impedance locus plots of human colon: **(A)** control; **(B)** ulcerative colitis. Z_{real} gives the ohmic component and $Z_{imaginary}$ gives the reactive component of the complex impedance. Intersections between the semicircle and x-axis at low and high frequencies represent the total resistance (R^t) and subepithelial resistance (R^{sub}), respectively. R^t minus R^{sub} gives the resistance of the epithelium (R^e). (Reproduced with permission from Schmitz.[8])

TABLE 1. Alternating current impedance analysis of the human colon

	R^t	R^e	R^{sub}	n
Control	109 ± 5	95 ± 5	14 ± 1	10
UC (mild to moderate)	$56 \pm 4^{***}$	$20 \pm 3^{***}$	$36 \pm 3^{***}$	11
UC (severe)	$77 \pm 8^{\#}$	22 ± 4	$54 \pm 4^{\#\#}$	7

NOTE: Transmural electrical resistance (R^t, $\Omega \cdot cm^2$), epithelial resistance (R^e, $\Omega \cdot cm^2$), and sub-epithelial resistance (R^{sub}, $\Omega \cdot cm^2$) were measured by alternating current impedance analysis in controls and patients with UC. All values are means $\pm$ SEM; n = number of patients. $^{***}p < 0.001$ vs. control; $^{\#}p < 0.05$ and $^{\#\#}p < 0.01$ vs. mild to moderate UC.[8]

TABLE 2. Tight junction strands in UC

	TJ strand count		Strand discontinuities		
	Surface	Crypt	Surface	Crypt	n
Control	6.94 ± 0.25	7.26 ± 0.31	0.08 ± 0.08	0.10 ± 0.08	9
UC	$4.76 \pm 0.47^{***}$	$5.46 \pm 0.37^{**}$	0.16 ± 0.11	0.16 ± 0.09	9

NOTE: Number of horizontally oriented strands in the main compact meshwork of the tight junction (TJ strand count) and frequency of strand discontinuities in epithelial cells from sigmoid colon of controls and patients with UC. Strand discontinuities are defined as strand interruptions of >25-nm length. The number of strand discontinuities is given per 1000-nm strand length. All values are means $\pm$ SEM; n = number of patients. $^{**}p < 0.01$, $^{***}p < 0.001$ (compared to control).[8]

TABLE 3. Tight junction depth in UC

	TJ mesh depth		Total TJ depth		
	Surface	Crypt	Surface	Crypt	n
Control	307 ± 22 nm	330 ± 37 nm	402 ± 42 nm	409 ± 46 nm	9
UC	244 ± 19 nm*	308 ± 31 nm	405 ± 78 nm	425 ± 61 nm	9

NOTE: Depth of the main tight junctional meshwork (TJ mesh depth) and depth of the total tight junction including aberrant strands (total TJ depth) in the sigmoid colon of controls and patients with UC. All values are means $\pm$ SEM; n = number of patients. $^{*}p < 0.05$ (compared to control).[8]

± 228 μm in UC ($n = 7$, $p < 0.05$; FIG. 2). At the same time, surface geometry showed a crypt rarefaction in UC (with a reduction of the surface area), while the enterocyte count per 100-μm mucosal length was slightly decreased in UC compared to control (18.4 ± 0.7 in control vs. 15.9 ± 0.6 in UC, $p < 0.05$; $n = 7$).

Freeze-Fracture Electron Microscopy

FIGURE 3 shows typical epithelial cell tight junction areas from the crypt region in control sigmoid (A) and in UC (B). The numerical data and the statistical evaluation are listed in TABLES 2 and 3. Surface and crypt compartment did not differ in the

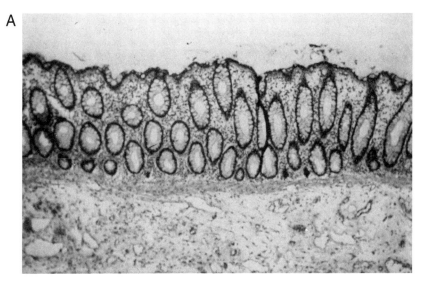

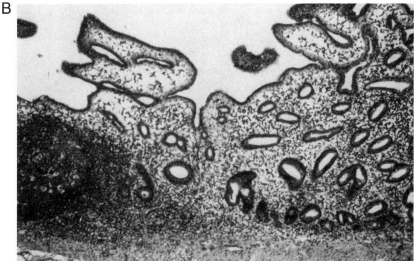

FIGURE 2. Conventional histology from colectomy specimens: **(A)** control sigmoid colon; **(B)** sigmoid colon from patients with ulcerative colitis (×53 magnification). (Reproduced with permission from Schmitz.[8])

number of horizontally oriented strands (TJ strand count, TABLE 2) or in the depth of the main tight junction meshwork (TJ depth, TABLE 3). By contrast, UC tissues demonstrated a decreased strand number at the surface as well as in crypt regions compared to control (TABLE 2). Further analysis of the strand distribution in the tight junctions revealed a significant number of tight junction regions with only 1 or 2 strands (FIG. 4), whereas tight junctions with high strand counts disappeared. How-

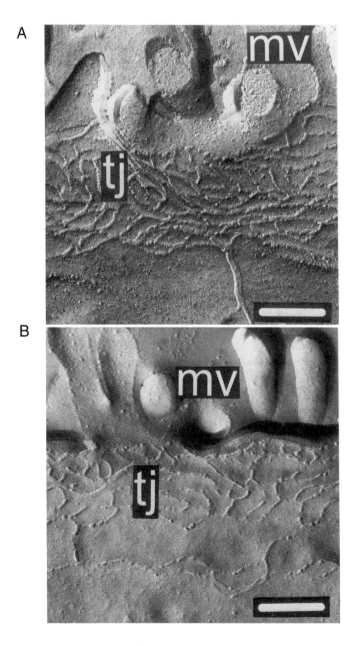

FIGURE 3. Freeze-fracture electron micrographs from the tight junctional region of colonic enterocytes: **(A)** control; **(B)** ulcerative colitis. The bar represents 200-nm length; mv = microvilli; tj = tight junction strands. (Reproduced with permission from Schmitz.[8])

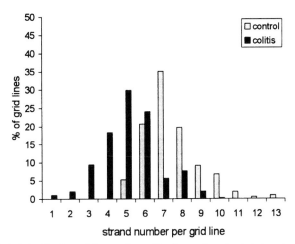

FIGURE 4. Distribution of the number of horizontally oriented strands along the tight junction of colonic enterocytes under control conditions and in ulcerative colitis. Values represent the percentage of grid lines with the respective strand count (related to the total number of grid lines analyzed). (Reproduced with permission from Schmitz.[8])

ever, in UC as in control, the strand distribution showed a Gaussian distribution, which suggests an orderly interference with the assembly/disassembly process of tight junction formation. In parallel, the depth of the main tight junction meshwork was reduced in UC, although this reduction reached statistical significance only at the surface (TABLE 3).

In contrast to the reduction in strand number, strand discontinuities (TABLE 2) were not frequent either in UC or in control, and statistical analysis revealed no difference between both groups. Furthermore, aberrant strands appeared below the main tight junction meshwork of surface and crypt tight junctions (FIG. 3), which explains why the total tight junction depth, despite the decreased depth of the main tight junction meshwork, showed a tendency towards higher values in UC when compared to control (TABLE 3). Aberrant strands below the main tight junction meshwork have been attributed to tight junction assembly and may be correlated to the increase in cell proliferation in UC.

Active Electrogenic Chloride Secretion

Before the results of these experiments can be interpreted, theoretical considerations need to be addressed. The subepithelial layer acts (like the bathing solution) as a resistor in series to the epithelium, which leads to underestimation of real transport rates.[13] Inasmuch as in our study the contribution of R^{sub} to R^t is higher in UC than in controls, the underestimation of transport rates is higher in UC than in controls; as a consequence of this, transport values need correction before comparison. This correction factor can be calculated by the ratio R^t/R^e and was 1.16 ± 0.02 ($n = 10$) for control and 3.09 ± 0.27 ($n = 11$, $p < 0.001$) for UC,[14] demonstrating only a minor underestimation of real transport rates in control, but a considerably greater one in UC.

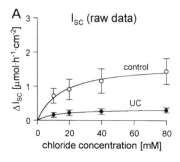

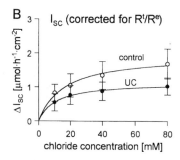

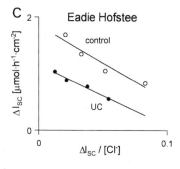

FIGURE 5. Electrogenic Cl⁻ secretory system in ulcerative colitis. Increase in short-circuit current (ΔI_{sc}) as a function of the chloride concentration in the bathing medium after stimulation with 0.1 mM carbachol, 10 μM PGE$_1$, and 10 mM theophylline. Data are given as (**A**) rough data and (**B**) corrected for subepithelial resistance contributions. (**C**) The Eadie-Hofstee plot of the data in panel **B**. After correction for the subepithelial resistance, the calculated maximum of ΔI_{sc} (ΔI_{sc}^{max}) was 1.80 ± 0.48 μmol·h⁻^{1}·cm⁻^{2} ($n = 6$) in control and 1.17 ± 0.28 μmol·h⁻^{1}·cm⁻^{2} ($n = 5$; n.s.) in ulcerative colitis. K_m was 13 ± 4 mM in control and 16 ± 6 mM (n.s.) in ulcerative colitis. (Reproduced with permission from Schmitz.[8])

In FIGURE 5A, raw data of the chloride kinetic experiments are shown, suggesting a decreased active electrogenic chloride secretion in UC compared to control. However, after correction for their different R^t/R^e ratios in UC, the real active chloride transport capacity was not different from control (FIG. 5B). The tendency towards lower values might reflect the altered mucosal architecture, leading to a decreased surface area in UC.

Conductance Scanning

In FIGURE 6, cross-sections and graphic illustrations of conductance scanning measurements are demonstrated for control (A, B) and for UC (C, D). In control, conductance scanning showed an even distribution of conductivity, suggesting a normal

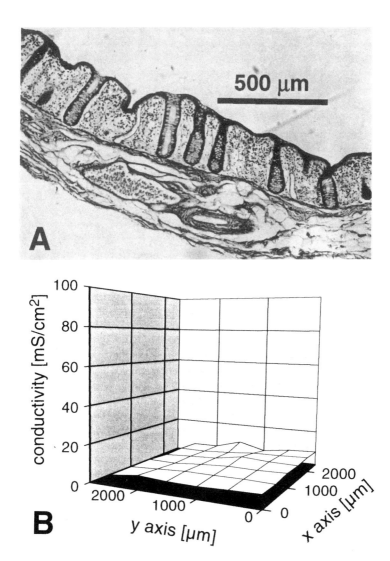

FIGURE 6. Conductance scanning and respective histology: control (**A**, **B**) and ulcerative colitis (**C**, **D**). (**A**) Homogeneous distribution of conductivity demonstrated in a 3D Cartesian mesh grid. (**B**) Respective histology of the tissue measured in panel **A** with normal histology findings.

intact epithelium, which was confirmed by subsequent histological examination. In UC, however, two different types of impaired epithelial conductivity have been found. First, in the area marked with an "I", an uneven increase in basal conductivity was found, without histological changes of the epithelial layer, possibly corresponding to

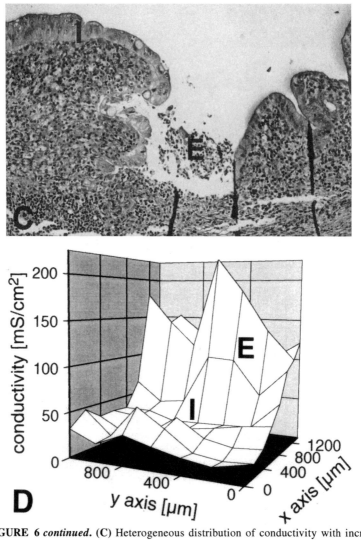

FIGURE 6 *continued*. **(C)** Heterogeneous distribution of conductivity with increased basal conductivity in an area marked with an "I" and a large conductivity spot in a second area ("E"). **(D)** Respective histology of the tissue measured in panel **C**. The increased conductivity in area I was histologically identified as intact surface epithelium, whereas the conductivity spot in area E was histologically identified as an erosion.

regions with decreased tight junction strand count. Second, a hot spot of conductivity was found in another area (marked with an "E"), which was histologically identified as an erosion, suggesting that erosions are structures of high epithelial leakiness.

DISCUSSION

The aim of the study was to investigate in more detail the pathomechanisms leading to intestinal barrier dysfunction with subsequent diarrhea in UC. In this respect, the role of tight junction alteration (which could lead to a leak flux diarrhea) and the influence of mucosal damage in the sense of erosions were especially investigated. For this purpose, various electrophysiological and morphological methods were applied, each of them giving more insight to different aspects of transport or barrier function in UC.

Alternating Current Impedance Analysis

Previous investigations in the Ussing chamber demonstrated that transepithelial resistance was reduced in UC, which corresponded to our results too.[15] However, due to the technique applied (conventional Ussing experiments), the real dimensions of barrier dysfunction did not become evident in previous studies. The intestinal barrier function relies only on the tightness of the epithelial layer because blood capillaries are located closely to the enterocytes. The subepithelium, however, is only of minor importance in this respect and even leads astray in the interpretation of conventional Ussing data. Only with means of alternating current impedance analysis did the real magnitude of the epithelial barrier defect in UC become evident, whereas in conventional Ussing experiments the reduction in R^e was masked by the increased subepithelial resistance. Taking geometrical factors into account (reduced surface area and reduced TJ area), which normally would lead to an increased electrical resistance, the barrier function was even more affected than expressed by the resistance values.

Freeze-Fracture Electron Microscopy

In order to clarify whether changes in tight junction architecture could explain the results obtained by impedance analysis, freeze-fracture EM was performed in the same specimen in which impedance measurements were performed. In this manner, correlation of tight junction morphology and epithelial resistance became possible. For interpretation of freeze-fracture analysis, it is important to consider that strand number and electrical resistance correlate not directly, but by a power function in which small changes in strand number have large effects on conductivities.[16,17]

When we determined the TJ strand count, we found a reduction in the number of horizontally oriented strands in UC compared to control in both the crypt and the surface epithelium. Additionally, TJ regions with very low strand numbers appeared in UC, which might be of special importance for the local barrier function of the epithelium. The decrease in strand count was accompanied by a reduction in the depth of the main tight junction meshwork. This reduction was only significant in the surface and not in the crypts, which might be due to the proliferation status of the epithelium with activation of TJ assembly in crypts.

By contrast, strand discontinuities were not more frequent in UC than in control. Nevertheless, they might be of importance in UC because, in UC, tight junction regions with strand counts of 1 or 2 appear. In these conditions, strand discontinuities would have dramatic effects on local conductivity and may enable even macromolecules or antigens to penetrate the intestinal barrier, whereas in control at least 5 strands form the tight junction meshwork.

Active Electrogenic Chloride Secretion

Besides the proof for viability of the tissues, the main result of this experimental set was that active electrogenic chloride secretion is not altered in UC. It is important to note that the intactness of this transport system was only demonstrable with the information from the impedance analysis experiments. In conventional Ussing experiments, real transport rates are generally underestimated by 10–15% due to the additional serial resistance between the voltage-sensing electrodes formed by the subepithelium. This is only of minor importance if groups are compared where the subepithelial contribution to R^t is equal. However, the results of the impedance analysis indicated that real transport rates in UC are much more underestimated than in controls due to the increased R^{sub}. Therefore, before comparison of transport rates was allowed, a correction of the transport values for their different R^t/R^e ratios was necessary. After this correction, the result of our chloride kinetic study was completely different. Whereas raw data suggested that active electrogenic chloride secretion was reduced in UC, correction for the different R^t/R^e ratios demonstrated only a shift towards lower electrogenic chloride transport capacity in UC, which might be explained by the reduced surface area per cm^2 in UC as mentioned above.

Conductance Scanning

With means of the conductance scanning technique, we were able to investigate UC tissues for their local distribution of conductivities. Whereas a homogeneous distribution of conductivity was detectable in control, we observed different types of increased conductivity in UC. In an area with intact surface epithelium, we observed an increased conductivity with inhomogeneous distribution. This area might reflect tight junction regions with decreased strand counts as demonstrated in the freeze-fracture analysis. However, regarding the results of Gitter *et al.*, we cannot exclude that some of the increased conductivity in this area arises from apoptotic regions with consecutive local conductivity increase. Conductance scanning experiments with the high resolution method are on the way to clarify this question. In a second area, a large peak of conductivity was detectable that was caused by an erosion as identified histologically. Until today, it is still unclear whether an erosion or ulcer is significant for local barrier function. The first results of our experiments suggest that erosions are areas of increased conductivity and that the local defect possibly allows invasion of antigens or noxious agents that could perpetuate inflammatory processes in the intestinal wall. However, as in the case of apoptosis, systematic experiments are warranted to answer this question.

Diarrheal Mechanisms in Ulcerative Colitis

With the exception of motility-driven mechanisms, all other forms of diarrhea are of osmotic nature (malabsorptive, secretory, or leak flux–induced). In the case of leak flux–induced diarrhea, solutes and water will enter the lumen depending on the electrochemical driving forces and permeability of the paracellular barrier. This mechanism was found to be exaggerated by *Clostridium difficile*.[18] This mechanism was also found to be involved in a further study by Fasano *et al.*, in which an attenuated *Vibrio cholerae* strain (which was depleted of the chloride secretion–inducing cholera toxin gene) still caused diarrhea in human volunteers.[19] This diarrhea was attributed to a zonula occludens toxin, which transiently decreased R^e of the ileum by altering tight junction structure. In our study, alternating current impedance analysis demonstrated that the epithelial resistance in UC is strongly impaired, which was paralleled by a decrease in tight junction strand count. From this, we conclude that a leak flux mechanism contributes to the diarrhea observed in UC.

In addition, preliminary results of conductance scanning experiments suggest that erosions and/or ulcers are relevant for barrier function in two respects. First, erosions or ulcers could be areas of highly impaired barrier function leading to an uncontrolled loss of solutes and water. Second, erosions or ulcers could represent areas of uncontrolled invasion of noxious agents or antigens, which could contribute to intestinal inflammation.

ACKNOWLEDGMENTS

We thank Anja Fromm, Sieglinde Lüderitz, and Ursula Lempart for assistance, and electronic engineer Detlef Sorgenfrei for support.

REFERENCES

1. RAMPTON, D.S. & G.E. SLADEN. 1984. Relationship between rectal mucosal prostaglandin production and water and electrolyte transport in ulcerative colitis. Digestion **30:** 13–22.
2. ANDUS, T. *et al.* 1991. Activation of monocytes during inflammatory bowel disease. Pathobiology **59:** 166–170.
3. HARRIS, J. & R. SHIELDS. 1970. Absorption and secretion of water and electrolytes by the intact human colon in diffuse untreated proctocolitis. Gut **11:** 27–33.
4. RASK-MADSEN, J. 1973. The relationship between sodium fluxes and electrical potentials across the normal and inflamed human rectal wall *in vivo*. Acta Med. Scand. **194:** 311–317.
5. RASK-MADSEN, J. *et al.* 1973. Electrolyte transport capacity and electrical potentials of the normal and the inflamed human rectum *in vivo*. Scand. J. Gastroenterol. **8:** 169–175.
6. SANDLE, G.I. *et al.* 1990. Cellular basis for defective electrolyte transport in inflamed human colon. Gastroenterology **99:** 97–105.
7. SCHMITZ, H. *et al.* 1996. Tumor necrosis factor-α induces Cl^- and K^+ secretion in human distal colon driven by prostaglandin E2. Am. J. Physiol. **271:** G669–G674.
8. SCHMITZ, H. *et al.* 1999. Altered tight junction structure contributes to the impaired epithelial barrier function in ulcerative colitis. Gastroenterology **116:** 301–309.
9. TRUELOVE, S.C. & W.C.D. RICHARDS. 1956. Biopsy studies in ulcerative colitis. Br. Med. J. **3:** 1315–1318.
10. GITTER, A.H. *et al.* 1998. Impedance analysis for the determination of epithelial and subepithelial resistance in intestinal tissues. J. Biochem. Biophys. Methods **37:** 35–46.
11. SCHULZKE, J.D. *et al.* 1992. Epithelial ion transport in the experimental short bowel syndrome of the rat. Gastroenterology **102:** 497–504.
12. GITTER, A.H. *et al.* 1997. Measurement of paracellular epithelial conductivity by conductance scanning. Pflüg. Arch. **434:** 830–840.
13. TAI, Y.H. & C.Y. TAI. 1981. The conventional short-circuiting technique under-short-circuits most epithelia. J. Membr. Biol. **59:** 173–177.

14. SCHULZKE, J.D. *et al.* 1987. Impaired intestinal sodium and chloride transport in the blind loop syndrome of the rat. Gastroenterology **92:** 693–698.
15. SANDLE, G.I. *et al.* 1990. Cellular basis for defective electrolyte transport in inflamed human colon. Gastroenterology **99:** 97–105.
16. CLAUDE, P. & D.A. GOODENOUGH. 1973. Fracture faces of zonulae occludentes from tight and leaky epithelia. J. Cell Biol. **58:** 390–400.
17. CLAUDE, P. 1978. Morphological factors influencing transepithelial permeability: a model for the resistance of the zonula occludens. J. Membr. Biol. **39:** 219–232.
18. MOORE, R. *et al.* 1990. *Clostridium difficile* toxin A increases intestinal permeability and induces Cl secretion. Am. J. Physiol. **259:** G165–G172.
19. FASANO, A. *et al.* 1991. *Vibrio cholerae* produces a second enterotoxin which affects intestinal tight junctions. Proc. Natl. Acad. Sci. USA **88:** 5242–5246.

Diarrhea in Ulcerative Colitis

The Role of Altered Colonic Sodium Transport

EMMA GREIG[a] AND GEOFFREY I. SANDLE[a,b,c]

[a]*Department of Medicine, University of Manchester, Hope Hospital, Salford M6 8HD, United Kingdom*

[b]*Molecular Medicine Unit, St. James's University Hospital, Leeds LS9 7TF, United Kingdom*

ABSTRACT: In normal human colon, water and sodium (Na^+) absorption are directly related. Defective Na^+ absorption may therefore be an important factor in the pathogenesis of diarrhea in ulcerative colitis (UC). Electrophysiological studies have revealed profound decreases in channel-mediated apical Na^+ entry and Na^+-K^+-ATPase–mediated basolateral Na^+ extrusion in surface epithelial cells in inflamed human distal colon. Recent molecular biological studies indicate that mucosal inflammation in UC leads to significant decreases in Na^+ channel β- and γ-subunit expression in the apical membrane of surface colonocytes, with a marked reduction in the levels of β- and γ-subunit–specific mRNAs. In addition, basolateral expression of the Na^+-K^+-ATPase α_1-isoform is reduced along the surface cell–crypt cell axis in UC, although there is no change in the level of the corresponding mRNA. Diarrhea in ulcerative colitis is therefore related, at least in part, to a major defect in electrogenic Na^+ absorption, which reflects changes in the levels of expression of critical subunits of both the apical Na^+ channel and basolateral Na^+-K^+-ATPase.

Na^+ ABSORPTION IN HEALTHY HUMAN COLON

In healthy individuals, 1.5–2 liters per day of Na^+- and Cl^--rich fluid pass from the small intestine into the colon. Of this, about 90% is absorbed. The main factor determining the extent of water movement out of the colonic lumen is the rate of Na^+ absorption,[1] but it should be emphasised that human colon possesses an array of Na^+ absorptive processes that are not distributed uniformly throughout its length. Thus, Cl^--dependent electroneutral Na^+ absorption is present throughout the large intestine (but is least evident in the rectum) and reflects either apical Na^+–Cl^- cotransport or an apical Na^+–H^+:Cl^-–HCO_3^- dual exchange mechanism.[2] Chloride-independent amiloride-insensitive Na^+ absorption, which may reflect both apical Na^+–H^+ exchange *and* amiloride-insensitive electrogenic Na^+ transport, appears to be a feature of the ascending (proximal) and transverse colonic segments.[2] In addition, bacterial fermentation of complex dietary carbohydrates in the ascending colon results in the production of short-chain fatty acids, all of which stimulate electroneutral Na^+ ab-

[c]Address for correspondence: Prof. G.I. Sandle, Molecular Medicine Unit, Level 6, Clinical Sciences Building, St. James's University Hospital, Leeds LS9 7TF, U.K. Voice: +44-113-206-5686; fax: +44-113-244-4475.

g.i.sandle@leeds.ac.uk

sorption to a progressively decreasing extent moving distally along the colon.[2] However, in the sigmoid colon and rectum (those segments that are invariably and usually maximally inflamed in acute UC), the dominant Na^+ absorptive mechanism is amiloride-sensitive electrogenic Na^+ transport.[2,3]

In recent years there has been a considerable increase in our understanding about the molecular structure, function, and regulation of apical Na^+ channels and basolateral Na^+-K^+-ATPase, which are the main integral membrane proteins involved in electrogenic Na^+ transport. The characteristics of these Na^+ transport proteins have been studied mainly in mammalian and amphibian renal epithelia. Little work has been done in mammalian (including human) colon using molecular techniques. A notable exception is the work of Rossier et al., who have cloned the α-, β- and γ-subunits of the amiloride-sensitive Na^+ channel and studied their epithelial distribution in rat distal colon.[4,5] As a consequence, we have been able to isolate a homologous α-subunit clone from a human distal colonic cDNA library,[6] which has a sequence identical to the Na^+ channel α-subunit found in human kidney and lung (G.I. Sandle, unpublished data). When overexpressed in Xenopus oocytes, this human colonic Na^+ channel subunit functions as an amiloride-sensitive Na^+ conductive pathway.[6] Although the Na^+ channel β- and γ-subunits in human distal colon have yet to be cloned, there is every likelihood that they will be similar if not identical to their counterparts in human kidney. The α-subunit represents the basic "building block" of the Na^+ channel, contains the amiloride binding site, and forms a heteromeric structure with the β- and γ-subunits, which are colocalized to the apical membrane of surface colonocytes.[7] The β- and γ-subunits are physiologically important because they ensure maximal activity of the heteromeric Na^+ channel and are upregulated by aldosterone.[5,7]

IMPAIRED COLONIC Na^+ ABSORPTION IN ULCERATIVE COLITIS

The large reserve capacity of the human colon for water transport allows it to absorb up to about five liters per day. In the presence of normal colonic transport function, diarrhea ensues only when the volume originating from the small intestine exceeds this level (for example, in patients with severe enteric infections or certain neuroendocrine tumors). A rather different situation exists in acute UC, where small intestinal fluid and electrolyte transport is normal but colonic mucosal inflammation is associated with altered electrolyte and water transport. It is generally accepted that diarrhea in these patients reflects, at least in part, leakage of a plasma-like solution and blood across the damaged mucosa, perhaps combined with altered colonic motility. In addition, increased levels of a variety of soluble inflammatory mediators are present in the inflamed colonic mucosa, all of which are capable of triggering electrogenic Cl^- secretion.[8,9] The tacit assumption has been that Cl^- secretion and the accompanying secretion of water are important transport abnormalities contributing to the pathogenesis of diarrhea in UC. However, Cl^- secretion probably plays an insignificant role in the production of diarrhea in this disease because (1) the high lumen-negative transmucosal electrical potential difference (pd) present in normal colonic and rectal mucosa is decreased or absent in patients with UC, which suggests loss of electrogenic Na^+ absorption rather than enhanced electrogenic Cl^- secretion;[10,11] (2) zero net Cl^- transport rather than net Cl^- secretion occurs in the proximal rectum of patients with acute UC;[10] (3) substantial decreases in Na^+ and Cl^- absorption, rather than Cl^- secretion, are seen in inflamed human colonic mucosa *in vitro;*[12] and (4) basolateral membrane Na^+-K^+-ATPase activity, a prerequisite for Cl^- secretion (in addition to electrogenic Na^+ absorption), is profoundly decreased in inflamed human colon.[11]

The overall picture of altered transport in UC is therefore one of diminished or negligible Na^+, Cl^-, and water absorption. The high pd normally present in human distal colon and rectum reflects electrogenic Na^+ absorption, which accounts for at least 60% of the total Na^+ absorbed in these segments, the remainder reflecting electroneutral NaCl absorption.[3] Net Na^+ and Cl^- absorption together provide the driving force for water absorption. Mucosal inflammation is associated with a marked decrease in pd, which is secondary to a substantial reduction in electrogenic Na^+ absorption, and a decrease in net Cl^- absorption, which probably reflects loss of passive (pd-dependent) Cl^- absorption and defective electroneutral Cl^-–HCO_3^- exchange.[10] These transport abnormalities have focused our attention on the cellular basis of defective Na^+ absorption and its role in the pathogenesis of diarrhea in patients with UC. Thus, previous electrophysiological studies have pinpointed defective amiloride-sensitive apical Na^+ entry in surface colonic epithelial cells and a 75% decrease in the maximal activity of basolateral membrane Na^+-K^+-ATPase as important factors leading to reduced electrogenic Na^+ absorption in UC.[11] Recent advances in the molecular biology of the amiloride-sensitive epithelial Na^+ channel and Na^+-K^+-ATPase have enabled us to explore the expression of these critical Na^+ transport proteins in the distal colon of patients with this disease.

MOLECULAR BASIS OF DEFECTIVE Na^+ ABSORPTION IN ULCERATIVE COLITIS

After written consent was obtained, four or five mucosal biopsies were taken from the sigmoid colon or proximal rectum of patients with UC who were undergoing routine colonoscopy or flexible sigmoidoscopy to assess the severity and extent of the disease. In all cases, patients were not receiving treatment either orally or in the form of enemas, and the degree of mucosal inflammation was judged to be of mild to moderate (grade II/III) severity. Biopsies of noninflamed mucosa (confirmed by routine histology) were obtained from the same regions in patients undergoing colonoscopy as part of the evaluation of functional abdominal pain.

The effect of UC on the expression of Na^+ channel subunit proteins was evaluated using polyclonal antirabbit antibodies that were specific for each subunit. The α-subunit was detected with an antibody raised against a full-length fusion protein generated from the α-subunit of the bovine renal epithelial Na^+ channel, the β-subunit with an antibody raised against a synthetic peptide corresponding to amino acids 411-420 of the human epithelial Na^+ channel β-subunit, and the γ-subunit with an antibody raised against an amiloride-sensitive Na^+ channel protein in bovine renal papilla (D. Benos, personal communication). Preliminary Western blot analyses revealed similar levels of α-subunit protein in patients with noninflamed mucosa and patients with UC. By contrast, the levels of β-subunit protein and γ-subunit protein appeared to be substantially lower in colitic patients compared with those in patients with noninflamed mucosa. Immunocytochemical evaluation of biopsy material from additional patients confirmed the presence of similar levels of Na^+ channel α-subunit protein in the apical membrane of surface (but not crypt) epithelial cells in both groups of patients, whereas levels of β- and γ-subunit proteins were greatly reduced or absent in patients with UC when compared with patients with noninflamed mucosa. Taken together, these findings indicate that in UC even mild to moderate mucosal inflammation is associated with decreased membrane expression of Na^+ channel β- and γ-subunits, which would be expected to reduce the overall level of Na^+ channel activity. In addition, using full-length human renal Na^+ channel α-, β-, and γ-subunit cDNAs as probes, *in situ* hybridization studies performed on noninflamed and inflamed mucosal biopsies suggested that levels of all three mRNAs were decreased in surface colonocytes in patients with UC. Thus, in UC, the levels of Na^+ channel β-and γ-subunit proteins appear to decrease in parallel with the levels of their specific mRNAs, suggesting a defect at the point of gene transcription and/or increased mRNA turnover. At present we have no explanation for the persistence of apparently normal levels of α-subunit protein in the face of reduced levels of α-subunit mRNA.

Similar experimental approaches were used to assess the effect of mucosal inflammation on the expression of the α_1-isoform of Na^+-K^+-ATPase, which, together with the β_1-isoform, constitute critical functional components of the Na^+-K^+-ATPase complex in ion-transporting epithelial cells.[13] Western blot analyses using a monoclonal antibody raised against rat kidney revealed a decrease of about 50% in α_1-isoform levels in biopsies from patients with UC compared with patients with noninflamed mucosa. Immunocytochemical studies also showed reduced expression of the α_1-isoform at the basolateral membrane of cells along the surface cell–crypt cell axis in biopsies from patients with UC. However, in contrast to our *in situ* hybridization evaluation of the Na^+ channel subunits, Northern blot analyses failed to reveal a difference in Na^+-K^+-ATPase α_1-isoform mRNA levels between the noninflamed and colitic biopsies, which suggests that decreased membrane expression of the Na^+-K^+-ATPase α_1-isoform in UC reflects a defect at the point of translation or impaired trafficking of the isoform to the basolateral domain.

IMPLICATIONS AND CONCLUSIONS

A more detailed picture of the ion transport abnormalities that occur in the distal colon and rectum in patients with UC is beginning to emerge. One important change seems to be a reduction or loss of electrogenic Na^+ transport, which in turn leads to a decrease in the passive (pd-dependent) absorption of Cl^- ions. The concomitant decrease in net water absorption is, therefore, likely to make a major contribution to diarrhea in patients with acute colitis. One drawback of our electrophysiological studies,[11] which involved transepithelial and intracellular measurements, was the inability to identify the individual contributions made by potentially defective apical Na^+ entry (mediated by amiloride-sensitive Na^+ channels) and impaired basolateral Na^+ extrusion (mediated by Na^+-K^+-ATPase) to the overall decrease in amiloride-sensitive Na^+ absorption. Another drawback arose from the fact that, because epithelial cell turnover is more rapid in inflamed colonic mucosa, "surface" colonocytes may have been undifferentiated crypt cells that are devoid of amiloride-sensitive apical Na^+ channels.[7,14] However, it would seem that this is not the case, because we have shown that the apical membrane of surface colonocytes in patients with UC contains normal levels of the Na^+ channel α-subunit, despite reduced levels of the β- and γ-subunits. Because coexpression of all three Na^+ channel subunits is required for full activity of the heteromeric channel,[5] it appears that defective electrogenic Na^+ absorption in UC reflects a combination of both reduced expression of the apical Na^+ channel in its heteromeric form and reduced expression of basolateral Na^+-K^+-ATPase (at least its α_1-isoform). It may also be that mucosal inflammation results in a decrease in distal colonic electroneutral NaCl absorption, as a consequence of reduced levels of the apical Na^+-H^+ exchanger, the apical Cl^--HCO_3^- exchanger, or a combination of both possibilities. This is clearly an important area for further study.

The abnormalities in the electrogenic Na^+ absorptive process that we have identified are particularly interesting from the point of view of understanding the antidiarrheal effect of corticosteroids in patients with UC. Despite the aforementioned membrane transport defects, Na^+ absorption across inflamed mucosa remains corticosteroid responsive. Pharmacological doses of hydrocortisone or methylprednisolone administered intravenously stimulate rectal Na^+ and water absorption and transmucosal pd after five hours to the same extent in healthy subjects and patients with acute UC.[10] These transport effects may reflect genomic events following crossover binding of the corticosteroids to mineralocorticoid receptors as well as glucocorticoid-receptor activation,[15,16] with the implication that corticosteroids stimulate transcription of Na^+ channel β- and/or γ-subunits,[7,17] even in the presence of mucosal inflammation. Alternatively, given the ability of the Na^+ channel α-subunit alone to function as a Na^+ conductive pathway,[4–6] the rapid stimulatory effect of corticosteroids on electrogenic Na^+ absorption across inflamed as well as noninflamed colon[10] may reflect largely nongenomic activation of basolateral Na^+-K^+-ATPase.[18] In any event, corticosteroids reduce diarrhea in UC by stimulating colonic Na^+ and water transport directly, quite apart from their ability to suppress inflammation. Additional studies are required to identify their effects at the molecular level.

ACKNOWLEDGMENTS

Antibodies to the Na^+ channel subunits and the Na^+-K^+-ATPase α_1-isoform were kindly donated by Dr. Dale Benos (University of Alabama at Birmingham, USA) and Dr. Kathleen Sweadner (Harvard University, USA), respectively. This work was supported by a Medical Research Council project grant and a Digestive Diseases Foundation/British Society of Gastroenterology Jubilee Fellowship awarded to E. Greig.

REFERENCES

1. BINDER, H.J. & G.I. SANDLE. 1994. Electrolyte transport in the mammalian colon. *In* Physiology of the Gastrointestinal Tract. L.R. Johnson, Ed.: 2133–2171. Raven Press. New York.
2. SANDLE, G.I. 1998. Salt and water absorption in the human colon: a modern appraisal. Gut **43**: 294–299.
3. SANDLE, G.I. 1994. Segmental differences in colonic function. *In* Short Chain Fatty Acids, Falk Symposium 73. H.J. Binder, J. Cummings & K. Soergel, Eds.: 29–43. Kluwer Academic Publishers. Lancaster, UK.

4. CANESSA, C.M., J-D. HORISBERGER & B.C. ROSSIER. 1993. Epithelial sodium channel related to proteins involved in neurodegeneration. Nature **361**: 467–470.
5. CANESSA, C.M., L. SCHILD, G. BUELL *et al.* 1994. Amiloride-sensitive epithelial Na$^+$ channel is made of three homologous subunits. Nature **367**: 463–467.
6. BAKER, E.H., R.P. BOOT-HANDFORD & G.I. SANDLE. 1996. Expression of the human colonic Na$^+$ channel α-subunit in *Xenopus* oocytes [abstract]. Clin. Sci. **90** (suppl. 34): 6P–7P.
7. RENARD, S., N. VOILLEY, F. BASSILANA *et al.* 1995. Localization and regulation by steroids of the α, β and γ subunits of the amiloride-sensitive Na$^+$ channel in colon, lung and kidney. Pflügers Arch. **430**: 299–307.
8. WARDLE, T.D., L. HALL & L.A. TURNBERG. 1993. Inter-relationships between inflammatory mediators, released from colonic mucosa in ulcerative colitis, and their effects on colonic secretion. Gut **34**: 503–508.
9. WARDLE, T.D. & L.A. TURNBERG. 1994. Potential role for interleukin-1 in the pathophysiology of ulcerative colitis. Clin. Sci. **86**: 619–626.
10. SANDLE, G.I., J.P. HAYSLETT & H.J. BINDER. 1986. Effect of glucocorticoids on rectal transport in normal subjects and patients with ulcerative colitis. Gut **27**: 309–316.
11. SANDLE, G.I., N. HIGGS, P. CROWE *et al.* 1990. Cellular basis for defective electrolyte transport in inflamed human colon. Gastroenterology **99**: 97–105.
12. HAWKER, P.C., J.S. MCKAY & L.A. TURNBERG. 1980. Electrolyte transport properties of colonic epithelium in patients with inflammatory bowel disease. Gastroenterology **79**: 508–511.
13. SWEADNER, K.J. 1989. Isozymes of the Na$^+$/K$^+$-ATPase. Biochim. Biophys. Acta **988**: 185–220.
14. LOMAX, R.B., C.M. MCNICHOLAS, M. LOMBÈS & G.I. SANDLE. 1994. Aldosterone-induced apical Na$^+$ and K$^+$ conductances are located predominantly in surface cells in rat distal colon. Am. J. Physiol. **266**: G71–G82.
15. MARUSIC, E.T., J.P. HAYSLETT & H.J. BINDER. 1981. Corticosteroid-binding studies in cytosol of colonic mucosa of the rat.. Am. J. Physiol. **240**: G417–G423.
16. TURNAMIAN, S.G. & H.J. BINDER. 1989. Regulation of active sodium and potassium transport in the distal colon of the rat. J. Clin. Invest. **84**: 1924–1929.
17. LINGUEGLIA, E., S. RENARD, R. WALDMANN *et al.* 1994. Different homologous subunits of the amiloride-sensitive Na$^+$ channel are differently regulated by aldosterone. J. Biol. Chem. **269**: 13736–13739.
18. EWART, H.S. & A. KLIP. 1995. Hormonal regulation of the Na$^+$-K$^+$-ATPase: mechanisms underlying rapid and sustained changes in pump activity. Am. J. Physiol. **269**: C295–C311.

Barrier Dysfunction and Crohn's Disease

JON MEDDINGS[a]

GI Research Group, University of Calgary, Calgary, Alberta, Canada

ABSTRACT: Crohn's disease is a debilitating illness of unknown etiology. A current hypothesis of disease pathogenesis suggests that this illness represents an abnormal immunological reaction to a luminal antigen. As part of this theory it is suggested that the luminal antigen is delivered to the mucosal immune system by a paracellular route. If this theory is correct several testable predictions can be made. In this manuscript these predictions are presented and the evidence to support or refute them is presented.

INTRODUCTION

The etiology of Crohn's disease is currently unknown. However, one major hypothesis contends that this disease represents an abnormal reaction of the mucosal immune system to a luminal antigen.[1] Data in support of this hypothesis are strong and comes from a variety of sources. The strongest support is from the clinical arena. Altering the flow of luminal contents to avoid areas of inflammation almost uniformly induces a remission of disease even in the absence of immunosuppressive therapy. Furthermore, because luminal antigens predominantly arise from either the luminal flora or ingested nutrients, it is also important to recognize that manipulations designed to alter these also improve disease outcome. Administration of antibiotics to alter intestinal flora is often used as a therapeutic strategy. Providing enteral nutrition in a predigested form is also useful in some clinical situations.[2] However, perhaps the strongest data along these lines comes from animal models of Crohn's disease. In the genetic knockout animals disease expression is critically dependent upon the luminal flora. Raising animals in a germ-free environment dramatically ameliorates disease expression.[1]

Given these findings a reasonable hypothesis regarding the genesis of Crohn's disease involves three separate events. The first is a genetic predisposition to react to a luminal antigen. The second is of course the presence of the offending antigen or inflammatory stimulus. The third appears to be the abnormal delivery of this agent to the mucosal immune system in a fashion that results in inflammation. Within the intestine most luminal antigens are presented to the mucosal immune system in a fashion destined to invoke oral tolerance rather than inflammation. This pathway involves presentation by either mucosal macrophages or enterocytes themselves to T cells in a fashion that suppresses the immune response. However, recent data suggest that abnormal delivery of luminal antigens may involve a paracellular pathway and that this mode of antigen delivery may invoke more of an inflammatory reaction.[3]

[a]Address for correspondence: Dr. Jon Meddings, 1705 Health Sciences Center, 3330 Hospital Dr. NW, Calgary, AB Canada T2N 4N1. Voice: 403-220-4557; fax: 403-220-8747.
meddings@ucalgary.ca

Under this scenario increased intestinal permeability is viewed as a means of presenting luminal antigens, or proinflammatory compounds, to the mucosal immune system and initiating mucosal inflammation. Of the three components involved in this hypothesis of disease generation, it is only this one that appears to vary in a physiological manner. Presumably, if we have a genetic predisposition to react to a luminal agent this remains with us for life. Furthermore, if the agent is a constituent of either the diet or the luminal flora then these will change only slowly with time. The regulatory step would primarily be alterations in delivery of the agent to the immune system. Inasmuch as the clinical course of Crohn's disease involves spontaneous fluctuating periods of active disease and remissions, it is hypothesized that it is the increased intestinal permeability that varies over time.

If this hypothesis of disease etiology is correct, several predictions about Crohn's disease can be made to test the hypothesis. These predictions are:

- Individuals destined to develop Crohn's disease will have increased intestinal permeability.
- Increased permeability will normalize with disease remission.
- Permeability will increase prior to relapses.
- Events that precipitate relapses will also increase intestinal permeability.

The focus of this manuscript will be to examine the evidence that supports these predictions, primarily in humans but to a lesser extent in animal models of Crohn's disease.

TESTING OF PREDICTIONS

Does Increased Intestinal Permeability Predate Disease Onset?

This is a very difficult question to examine in humans. The major problem is that we cannot identify patients prior to their having disease. The approach that has been used is to identify groups of individuals at high risk of developing Crohn's disease and determining whether they contain a subgroup that exhibits increased permeability. Furthermore, because we also expect that individuals who will develop Crohn's disease will have increases in permeability that normalize over time, that is, they may have cyclic variations in permeability, we would also predict that at any point in time permeability might be increased or appear normal and the individual may still go on and develop Crohn's disease. However, the fact still remains that if this hypothesis is correct then at least a subset of individuals who will subsequently develop Crohn's disease should have increased intestinal permeability prior to disease onset.

The best studied high risk group are first degree relatives of patients who already have Crohn's disease. Epidemiological data would suggest that up to 10% of this group will go on and develop Crohn's disease at some time during their life. The original experiments evaluating this risk group were presented in the late 1980s and suggested that a surprising number of relatives had increased permeability.[4] These studies have been criticized for the choice of probes to determine permeability, and subsequent work failed to corroborate the early data. However, using more conventional methods to determine intestinal permeability, it has now been reported that on the order of 10% of first degree relatives have increased intestinal permeability. These data were first reported from our group but have now been replicated by numerous groups around the world.[5-9]

It is difficult to understand what this finding represents. No longtitudinal studies have been presented to suggest that the individuals with increased permeability are the ones that will later develop Crohn's disease. Clearly this would make the argument more compelling. However, it has been demonstrated that the relatives with a permeability abnormality also demonstrate an increased fraction of circulating B cells that express the CD45RO phenotype.[6] This marker suggests that these cells are immunological memory cells and suggests increased antigen delivery in these individuals. Patients with Crohn's disease also express an increased fraction of these cells in the peripheral circulation. Therefore, a group of individuals at high risk for the development of Crohn's disease clearly has increased permeability, and in these people it is associated with a marker suggesting activation of the immune system.

An interesting extension of this concept has also been reported. Although increased permeability in relatives may have its origin in genetics, it is equally possible that this represents a reaction to something in the environment, which is of course shared by close relatives. To address this Peeters *et al.* have studied both families of patients with Crohn's disease as well as spouses of patients with Crohn's disease.[8] In addition to demonstrating that a subgroup of relatives had increased permeability, this group also demonstrated that about a third of spouses have increased intestinal permeability. Although the number of spouses studied was small, this study has offered the important suggestion that perhaps increased permeability in relatives may be induced by an environmental agent.

The possible involvement of environmental factors in regulation of intestinal permeability also opens the conceptual door for other considerations. Although fluctuating increases in permeability may occur intrinsically, we could also postulate that individuals at risk of developing Crohn's disease may be inordinately sensitive to environmental agents that increase small intestinal permeability. Our group has recently tested this hypothesis. By giving relatives aspirin and testing permeability before and after, we have identified that a subgroup of relatives are exquisitely sensitive to the permeability-enhancing effects of this NSAID. Of interest was the observation that this sensitivity was only apparent in the small intestine and not evident in the stomach.[7]

The conclusions of these studies are readily apparent. Individuals at risk of developing Crohn's disease are more likely to have increased small intestinal permeability or increased sensitivity to the permeability-enhancing effects of aspirin than the general population. Furthermore, the same individuals have evidence of an immunological correlate for these findings. Whether this represents a genetic predisposition or the effect of an environmental factor is unclear. However, the data obtained from spousal studies would support the latter interpretation.

Increased Intestinal Permeability Will Normalize during Remission

The evidence on this point is clear. Small intestinal disease, when active, is characterized by increased permeability. However, when the disease is brought into remission the abnormal permeability returns to the normal range.[10,11] This has been demonstrated in multiple studies, and determinations of small intestinal permeability are used in some centers to follow disease activity.

Do Increases in Intestinal Permeability Precede Disease Relapse?

This is a critical question and one that has been relatively poorly studied. It is important not only from the perspective of disease mechanisms but also for the treatment of disease. In treating patients with Crohn's disease it would be useful to be able to identify imminent relapses and start therapy immediately prior to disease recurrence. There is a belief that such a strategy might reduce the number of drugs consumed by these patients and improve patient well-being.

The first study examining this question examined a group of patients that had entered remission following surgical therapy.[12] Small intestinal permeability was determined and patients stratified into two groups: normal or high permeability. Over a one year follow-up 70% of the patients with high permeability had a relapse as compared to only 17% of those with a low permeability.

In a follow-up study, Hilsden *et al.* sequentially determined small intestinal permeability every four months. Once again, over 70% of small intestinal disease relapses were preceeded by increased permeability. Of interest was the observation that those relapses not preceded by increased permeability occurred more than 100 days following the most recent permeability determination. This raises the important question of whether more frequent determinations would increase the yield.

Do Events That Precipitate Relapse Also Increase Permeability?

Most relapses appear to be random events with no clear precipitating factors. This makes the answer to this question difficult. However, a small proportion of disease relapses are temporally linked to either NSAID ingestion or psychological stress. Both factors are amenable to study.

It is clear that both in animal models and humans the ingestion of NSAID's leads to an increase in small intestinal permeability.[13–15] There is evidence that certain compounds preferentially damage the small intestine as compared to the stomach. In fact certain formulations that appear to decrease gastric damage appear to shift proximal damage to more distal sites within the small intestine.[16] As previously mentioned it is also apparent that some relatives of patients with Crohn's disease and patients themselves appear to have an exaggerated response to orally ingested NSAIDs.[7] Whether this is the same subgroup that develops NSAID-associated relapses has not been studied.

There are now also convincing data that environmental stress increases intestinal permeability for a variety of molecular sizes.[17,18] It is also clear that much of this increase is cholinergically mediated in animal models. However, there is also data that the hypothalamic-pituitary axis may be involved.[19] Regardless of the mechanism, it is clear that psychological stress can increase gastrointestinal permeability, and this may increase delivery of luminal antigens to the mucosal immune system.

CONCLUSIONS

Inflammatory bowel disease, that resembles Crohn's disease, can be initiated in animal models by increasing the delivery of luminal antigens to the mucosal immune system. This has led to the hypothesis that a defective mucosal barrier may be a very early event in human Crohn's disease and that increases in intestinal permeability may underlie the development of the disease and the relapses that characterize it clinically. The predictions of this model, outlined above, are all supported by experimental data in humans and experimental animals.

At the very least a subgroup of patients at high risk for the development of Crohn's disease has increased permeability. Whether this group ultimately develops disease is a critical question that must be addressed by future prospective studies. However, there is a strong suggestion that this may represent a significant risk factor. What remains unclear is whether this represents a structural defect in tight junction formation, or regulation or damage to the enterocyte itself. These questions all represent future questions that will be important in further understanding the etiology of this debilitating disease.

REFERENCES

1. ELSON, C.O., R.B. SARTOR, G.S. TENNYSON & R.H. RIDDELL. 1995. Experimental models of inflammatory bowel disease. Gastroenterology **109:** 1344–1367.
2. SANDERSON, I.R., P. BOULTON, I. MENZIES & J.A. WALKER-SMITH. 1987. Improvement of abnormal lactulose/rhamnose permeability in active Crohn's disease of the small bowel by an elemental diet. Gut **28:** 1073–1076.
3. TERATO, K., X.J. YE, H. MIYAHARA et al. 1996. Induction of chronic autoimmune arthritis in DBA/1 mice by oral administration of type II collagen and *Escherichia coli* lipopolysaccharide. Br. J. Rheumatol. **35:** 828–838.
4. HOLLANDER, D., C.M. VADHEIM, E. BRETTHOLZ, et al. 1986. Increased intestinal permeability in patients with Crohn's disease and their relatives. Ann. Int. Med. **105:** 883–885.
5. MAY, G.R., L.R. SUTHERLAND & J.B. MEDDINGS. 1993. Is small intestinal permeability really increased in relatives of patients with Crohn's disease? Gastroenterology **104:** 1627–1632.
6. YACYSHYN, B.R. & J.B. MEDDINGS. 1995. CD45RO expression on circulating CD19+ B cells in Crohn's disease correlates with intestinal permeability. Gastroenterology **108:** 132–137.
7. HILSDEN, R.J., J.B. MEDDINGS & L.R. SUTHERLAND. 1996. Intestinal permeability changes in response to acetylsalicylic acid in relatives of patients with Crohn's disease. Gastroenterology **110:** 1395–1403.
8. PEETERS, M., B. GEYPENS, D. CLAUS et al. 1997. Clustering of increased small intestinal permeability in families with Crohn's disease. Gastroenterology **113:** 802–807.
9. MUNKHOLM, P., E. LANGHOLZ, D. HOLLANDER et al. 1994. Intestinal permeability in patients with Crohn's disease and ulcerative colitis and their first degree relatives. Gut **35:** 68–72.
10. TEAHON, K., P. SMETHURST, M. PEARSON et al. 1991. The effect of elemental diet on intestinal permeability and inflammation in Crohn's disease. Gastroenterology **101:** 84–89.
11. MURPHY, M.S., E.J. EASTHAM, R. NELSON et al. 1989. Intestinal permeability in Crohn's disease. Arch Dis. Child **64:** 321–325.
12. WYATT, J., H. VOGELSANG, W. HUBL et al. 1993. Intestinal permeability and the prediction of relapse in Crohn's disease. Lancet **341:** 1437–1439.
13. SIGTHORSSON, G., J. TIBBLE, J. HAYLLAR et al. 1998. Intestinal permeability and inflammation in patients on NSAIDs. Gut **43:** 506–511.
14. MIELANTS, H. & E.M. VEYS. 1985. NSAID and the leaky gut. Lancet **1:** 218
15. BJARNASON, I., P. SMETHURST, C.G. FENN et al. 1989. NSAID small bowel injury and cytoprotection. Gastroenterology **97(5):** 1344–1345.
16. DAVIES, N.M. & J.L. WALLACE. 1997. Nonsteroidal anti-inflammatory drug-induced gastrointestinal toxicity: new insights into an old problem. [review] [89 refs]. J. Gastroenterol. **32:** 127–133.

17. KILIAAN, A.J., P.R. SAUNDERS, P.B. BIJLSMA *et al.* 1998. Stress stimulates transepithelial macromolecular uptake in rat jejunum. Am. J. Physiol. Gastrointest. Liver Physiol. **275:** G1037–G1044.
18. SAUNDERS, P.R., N.P.M. HANSSEN & M.H. PERDUE. 1997. Cholinergic nerves mediate stress-induced intestinal transport abnormalities in Wistar-Kyoto rats. Am. J. Physiol. Gastrointest. Liver Physiol. **273:** G486–G490.
19. SPITZ, J., G. HECHT, M. TAVERAS *et al.* 1994. The effect of dexamethasone administration on rat intestinal permeability: the role of bacterial adherence. Gastroenterology **106:** 35–41.

Mechanisms of Cholera Toxin–Induced Diarrhea

E. BEUBLER[a] AND R. SCHULIGOI

Department of Experimental and Clinical Pharmacology,
Karl-Franzens-University of Graz, 8010 Graz, Austria

ABSTRACT: In the pathogenesis of cholera, cyclic adenosine monophosphate, 5-hydroxytryptamine, prostaglandins, and the function of neuronal structures have been implicated. To elucidate the role of different isoforms of cyclooxygenase (COX)-1 and COX-2, selective COX-2 inhibitors were used. The selective COX-2 inhibitors NS-398 and DFU completely suppressed cholera toxin–induced prostaglandin E_2 biosysthesis and caused a dose-dependent inhibition of cholera toxin–induced fluid secretion in the rat jejunum *in vivo*. Constitutive expression of COX-1 but also of COX-2 mRNA was found in mucosal scrapings of the rat jejunum. Cholera toxin had no effect on COX-1 as well as COX-2 mRNA expression. Treatment of rats with dexamethasone did not effect cholera toxin–induced prostaglandin E_2 biosynthesis and did not influence the expression of COX-2 mRNA, further substantiating that cholera toxin does not cause an induction of COX-2 mRNA. Treatment of rats with *E. coli* lipopolysaccharide caused a marked increase in COX-2 mRNA expression that was inhibited by dexamethasone. In conclusion, the results provide evidence that cholera toxin, in addition to other mediators, uses prostaglandin E_2 to exert its secretory effect and that in the case of cholera toxin prostaglandins are metabolized via COX-2.

INTRODUCTION

The diarrhea of cholera is commonly considered to depend on a cyclic adenosine monophosphate (cAMP)–mediated active secretory mechanism. However, several other mediators have been implicated in the mediation of cholera toxin–induced intestinal fluid secretion. In 1970, Bhide and coworkers[1] demonstrated an increase in 5-HT blood levels in choleraic rabbits. Cholera toxin administered into the duodenum of rabbits caused severe degranulation of enterochromaffin cells, as revealed by electron microscopy.[2] From these experiments, a hypothesis was proposed that cholera toxin stimulates an apical receptor on the enterochromaffin cells and that serotonin and a polypeptide released by the stimulus may mediate the diarrheagenic action of cholera toxin. After these observations, much evidence accumulated to prove the involvement of 5-HT in the genesis of cholera toxin–induced fluid and electrolyte secretion. The involvement of 5-HT in choleraic secretion was proved by inducing tachyphylaxis against 5-HT in the experimental animals, cats and rats, by intravenous infusion of increasing doses of 5-HT. In these animals, cholera toxin–induced

[a]Address for correspondence: E. Beubler, Department of Experimental and Clinical Pharmacology, Karl-Franzens-University of Graz, 8010 Graz, Austria. Fax: +43-316-380-9645.
eckhard.beubler@kfunigraz.ac.at

secretion was inhibited.[3] In a histochemical study, cholera toxin was shown to cause a significant depletion of 5-HT from enterochromaffin cells in the feline small intestine.[4]

It has been further suggested that cholera toxin may cause diarrhea by stimulating prostaglandin (PG) synthesis.[5] This concept is supported by the observations that cholera toxin is apparently associated with increased local PG synthesis[6–9] and that PG-synthetase inhibitors impair the secretory effect of cholera toxin.[10–13] The finding that indomethacin in some studies has been reported to inhibit cholera toxin–induced secretion but not mucosal cAMP accumulation[13,14] favors the notion that PGs may play a primary role in the secretory mechanism. This view is supported by the observation that low concentrations of PGs exert a secretory effect by facilitating the entry of calcium into the cell, rather than by stimulating the adenylate cyclase-cAMP system.[15] On the other hand, PGE_2 has been shown to be an important intermediate in the transduction mechanism that leads to 5-HT–induced intestinal secretion.[15] The present study was performed to elucidate the role of the different isoforms of cyclooxygenase, COX-1 and COX-2, in cholera toxin–induced fluid secretion by using specific COX-2 inhibitors and determination of mRNA expression of COX-1 and COX-2 in mucosal scrapings.

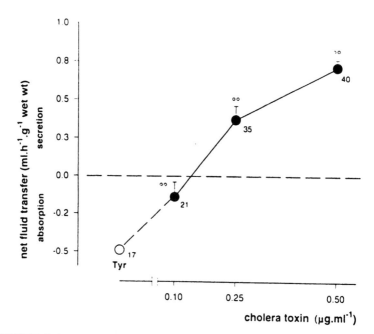

FIGURE 1. Dose-response relationship for the effect of cholera toxin (4 h) on net fluid transport in the rat jejunum *in vivo*. Each *point* represents the mean ± SEM. The figures indicate the number of experiments; $°°p < 0.01$ compared with control. (Reproduced with permission from Beubler *et al.*[19])

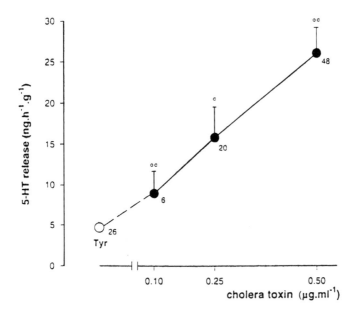

FIGURE 2. Dose-response relationship for the effect of cholera toxin (4 h) on luminal 5-HT output in the rat jejunum *in vivo*. Each *point* represents the mean ± SEM. The figures indicate the number of experiments; $^{\circ}p < 0.05$, $^{\circ\circ}p < 0.01$ compared with control. (Reproduced with permission from Beubler & Horina.[20])

METHODS

The experiments were performed in a tied off loop model in the rat jejunum *in vivo*. Net fluid transfer rates were determined gravimetrically after the instillation of Tyrode's solution into the gut lumen. Cholera toxin was administered intraluminally (0.1–0.5 g/mL^{-1}, 4 h). The specific COX-2 antagonists NS-398 ([*N*-(2-cyclo-hexaloxy-4-nitrophenyl) methanesulfonamide])[16] and DFU (5,5-dimethyl-3-(3-flu-orophenyl)-4-(4-methylsulphonyl)phenyl-2(5H)-furanone)[17] were administered subcutaneously. *E. coli* lipopolysaccharide (LPS) (5 mg/kg) was administered intra-peritoneally. PGE_2 was measured in the intraluminal fluid by radioimmunoassay. Reverse tanscriptase polymerase chain reaction (RT-PCR) was performed using specific primers to determine COX-1 and COX-2 mRNA expression using glyceralde-hyde-3-phosphatate dehydrogenase (GAPDH) as internal control.[18]

RESULTS AND DISCUSSION

In the rat jejunum *in vivo*, cholera toxin (0.1–0.5 μg/mL) dose dependently in-creased intestinal fluid (FIG. 1) and electrolyte secretion as well as luminal 5-HT (FIG. 2) and prostaglandin E_2 output (FIG. 3).[16,17]

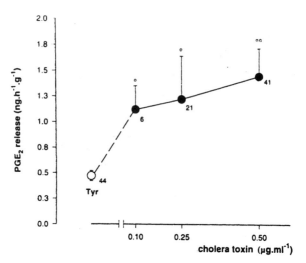

FIGURE 3. Dose-response relationship for the effect of cholera toxin (4 h) on luminal PGE_2 output in the rat jejunum *in vivo*. Each *point* represents the mean ± SEM. The figures indicate the number of experiments; *$p < 0.05$, **$p < 0.01$ compared with control. (Reproduced with permission from Beubler *et al.*[19])

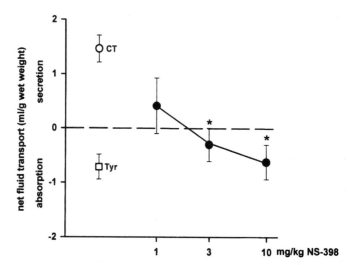

FIGURE 4. Effect of NS-398 on cholera toxin–induced net fluid secretion in the rat jejunum *in vivo*. Each *point* represents the mean ± SEM of 10–15 experiments. *$p < 0.05$ compared with cholera toxin.

Cholera toxin–induced fluid secretion (0.5 µg/mL) was dose dependently inhibited by the specific COX-2 inhibitors NS-398, at doses from 1 to 10 mg/kg (FIG. 4), and DFU, at doses from 0.125 to 2 mg/kg (FIG. 5). Cholera toxin also stimulated PGE_2 release into the gut lumen in these experiments. Both NS-398 (3 mg/kg) and

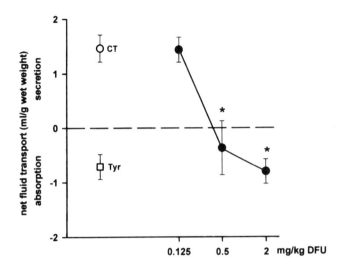

FIGURE 5. Effect of DFU on cholera toxin–induced net fluid secretion in the rat jejunum *in vivo*. Each *point* represents the mean ± SEM of 10–15 experiments. *$p < 0.05$ compared with cholera toxin.

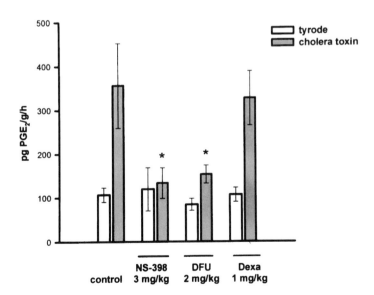

FIGURE 6. Effect of cholera toxin (0.5 µg/mL) on PGE_2 release into the rat jejunum in the absence and presence of NS-398 (3 mg/kg), DFU (2 mg/kg), and dexamethasone. Each *column* represents the mean ± SEM of 6–20 experiments. *$p < 0.05$ compared with cholera toxin.

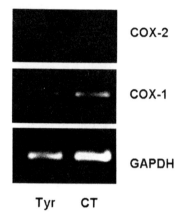

COX-2

COX-1

GAPDH

Tyr   CT

FIGURE 7. Expression of COX-1 and COX-2 mRNA in mucosal scrapings of the rat jejunum, treated with cholera toxin (CT) (0.5 µg/mL, 4 h) versus controls (Tyr). The agarose gel shows a representative RT-PCR with ethidium bromide–stained DNA bands; GAPDH was used as internal standard.

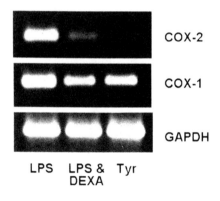

COX-2

COX-1

GAPDH

LPS   LPS &   Tyr
       DEXA

FIGURE 8. Expression of COX-2 mRNA in mucosal scrapings of the rat jejunum, treated with LPS (5 mg/kg sc) without and with pretreatment with dexamethasone (LPS & Dexa) (1 mg/kg) versus controls (Tyr). The agarose gel shows a representative RT-PCR with ethidium bromide–stained DNA bands; GAPDH was used as internal standard.

DFU (2 mg/kg) but not dexamethasone (1 mg/kg) inhibited cholera toxin–induced PGE_2 formation (FIG. 6). These data suggest that cholera toxin stimulates PGE_2 formation by stimulation primarily of COX- 2..

Because dexamethason (1 mg/kg sc) is without any effect on cholera toxin–induced PGE_2 formation, the cyclooxygenase involved is likely to be a constitutive one, but not an inducible one. This hypothesis is further substantiated by the following RT-PCR experiments. RT-PCR experiments show that COX-2 as well as COX-1 mRNA is present in mucosal scrapings (FIG. 7). Cholera toxin does not influence COX-2 mRNA expression, again indicating that COX-2 is constitutively present in mucosal scrapings. This is further supported by the failure of dexamethasone to influence COX-2 mRNA expression in untreated controls and in cholera toxin–treated animals (data not shown). To evaluate the methods used, it is shown that LPS, (5 mg/kg ip) as expected, increases COX-2 mRNA in mucosal scrapings and that this increase is inhibited by pretreatment of the rats with dexamethasone (1 mg/kg sc) (FIG. 8).

CONCLUSION

Our results provide further evidence that cholera toxin uses serotonin to induce its effect. Serotonin, in a second step, stimulates via 5-HT_2 receptors a predominantly constitutive COX-2 that is responsible for PGE_2 formation. COX-2 is obviously not induced by cholera toxin as it is by LPS. The role of COX-1 in cholera toxin–induced secretion remains to be elucidated.

ACKNOWLEDGMENTS

This work was supported by Austrian Scientific Research Fund Grant Nos. 10007 and 12158 and by the Jublilee Foundation of the Austrian National Bank Grant No. 6505.

REFERENCES

1. BHIDE, M.B., V.A. AROSKAR & N.K. DUTTA. 1970. Release of active substances by cholera toxin. Indian J. Med. Res. **58:** 548–550.
2. OSAKA, M., T. FUJITA & Y. YANATORI. 1975. On the possible role of intestinal hormones as the diarrhoeagenic messenger in cholera. Virchows Arch. B. Cell Pathol. **18:** 287–296.
3. CASSUTO, J., M. JODAL, R. TUTTLE & O. LUNDGREN. 1982. 5-Hydroxytryptamine and cholera secretion. Physiological and pharmacological studies in cats and rats. Scand. J. Gastroenterol. **17:** 695–703.
4. NILSSON, O., J. CASSUTO, P.A. LARSSON et al. 1983. 5-Hydroxytryptamine and cholera secretion: a histochemical and physiological study in cats. Gut **24:** 542–548.
5. BENNETT, A. 1971. Cholera and prostaglandins. Nature **231:** 536.
6. OKPAKO, D. 1975. Prostaglandins and cholera: the occurrence of prostaglandin-like smooth muscle contracting substances in cholera diarrhoea. Prostaglandins **10:** 769–777.
7. SPEELMANN, P., G.H. RABBANI, K. BUKHAVE & J. RASK-MADSEN. 1985. Increased jejunal prostaglandin E_2 concentrations in patients with acute cholera. Gut **26:** 188–193.
8. BEDWANI, J.R. & D. OKPAKO. 1975. Effects of crude and pure cholera toxin on prostaglandin release from the rabbit ileum. Prostaglandins **10:** 117–127.
9. TOTHILL, A. 1976. Prostaglandin E_2: a factor in the pathogenesis of cholera. Prostaglandins **11:** 925–933.
10. JACOBY, H.I. & C.H. MARSHALL. 1972. Antagonism of cholera enterotoxin by antiinflammatory agents in the rat. Nature **235:** 163–164.
11. FINCK, A.D. & R.L. KATZ. 1972. Prevention of cholera-induced intestinal secretion in the cat by aspirin. Nature **238:** 273–274.
12. GOTS, R.E., S.B. DORMAL & R.A. GIANNELLA. 1974. Indomethacin inhibition of *Salmonella typhimurium, Shigella flexneri* and cholera-mediated rabbit ileal secretion. J. Infect. Dis. **130:** 280–284.
13. WALD, A., G.S. GOTTERER, G.R. RAJENDRA et al. 1977. Effect of indomethacin on cholera-induced fluid movements, unidirectional sodium fluxes and intestinal cAMP. Gastroenterology **72:** 106–110.
14. WILSON, E.E., S. EL HINDI, P. TAO & L. POPPE. 1975. Effects of indomethacin on intestinal secretion, prostaglandin E and cyclic cAMP. Evidence against a role for prostaglandins in cholera toxin–induced secretion. Prostaglandins **10:** 581–587.
15. BEUBLER, E., K. BUKHAVE & J. RASK-MADSEN. 1986. Significance of calcium for the prostaglandin E_2–mediated secretory response to 5-hydroxytryptamine in the small intestine of the rat *in vivo.* Gastroenterology **90:** 1972–1977.

16. GILROY, D.W., A. TOMLINSON & D.A. WILLOUGHBY. 1998. Differential effects of inhibitors of cyclooxygenase (cyclooxygenase 1 and cyclooxygenase 2) in acute inflammation. Eur. J. Pharmacol. **355:** 211–217.
17. RIENDEAU, D., M.D. PERCIVAL, S. BOYCE *et al.* 1997. Biochemical and pharmacological profile of a tetrasubstituted furanone as a highly selective COX-2 inhibitor. Br. J. Pharmacol. **121:** 105–117.
18. BEICHE, F., K. BRUNE, G. GEISSLINGER & M. GOPPELT-STRUEBE. 1998. Expression of cyclooxygenase isoforms in the rat spinal cord and their regulation during adjuvant-induced arthritis. Inflamm. Res. **47:** 482–487.
19. BEUBLER, E., G. KOLLAR, A. SARIA *et al.* 1989. Involvement of 5-hydroxytryptamine, prostaglandin E_2, and cyclic adenosine monophosphate in cholera toxin–induced fluid secretion in the small intestine of the rat *in vivo*. Gastroenterology **96:** 368–376.
20. BEUBLER, E. & G. HORINA. 1990. 5-HT_2 and 5-HT_3 receptor subtypes mediate cholera toxin–induced intestinal fluid secretion in the rat. Gastroenterology **99:** 83–89.

Effects of *Clostridium difficile* Toxins on Epithelial Cell Barrier

CHARALABOS POTHOULAKIS[a]

Division of Gastroenterology, Beth Israel Deaconess Medical Center,
Harvard Medical School, Boston, Massachusetts 02215, USA

ABSTRACT: *Clostridium difficile* is the primary agent responsible for many patients with antibiotic-associated diarrhea and almost all patients with pseudomembranous colitis following antibiotic therapy. *C. difficile* infection is the most frequent form of colitis in hospitals and nursing homes and affects millions of patients in the United States and abroad. The first event in the pathogenesis of *C. difficile* infection involves alterations of the indigenous colonic microflora by antibiotics, followed by colonization with *C. difficile*. *C. difficile* causes diarrhea and colitis by releasing two high molecular weight protein exotoxins, toxin A and toxin B, with potent cytotoxic and enterotoxic properties. Evidence presented here indicates that *C. difficile* toxins compromise the epithelial cell barrier by at least two pathophysiologic pathways, one involving disaggregation of actin microfilaments in colonocytes via glucosylation of the Rho family of proteins leading to epithelial cell destruction and opening of the tight junctions, whereas the other appears to involve early release of proinflammatory cytokines from intestinal epithelial cells probably via activation of MAP kinases. We speculate that cytokines released from intestinal epithelial cells in response to toxin A exposure will diffuse into the lamina propria and activate macrophages, enteric nerves, and sensory neurons to release SP, CGRP, and NT, which, in turn, interact with immune and inflammatory cells and amplify the inflammatory response. Dissection of this inflammatory cascade may help us understand the pathophysiology of inflammatory diarrhea caused by this important pathogen.

INTRODUCTION

Clostridium difficile is the primary agent responsible for many patients with antibiotic-associated diarrhea and almost all patients with pseudomembranous colitis following antibiotic therapy.[1] *C. difficile* infection is the most frequent form of colitis in hospitals and nursing homes and affects millions of patients in the United States and abroad.[2] The first event in the pathogenesis of *C. difficile* infection involves alterations of the indigenous colonic microflora by antibiotics, followed by colonization with *C. difficile*.[3] *C. difficile* causes diarrhea and colitis by releasing two high molecular weight protein exotoxins, toxin A and toxin B, with potent cytotoxic and enterotoxic properties. One of the features of *C. difficile* colitis is the acute inflammatory infiltrate in the colonic mucosa associated with destruction of

[a]Address for correspondence: Charalabos Pothoulakis, M.D., Division of Gastroenterology, Beth Israel Deaconess Medical Center, 330 Brookline Avenue, Boston, MA 02215. Voice: 617-667-1246; fax: 617-975-5071.
cpothoul@caregroup.harvard.edu

epithelial cells.[1] Studies with colonic cell lines and native human colon have shown the ability of these clostridial toxins to act on intestinal epithelial cells, alter the actin cytoskeleton, and cause epithelial cell damage and increased permeability of the tight junctions.[3] Animal studies demonstrated that *C. difficile* toxin A causes inflammatory diarrhea and increases intestinal permeability by a mechanism involving interactions between sensory neurons, neuropeptides, and inflammatory cells.[4] Thus, *C. difficile* toxins can activate several different cell types leading to alterations of the epithelial cell barrier and ultimately to intestinal secretion, colonocyte damage, and intestinal inflammation.

BIOLOGY OF *C. DIFFICILE* TOXINS

C. difficile is a gram-positive spore-forming anaerobic pathogen that mediates its intestinal effects by releasing the toxins A and B.[1] Cloning and sequence analysis predicts molecular masses of 308 and 270 kDa for toxin A and B, respectively, and amino acid analysis revealed significant homology between the two toxins.[5,6] This homology probably accounts for the similar biologic actions of these toxins in both intestinal and nonintestinal cells. Toxin A is an enterotoxin that stimulates fluid secretion and intestinal inflammation when administered in animal intestine.[7,8] Toxin A also possesses cytotoxic activity against cultured cells, as evidenced by rounding of intestinal and nonintestinal cells in culture[9,10] and hemagglutinating activity against rabbit erythrocytes.[11] Toxin A is a neurotoxin causing paralysis and death when injected parenterally to laboratory animals[12] and possesses excitatory action on submucosal secretomotor neurons *in vitro*.[13] By contrast, toxin B does not cause intestinal effects in animals,[8] probably owing to the absence of receptors for this toxin in animal intestine. Toxin B, however, is a potent cytotoxin causing rounding of cells at very low concentrations[14] and, like toxin A, is lethal when injected to animals.[12] Although toxin B is inactive in animal intestine, it is cytotoxic in human intestinal cell lines.[15,16] Both toxins can also directly activate human monocytes to release proinflammatory cytokines, such as IL-1β, TNF-α, and IL-6,[17,18] and toxin A stimulates human neutrophils, as evidenced by increased neutrophil calcium levels and chemotaxis.[19]

CELLULAR MECHANISM OF ACTION

Binding of *C. difficile* toxins to specific surface receptors appears to be an important step in the expression of the biologic actions of *C. difficile* toxins (reviewed in Ref. 20). After receptor binding *C. difficile* toxins are internalized into the cytosol by endocytosis via coated pits.[9] Once into the cell the catalytic action of these toxins is identical. The main toxin effect is cell rounding caused by disaggregation of filamentous action. Early studies showed that exposure of fibroblast monolayers to *C. difficile* toxin B resulted in increased levels of soluble or cytosolic actin, whereas filamentous actin was decreased.[14,21] This significant increase in the ratio of cytosolic actin was nearly completed prior to the onset of cell rounding.[14] Purified toxin B had no effect on either purified cytosolic actin or on polymerization of pure actin,[14] indicating that it may interact with another factor(s) important for actin polymerization and microfilament formation. Recent exciting results demonstrated that this actin effect is caused by a direct enzymatic modification of the Rho family of proteins,[22–24] small GTPases of the Ras superfamily, which regulate assembly of actin microfilaments. *C. difficile* toxins possess glucosyltransferase activity against Rho proteins as evidenced by their ability to enzymatically transfer a glucosyl residue from UDP glucose to threonine 37 of Rho, Rac, and Cdc-42.[22,23,25] This modification leads to disassembly of actin stress fibers, disruption of the actin-associated adhesion plaque proteins, and cell detachment and rounding.

EFFECTS ON INTESTINAL ELECTROPHYSIOLOGY AND MORPHOLOGY

Disaggregation of filamentous actin may be responsible for the dysfunction of tight junctions in animal and human intestinal epithelium following exposure to *C. difficile* toxins. Exposure of cultured human intestinal epithelial (T84) cell monolayers to toxins A and B diminished transepithelial resistance and increased epithelial cell permeability.[15,26] Flux studies demonstrated that the permeability defect is at the level of the intercellular tight junctions.[15,26] Both toxins altered the F actin comprising the perijunctional ring as evidenced by F actin condensation into discrete plaques in toxin A– and B–exposed cells.[15,26] Because the integrity of cytoskeletal actin is important in the regulation of tight junctional permeability, these studies indicate that the effects of toxins A and B on epithelial barrier function may be related to cytoskeletal alterations resulting from modification of the Rho family of proteins. Indeed, Nusrat *et al.*[27] provided strong evidence that the rho protein regulates tight junctions and organization of actin microfilaments in polarized T84 colonic cell monolayers.

Electrophysiologic studies using animal and human mucosa strips placed in Ussing chambers also pointed to a similar toxin effect in human colon and confirmed the studies with the colonic cell lines. Exposure of guinea pig ileum to toxin A resulted in increased permeability to radiolabeled mannitol and inulin and a substantial drop in transepithelial resistance.[28] Toxin A exposure also caused substantial structural alterations of villus tip cells paralleled by destruction of tight junctions and alterations of actin microfilaments.[28] Riegler *et al.*[29] showed that luminal exposure of colonic mucosa to toxins A and B significantly decreased potential difference, short-circuit current and mucosal resistance and stimulated a significant increase on serosal to mucosal permeability to [^{3}H]mannitol. Light and scanning electron microscopy demonstrated that *C. difficile* toxins caused a dramatic destruction of surface human colonocytes, while crypt colonocytes remained intact,[29] in agreement with previous studies showing binding of toxin A only on the villus tip and not on the crypt cells of the rabbit brush border.[30] Interestingly, lower toxin doses caused a patchy distribution of the colonocyte cell damage and, in some cases, single epithelial cell lesions were evident, reminiscent of cells undergoing apoptosis.[29] These findings may be related to the some *C. difficile*–associated pseudomembranous colitis cases characterized by a patchy distribution of pseudomembranes in the diseased colon.[1] Recent results also demonstrated that toxin A induces apoptosis and cell de-

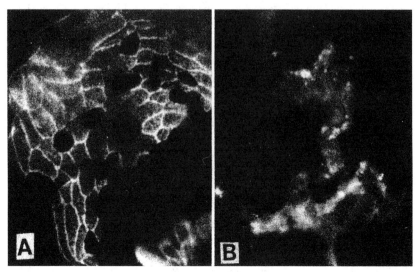

FIGURE 1. Toxin B causes disorganization of cellular F-actin. Fluorescent photomicrograph of human colonic mucosal sheets exposed to either buffer alone (**A**) or buffer containing 3 nM of purified toxin B (**B**). Cells were fixed and F-actin was stained with rhodamine-labeled phalloidin. Note the complete disorganization of F-actin and the clumping of F-actin within cells in toxin B–exposed tissues (**B**), as compared to the polygonal shape of F-actin in buffer-exposed tissues (**A**). magnification ×400. (Reproduced with permission from Riegler *et al.*[29])

tachment in human colonocyte cell lines and human colonic mucosal biopsies placed in organ culture.[31] Fluorescent microscopy studies demonstrated disorganization of the F actin fibers, as evidenced by loss of normal actin microfilament staining and condensation of F actin in toxin A– and B–exposed human colonic mucosa (FIG. 1).[29] Interestingly, and in agreement with studies using colonic cell lines, toxin B was more potent that toxin A in inducing these effects,[29] indicating that, in contrast to animal enterocolitis, toxin B may be involved in the pathogenesis of human *C. difficile* infection.

PROTECTIVE EFFECTS OF EPIDERMAL GROWTH FACTOR AND TREFOIL PEPTIDES

Based on previous studies suggesting the importance of epidermal growth factor (EGF) for the integrity and maintenance of the epithelial barrier function, several laboratories examined its role in epithelial cell damage in response to *C. difficile* toxins. Using colonic adenocarcinoma Caco-2 cells Lawrence *et al.*[16] showed that EGF altered the decline in transepithelial resistance caused by toxin B exposure. Studies by Riegler *et al.*[32] also demonstrated that the electrophysiologic changes following toxin A and B administration on human colonic mucosa were dramatically reduced by prior serosal application of EGF. EGF was also effective in diminishing epithelial cell damage and disruption of cytoskeletal F actin in response to toxins A and B.[32]

Because EGF participates in F actin polymerization and assembly of focal adhesions, the protective effects of EGF in toxin A– and B–induced epithelial cell damage and F actin disorganization may be related to the stabilization of cytoskeleton. Trefoil peptides, known to protect the intestinal epithelial cell barrier from various insults, were also effective in epithelial cell damage caused by toxin A. Kindon *et al.*[33] showed that addition of recombinant human intestinal trefoil factor in human colonic T84 cell monolayers inhibited toxin A–mediated increases in mannitol permeability in these cells. Interestingly, administration of the trefoil protein together with mucin glycoprotein further enhanced this protective effect.[33]

ROLE OF SENSORY NEUROPEPTIDES IN *C. DIFFICILE* TOXIN A–MEDIATED INTESTINAL INFLAMMATION AND MUCOSAL DAMAGE

One of the major characteristics in animal models of *C. difficile* toxin A–induced enterocolitis is the acute inflammatory infiltrate characterized by transmigration of neutrophils in the intestinal mucosa and enterocyte necrosis. Injection of toxin A into ileal or colonic loops of anesthetized animals caused mucosal neutrophil infiltration and increased intestinal secretion and mucosal permeability 1–4 h after toxin administration.[8,34–36] Neutrophil recruitment is a key event for the expression of full-blown enterocolitis in response to toxin A. For example, administration of a monoclonal antibody directed against the adhesion molecule CD18 to rabbits inhibited inflammation and epithelial cell damage in response to toxin A.[37] Although the mechanism(s) leading to this acute inflammatory response is not completely understood, it appears that activation of sensory nerves and release of sensory neuropeptides, such as substance P (SP) and calcitonin gene–related peptide (CGRP), are important in the mediation and amplification of the inflammatory signal. We recently showed that 30–60 min after toxin A administration into rat ileum SP and CGRP content are elevated in the cell bodies of the dorsal root ganglia in the spinal cord followed by increased levels of these peptides in the intestinal mucosa.[35,38] Administration of capsaicin, a neurotoxin that depletes the nerve endings of sensory neurons from SP and CGRP, or peripheral injection of either SP receptor or CGRP antagonists substantially reduced toxin A–induced enterocyte necrosis and mucosal permeability in rats.[38–41] Furthermore, SP (neurokinin-1) receptor expression was increased in the intestinal epithelium shortly after exposure to toxin A, and before mucosal inflammation and intestinal secretion to this toxin were evident.[42] Further support for the importance of SP and its receptor in the intestinal effects of *C. difficile* toxin A was provided by recent results showing that mice that genetically lack the SP (neurokinin-1) receptor have dramatically reduced intestinal responses to toxin A (FIG. 2).[43] Taken together, these studies suggest that SP-containing sensory nerves and the SP (neurokinin-1) receptor represent a major amplification system in toxin A–mediated acute intestinal inflammation and destruction of the epithelial cell barrier. Interestingly, increased SP receptor expression occurs also in the intestine of patients with *C. difficile* pseudomembranous colitis,[44] indicating that SP receptors may play an important role in the pathophysiology of *C. difficile* infection.

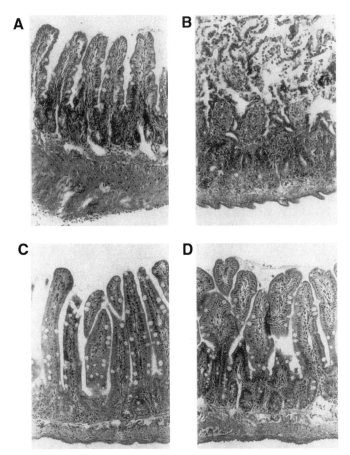

FIGURE 2. SP (NK-1) receptor–deficient mice are protected from histologic damage and ileal inflammation in response to toxin A. Toxin A or buffer (control) were injected into ileal loops of either NK receptor–deficient or wild-type mice. (**A**) Control wild-type mouse ileum 4 h after injection of buffer, showing normal villus epithelium. (**B**) Wild-type mouse ileum 4 h after injection of toxin A; there is severe necrosis and destruction of mucosal architecture and infiltration of lamina propria with inflammatory cells. (**C**) NK-1 receptor–deficient mouse ileum 4 hours after injection of buffer showing normal mucosal architecture. (**D**) NK-1 receptor–deficient ileum exposed to toxin A for 4 h shows lack of toxin-associated necroinflammatory changes. All sections were stained with hematoxylin and eosin. Original magnification: ×160. (Reproduced with permission from Castagliuolo *et al.*[43])

NEUROPEPTIDE–MAST CELL INTERACTIONS IN INTESTINAL INFLAMMATION

Several observations support a critical role for mucosal mast cells in the pathophysiology of toxin A enteritis. Shortly after toxin A exposure there is substantial degranulation of mucosal mast cells and release of the mast cell products rat mast cell protease II,[39,40] and leukotriene C4 and PAF.[45] Administration of the mast cell stabilizer ketotifen to rats substantially reduced the intestinal responses to toxin A.[45] Wershil *et al.*[46] showed that mast cell–deficient mice have diminished neutrophil recruitment and intestinal secretion after toxin A administration, providing direct evidence for involvement of mast cells in the intestinal effects of toxin A. Studies from our laboratory also supported a SP-dependent pathway in mast cell activation during the course of toxin A enteritis. For example, administration of a specific SP receptor antagonist to rats diminished mucosal mast cell activation in response to toxin A.[40] Results from Wershil *et al.*[46] also suggested a SP–mast cell interaction involved in neutrophil recruitment during toxin A enteritis. These results are in line with previous studies that provided anatomical and functional evidence for nerve–mast cell interactions in the intestinal mucosa.[47,48]

Recent results suggest that neurotensin (NT), a peptide primarily distributed in the brain and the GI tract, may be important in mast cell activation in different intestinal pathophysiologies. For example, intravenous injection of NT to rats results in elevated plasma histamine levels,[49] and exposure of serosal mast cells to NT *in vitro* stimulates histamine release via a receptor-mediated mechanism.[50] Furthermore, administration of a NT receptor antagonist to rats dramatically inhibited degranulation of colonic mast cells in response to immobilization stress.[51] We have recently reported that NT and its receptor were elevated in the rat colonic mucosa following toxin A administration, and pretreatment of rats with a specific NT receptor antagonist inhibited toxin A–induced changes in colonic secretion, mucosal permeability, and mast cell activation.[52] NT itself induced mast cell degranulation in colonic explants *in vitro,* and this effect was inhibited by the SP (neurokinin-1) receptor antagonist CP-96345.[52] These results suggest that neurotensin can serve as a proinflammatory peptide in the GI tract and activate mast cells via a SP-dependent mechanism.

ROLE OF SUBSTANCE P IN ACTIVATION OF INTESTINAL MACROPHAGES IN TOXIN A ENTERITIS

SP plays a critical role in activation of lamina propria macrophages during acute intestinal inflammation in response to toxin A. Thus, administration of a SP (neurokinin-1) receptor antagonist to rats resulted in diminished TNF-α levels released from lamina propria macrophages isolated from toxin A–injected ileal loops.[35] Rat intestinal macrophages are a source of SP during toxin A–mediated enteritis, and SP can also stimulate intestinal macrophages to release TNF-α,[35] a key cytokine involved in various types of intestinal inflammation. The presence of functional receptors for SP on intestinal macrophages suggests an autocrine and/or paracrine pathway in the regulation of TNF-α secretion in response to SP. Recent results from our laboratory also demonstrated that the intestinal levels of TNF-α mRNA and protein are diminished in SP (neurokinin-1) receptor–deficient mice as compared to controls following ileal toxin A administration,[43] providing direct evidence for the SP dependency of TNF-α secretion in the inflamed gut.

A critical question in the pathophysiology of toxin A enterocolitis involves the mechanism by which toxin A initiates inflammation leading to neuroimmune cell activation, release of proinflammatory cytokines, and epithelial cell destruction. One mechanism is by stimulating release of proinflammatory cytokines upon binding to the enterocyte. For example, we and others showed that MIP-2, and IL-8 are released from intestinal epithelial cells in response to toxin A administration.[31,53] In addition, toxin A can stimulate several MAP kinases 1–2 min after exposure of human macrophage/monocyte (THP-1) cells to toxin A cells.[54] Activation of MAP kinases was also associated with IL-8 release and expression of toxin A enterotoxicity in mouse ileum[54] and preceded Rho glucosylation, suggesting that MAP kinase activation may be independent of the Rho-glucosylating activity of toxin A.

In summary, the evidence presented here indicates that *C. difficile* toxins compromise the epithelial cell barrier by at least two pathophysiologic pathways. One pathway involves disaggregation of actin microfilaments in colonocytes via glucosylation of the Rho family of proteins leading to epithelial cell destruction and opening of the tight junctions. The other appears to involve early release of proinflammatory cytokines from intestinal epithelial cells probably via activation of MAP kinases. We speculate that cytokines released from intestinal epithelial cells in response to toxin A exposure will diffuse into the lamina propria and activate macrophages, enteric nerves, and sensory neurons to release SP, CGRP, and NT, which, in turn, will interact with immune and inflammatory cells and amplify the inflammatory response. Dissection of this inflammatory cascade may help us understand the pathophysiology of inflammatory diarrhea caused by this important pathogen.

REFERENCES

1. KELLY, C.P., C. POTHOULAKIS & J.T. LAMONT. 1994. *Clostridium difficile* colitis. N. Engl. J. Med. **330:** 257–262.
2. MCFARLAND, L.V., M.E. MULLIGUN, R.Y.Y. KWOK & W.E. STAM. 1989. Nosocomial acquisition of *Clostridium difficile* infection. N. Engl. J. Med. **320:** 204–210.
3. POTHOULAKIS, C. 1996. Pathogenesis of *Clostridium difficile*-associated diarrhoea. Eur. J. Gastroenterol. Hepatol. **8:** 1041–1047.
4. POTHOULAKIS, C., I. CASTAGLIUOLO & J.T. LAMONT. 1998. Neurons and mast cells modulate secretory and inflammatory responses to enterotoxins. News Physiol. Sci. **13:** 58–63.
5. DOVE, C.H., S.Z. WANG, S.B. PRICE *et al.* 1990. Molecular characterization of the *Clostridium difficile* toxin A gene. Infect. Immun. **58:** 480–488.
6. JOHNSON, J.L., C. PHELPS, L. BARROSO *et al.* 1990. Cloning and expression of the toxin B gene of *Clostridium difficile*. Curr. Microbiol. **20:** 397–401.
7. MITCHELL, T.J., J.M. KETLEY, S.C. HASLAM *et al.* 1986. Effect of toxins A and B of *Clostridium difficile* on rabbit ileum and colon. Gut **27:** 78–85.
8. TRIADAFILOPOULOS, G., C. POTHOULAKIS, M. O'BRIEN & J.T. LAMONT. 1987. Differential effects of *Clostridium difficile* toxins A and B on rabbit ileum. Gastroenterology **93:** 273–279.
9. FIORENTINI, C. & M. THELESTAM. 1991. *Clostridium difficile* toxin A and its effects on cells. Toxicon **29:** 543–567.
10. TUCKER, K.D., P.F. CARRING & T.D. WILKINS. 1990. Toxin A *of Clostridium difficile* is a potent cytotoxin. J. Clin. Microbiol. **28:** 869–871.
11. KRIVAN, H., C.F. CLARK, D.F. SMITH & T.D. WILKINS. 1986. Cell surface binding site for *Clostridium difficile* enterotoxin: evidence for a glycoconjugate containing the sequence Gal alpha 1–3Gal beta-4GlcNAc. Infect. Immun. **53:** 573–581.
12. ARNON, S.S., D.C. MILLS, P.A. DAY *et al.* 1984. Rapid death of infant rhesus monkeys injected with *Clostridium difficile* toxins A and B: physiologic and pathophysiologic basis. J. Pediatr. **101:** 34–40.
13. XIA, Y., C. POTHOULAKIS & J.D. WOOD. 1997. *Clostridium difficile* toxin excites enteric neurons and suppresses inhibitory noradrenergic neurotransmission in the submucous plexus of guinea-pig small intestine. Gastroenterology **112:** A1122.

14. POTHOULAKIS, C., L.M. BARONE, R. ELY *et al.* 1986. Purification and properties of *Clostridium difficile* cytotoxin B. J. Biol. Chem. **261:** 1316–1321.
15. HECHT, G., A. KOUTSOURIS, C. POTHOULAKIS, *et al.* 1992. *Clostridium difficile* toxin B disrupts the barrier function of T_{84} monolayers. Gastroenterology **102:** 416–423.
16. LAWRENCE, J.P., L. BREVETTI, R.J. OBISO *et al.* 1997. Effects of epidermal growth factor and *Clostridium difficile* toxin B in a model of mucosal injury. J. Pediatr. Surg. **32:** 430–433.
17. LINEVSKY, J.K., C. POTHOULAKIS, S. KEATES *et al.* 1997. IL-8 release and neutrophil activation by *Clostridium difficile* toxin A–exposed human monocytes. Am. J. Physiol. **273:** G1333–G1340.
18. FLEGEL, W.A., F. MULLER, W. DAUBENER *et al.* 1991. Cytokine response by human monocytes to *Clostridium difficile* toxin A and toxin B. Infect. Immun. **59:** 3659–3666.
19. POTHOULAKIS, C., R. SULLIVAN, D. MELNICK *et al.* 1988. *Clostridium difficile* toxin A stimulates intracellular calcium release and chemotactic response in human granulocytes. J. Clin. Invest. **81:** 1741–1745.
20. POTHOULAKIS, C. 1999. Enterotoxin A of *Clostridium difficile* and α-Gal epitopes. *In* Subcellular Biochemistry: Vol. 32, α-Gal and Anti-Gal. U. Galili & C. Avila, Eds.: 215–227. Kluwer Academic/Plenum Publishers. New York.
21. OTTLINGER, M.E. & S. LIN. 1988. *Clostridium difficile* toxin B induces reorganization of actin, vinculin, and talin in cultured cells. Exp. Cell Res. **174:** 215–229.
22. JUST, I., G. FRITZ, K. AKTORIES *et al.* 1994. *Clostridium difficile* toxin B acts on the GTP-binding protein Rho. J. Biol. Chem. **269:** 10706–10712.
23. JUST, I., M. WILM, J. SELZER *et al.* 1995. The enterotoxin from *Clostridium difficile* (ToxA) monoglucosylates the Rho proteins. J. Biol. Chem. **270:** 13932–13936.
24. DILLON, S., E. RUBIN, M. YAKUBOVICH *et al.* 1995. Involvement of ras-related Rho proteins in the mechanism of action of *Clostridium difficile* toxin A and B. Infect. Immun. **63:** 1421–1426.
25. JUST, I., J. SELZER, M. WILM *et al.* 1995. Glucosylation of *Rho* proteins by *Clostridium difficile* toxin B. Nature **375:** 500–503.
26. HECHT, G., C. POTHOULAKIS, J.T. LAMONT & J.L. MADARA. 1988. *Clostridium difficile* toxin A perturbs cytoskeletal structure and junction permeability in cultured human epithelial cells. J. Clin. Invest. **82:** 1516–1524.
27. NAUSTRAT, A., M. GIRY, J.R. TURNER *et al.* 1995. Rho protein regulates tight junctions and perijunctional actin organization in polarized epithelia. Proc. Natl. Acad. Sci. USA **92:** 10629–10633.
28. MOORE, R., C. POTHOULAKIS, J.T. LAMONT *et al.* 1990. *C. difficile* toxin A increases intestinal permeability and induces Cl⁻ secretion. Am. J. Physiol. **259:** G165–G174.
29. RIEGLER, M., R. SEDIVY, C. POTHOULAKIS, *et al.* 1995. *Clostridium difficile* toxin B is more potent than toxin A in damaging human colonic epithelium *in vitro*. J. Clin. Invest. **95:** 2004–2011.
30. POTHOULAKIS, C., R.J. GILBERT, C. CLADARAS *et al.* 1996. Rabbit sucrase-isomaltase contains a functional receptor for *Clostridium difficile* toxin A. J. Clin. Invest. **98:** 641–649.
31. MAHIDA, Y.R., S. MAKH, S. HYDE *et al.* 1996. Effect of *Clostridium difficile* toxin A on human intestinal epithelial cells: induction of interleukin 8 production and apoptosis after cell detachment. Gut **38:** 337–347.
32. RIEGLER, M., R. SEDIVE, T. SOGUKOGLU *et al.* 1997. Epidermal growth factor attenuates *Clostridium difficile* toxin A- and B-induced damage of human colonic mucosa *in vitro*. Am. J. Physiol. **273:** G1014–G1022.
33. KINDON, H., C. POTHOULAKIS, L. THIM *et al.* 1995. Trefoil peptides protect intestinal epithelial monolayer/barrier function: cooperative interaction with mucin glycoprotein. Gastroenterology **109:** 516–523.
34. TRIADAFILOPOULOS, G., C. POTHOULAKIS, R. WEISS *et al.* 1989. Comparative study of *Clostridium difficile* toxin A and cholera toxin in rabbit ileum. Role of prostaglandins and leukotrienes. Gastroenterology **97:** 1186–1192.
35. CASTAGLIUOLO, I., A.C. KEATES, B. QIU *et al.* 1997. Substance P responses in dorsal root ganglia and intestinal macrophages during *Clostridium difficile* toxin A enteritis in rats. Proc. Natl. Acad. Sci. USA **94:** 4788–4793.

36. QIU, B., C. POTHOULAKIS, I. CASTAGLIUOLO et al. 1999. Participation of reactive oxygen metabolites in Clostridium difficile toxin A–induced enteritis in rats. Am. J. Physiol. **276:** G485–G490.
37. KELLY, C.P, S.D. BECKER, J.K. LINEVSKY et al. 1994. Neutrophil recruitment in Clostridium difficile toxin A enteritis. J. Clin. Invest. **93:** 1257–1265.
38. KEATES, A.C., I. CASTAGLIUOLO, B. QIU et al. 1988. CGRP upregulation in dorsal root ganglia and ileal mucosa during Clostridium difficile toxin A–induced enteritis. Am. J. Physiol. **274:** G196–G202.
39. CASTAGLIUOLO, I., J.T. LAMONT, R. LETOURNEAU et al. 1994. Neuronal involvement in the intestinal effects of Clostridium difficile toxin A and Vibrio cholera enterotoxin. Gastroenterology **107:** 657–665.
40. POTHOULAKIS, C., I. CASTAGLIUOLO, J.T. LAMONT et al. 1994. CP-96,345, a substance P antagonist, inhibits rat intestinal responses to toxin A but not cholera toxin. Proc. Natl. Acad. Sci. USA **91:** 947–951.
41. MANTYH, C.R., T.N. PAPPAS, J.A. LAPP et al. 1996. Substance P activation of enteric neurons in response to intraluminal Clostridium difficile toxin A in the rat ileum. Gastroenterology **111:** 1272–1280.
42. POTHOULAKIS, C., I. CASTAGLIUOLO, S.E. LEEMAN et al. 1998. Increased substance P receptor expression in intestinal epithelial cells during Clostridium difficile toxin A enteritis in rats. Am. J. Physiol. **275:** G68–G75.
43. CASTAGLIUOLO, I., M. RIEGLER, S. NIKULASSON et al 1998. NK-1 receptor is required in Clostridium difficile–induced enteritis. J. Clin. Invest. **101:** 1547–1550.
44. MANTYH, C.R., J.E. MAGGIO, P.W. MANTYH et al. 1996. Increased substance P receptor expression by blood vessels and lymphoid aggregates in Clostridium difficile–induced pseudomembranous colitis. Dig. Dis. Sci. **41:** 614–620.
45. POTHOULAKIS, C., F. KARMELI, C.P. KELLY et al. 1993. Ketotifen inhibits toxin A–induced enteritis in rat ileum. Gastroenterology **105:** 701–707.
46. WERSHIL, B., I. CASTAGLIUOLO & C. POTHOULAKIS. 1998. Mast cell involvement in Clostridium difficile toxin A–induced intestinal fluid secretion and neutrophil recruitment in mice. Gastroenterology **114:** 956–964.
47. STEAD, R.H., M. TOMOIKA, G. QUINONEZ et al. 1987. Intestinal mucosal mast cells in normal and nematode-infected rat intestines are in intimate contact with peptidergic nerves. Proc. Natl. Acad. Sci. USA **84:** 2975–2979.
48. STEAD, R.H., M.F. DIXON, N.H. BRAMWELL et al. 1989. Mast cells are closely apposed to nerves in the human gastrointestinal mucosa. Gastroenterology **97:** 575–585.
49. CARRAWAY, R.E., D.E. COCHRANE, R. SALMONSEN et al. 1991. Neurotensin elevates hematocrit and plasma levels of the leukotrienes LTB4, LTC4, LTD4, and LTE4, in anesthetized rats. Peptides **12:** 1105–1111.
50. FELBDERG, R.S., D.E. COCHRANE, R.E. CARRAWAY et al 1998. Evidence for a neurotensin receptor in rat serosal mast cells. Inflamm. Res. **47:** 245–250.
51. CASTAGLIUOLO, I., S.E. LEEMAN, E. BARTOLAK-SUKI et al. 1996. A neurotensin antagonist, SR 48692, inhibits colonic responses to immobilization stress in rats. Proc. Natl. Acad. Sci. USA **93:** 12611–12615.
52. CASTAGLIUOLO, I., C-C. WANG, L. VALENICK et al. 1999. Neurotensin is a proinflammatory peptide in colonic inflammation. J. Clin. Invest. **103:** 843–849.
53. CASTAGLIUOLO, I., A.C. KEATES, C.C. WANG et al. 1988. Clostridium difficile toxin A stimulates macrophage inflammatory protein-2 production in rat intestinal epithelial cells. J. Immunol. **160:** 6039–6045.
54. WARNY, M., S. KEATES, A.C. KEATES et al. 2000. p38 MAP kinase acitvation by Clostridium difficile toxin A mediates monocyte necrosis, IL-8 production and enteritis. J. Clin. Invest. **105:** 1147–1156.

Ion Transport during Growth and Differentiation

J. VENKATASUBRAMANIAN, J. SAHI, AND M. C. RAO[a]

Department of Physiology and Biophysics, University of Illinois at Chicago, Chicago, Illinois 60612-7342, USA

ABSTRACT: The major function of the adult colon is to reabsorb fluid from the chyme. This ability to conserve salt and water is especially important in newborns, where reserves are small and diarrhea is frequent. Although much is known about regulation of Cl^- transport in the adult colon, postnatal changes in electrolyte transport are not well characterized. We have established an *in vitro* model to study colonic epithelial cells (colonocytes) at different stages of development. Primary cultures were isolated from newborn, weanling, and adult rabbit colon and properties such as growth and Cl^- transport characterized. The isolation procedure yielded a crypt-enriched population of cells, and the cell yield per gram mucosa increased with age. The colonocytes also showed an age-related decrease in attachment to extracellular matrix, with maximum attachment seen with Matrigel and collagen IV. The crypt enrichment was confirmed by demonstrating that the cell population was capable of transporting Cl^-, which was stimulated by agents such as forskolin and phorbol esters at all ages. Agents that increased intracellular cGMP, however, did not increase Cl^- transport at any age. It was interesting to observe that the secondary bile acid, taurodeoxycholate, stimulated Cl^- transport only in the adult but not newborn or weanling distal colonocytes. We have demonstrated that rabbit distal colonocytes can be kept viable in culture and transport Cl^- at all ages. However, the regulation of Cl^- transport changes during ontogeny and depends on the signaling pathway.

INTRODUCTION

To accommodate the varying nutritional needs at different periods in the life of a mammal, the gastrointestinal tract undergoes structural/functional adaptation. The most dramatic of these in normal physiology are the transitions needed at parturition and weaning. In the former, the source of nutrition switches from the maternal circulation to colostrum and breast milk, and in the latter from milk to the adult form of nutrition, which varies with the species. An additional contributor to the luminal milieu is the bacterial colonization that begins soon after birth, and, depending on the species, takes different times to reach adult levels. The structures and functions of different regions of the intestine undergo modifications to meet these changing demands. Thus, much attention has been paid to the ontogeny of macromolecular hy-

[a]Address for correspondence: Prof. M. C. Rao, Department of Physiology and Biophysics, University of Illinois at Chicago, 835 S. Wolcott Ave, m/c 901, Chicago, IL 60612-7342. Voice: 312-996-7884; fax: 312-996-1414.
 meenarao@uic.edu

drolases and nutrient transport in the small intestine[1] where maturation occurs in concert with weaning. In addition, elegant studies using transgenic methodology have provided information on the factors contributing to the development of cell lineages in the crypt-villus and cephalocaudal axes of the gut.[2] The picture emerging from such studies is that developmental regulation is multifactorial, species dependent, and complex.

Other demands on the developing intestine include a maturation of its motility function, immune system, and ability to cope with the movement of large amounts of fluid. For example, in the healthy adult the intestine secretes 1 liter of fluid, processes as much as 9 liters per day, reabsorbing 8.8 liters, and losing <200 mL in the stool.[3] To maintain luminal osmolarity and adapt to the changing nutritional demands during development, the intestine also has to adjust its fluid transport properties. The ability to conserve water and salt are of greater importance in the newborn, where reserves are small and diarrhea is frequent. In contrast to nutrient digestion and absorption, the developmental changes in ion transport have been examined only in a few studies in the small intestine[4] and to a lesser extent in the colon.[5] As compared to the adult, the neonatal small intestine exhibits greater transepithelial conductance and permeability to ions. Whereas the adult mammalian colon is chiefly responsible for the conservation of water and electrolytes, the colon plays a role in nutrient absorption in the neonatal animal.[6] It has been well characterized in the adult intestine that the balance of fluid absorptive and secretory processes is maintained by cell-specific ion transporters and their modulation by specific neurohumoral regulatory cascades. It is not known whether the transport processes and their regulatory mechanisms mature in concert during development or if they follow different time courses.

The majority of studies on postnatal developmental changes in colonic ion transport have focused on Na^+ transport.[7–10] Higher rates of Na^+ absorption as compared to the adult are seen in the rectum of human infants[10] and in the distal and proximal colon of suckling rabbits.[9] In distal colon, this appeared to be regulated by high circulating levels of aldosterone.[9] In the few studies conducted on colonic anion transport, preterm and newborn infants have been observed to have poorly developed anion exchange processes as compared to adults.[7,8] Potter et al.[11] demonstrated that net flux of Na^+ and Cl^- was similar in the neonatal and adult rabbit colon, but in the neonate, unlike the adult, Cl^- transport is not linked to Na^+ transport.

Aberrations in the regulation of fluid transport have severe pathophysiological consequences in both the small and large intestine,[12,13] and neonatal mammals, including humans, are particularly susceptible to diarrhea. It has been postulated that the regulatory mechanisms governing the balance of absorption:secretion are not fully geared to causing maximal absorption in the neonatal epithelium.[14] As the animal matures through weaning into adulthood, the balance is shifted so that absorption predominates. A number of factors including increased tissue permeability, decreased absorptive processes, and increased tissue receptors for enterotoxins have been suggested to contribute to the susceptibility of the neonate. Conversely, it is conceivable that the developing colon may have its own "protective" mechanisms to prevent excessive fluid loss. The precise cellular basis for the susceptibility of the neonate or its ability to protect itself has not been investigated in detail. However, the relatively few ontogenic studies of ion transport have provided some interesting insights into regulatory mechanisms that cater to the unique demands on the intestine during development.[14–16] Another emerging factor to consider, based largely on studies in adult mammals, is that there are distinct differences in ion transport and its regulation in the various segments of the colon. Our laboratory has been investigating the ontogeny of Cl⁻ transport and its regulation in the rabbit colon.[17–21] As elaborated in the following sections, our findings suggest that the developing colon has evolved some protective mechanisms that make it refractory to the onslaught of humoral or microbe-derived secretagogues. This review focuses on our findings on the distal colon.

THE MODEL

The normal adult mammalian colon exhibits net absorption of Na^+, Cl^-, short-chain fatty acids, and fluid and net secretion of K^+ and HCO_3^-.[12] However, Cl^- secretion is important for fluid movement. A structure–function dogma applicable to all mammals has been that secretory and absorptive processes are spatially separated in the crypt and surface cells, respectively.[22,23] Recent evidence suggests that there may be a gradation of transporters from the crypt base to the surface[24] with intricate intracellular tuning of individual transporters to prevent absorption and secretion from occurring in the same cell simultaneously.[25] The integrity of epithelium and colonocyte function can be regulated by luminal factors, such as bacterial metabolites and waste products, and by submucosal factors, such as neurohumoral and immunomodulators.[12,26] Distal colonic ion transport has been extensively studied in a number of mammalian species, especially in the rabbit and in the rat. Although nutrient and Na^+ transport has been examined in the developing rat, some intriguing age-dependent changes in the regulation of Cl^- transport have been reported in the rabbit.[9,11] (See below and especially REGULATION OF COLONIC CHLORIDE TRANSPORT AT DIFFERENT AGES.)

Studies of the distal colon in a variety of adult mammals, including the rabbit, have contributed greatly to the current view of colonic Cl^- secretion. In the adult rabbit distal colon, the bulk of net Cl^- secretion occurs via conductive pathways.[27] In a prototypic distal colonic secretory cell, Cl^- enters via the bumetanide-sensitive Na^+-K^+-$2Cl^-$ cotransporter, NKCC-1, located on the basolateral membrane, and accumulates in the cell above its electrochemical equilibrium.[12] Recycling across the basolateral membrane of Na^+ by the Na^+/K^+-ATPase pump, and K^+ via the K^+ channel, maintains a favorable electrochemical gradient for Cl^- exit via the apical membrane.[12] In the unstimulated cell the apical membrane is relatively impermeable to Cl^-, but in response to secretagogues, apical membrane Cl^- channel(s) are activated, resulting in Cl^- secretion.[26,28] The major Cl^- channel on the apical membrane is the cystic fibrosis gene product CFTR.[29] It remains to be established if all these pathways are operative in the colonocyte during development.

To delineate the cellular basis for any age-related differences in Cl^- transport, we have developed and partially characterized a primary culture model of crypt-enriched colonocytes isolated from newborn (7 days), weanling (28 days), and adult (6 months) rabbits. The colonocytes are isolated by limited enzymatic digestion followed by a series of low-speed centrifugations. Morphological examination of the

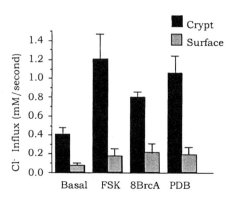

FIGURE 1. Heterogeneity in chloride permeability between the crypt and surface adult colonocytes. The colonocytes were isolated by enzymatic digestion and cultured for 24 h. Cl⁻ transport was measured by MQAE fluorescence (see text). Agents tested were forskolin (Fsk: 1 μM); 8-Br-cAMP (8BrcA: 100 μM); phorbol dibutyrate (PDB: 1 μM) ($n = 3$).

pellet fractions shows crypt-like structures, whereas the supernatant fractions comprise single cells, presumably of surface origin. To determine if these two cell fractions are functionally distinct, the "crypt-" and "surface-enriched" fractions from adult colonic mucosa were cultured separately for 24 hours. They maintained 90% viability over this period. Chloride transport was assessed in these cells in a suspension using the halide-sensitive fluorescent probe MQAE (6-methoxy-quinolyl acetoethyl ester),[30] in the presence and absence of the Cl⁻ channel blocker diphenylamine 2-carboxylate (DPC: 50 μM) and the Na^+-K^+-$2Cl^-$ cotransport inhibitor furosemide (10 μM). As shown in FIGURE 1, basal Cl⁻ transport in the surface cells was 50–70% lower than that found in crypt cells. Similarly, although both cell populations responded to the cAMP-dependent secretagogues, 8-Br-cAMP (10 μM), forskolin (1 μM), and the protein kinase C (PKC) activator phorbol dibutyrate (1 μM), the responses in the crypt population were much greater. These data suggest that our procedure allows us to isolate a crypt-enriched fraction and support the concept that crypt rather than surface cells are the major sites of Cl⁻ transport.[19,30,31] This reductionist cell model allows us initially to define Cl⁻ transport properties intrinsic to colonocytes, devoid of subepithelial influences. Later approaches to study increasingly complex subepithelial interactions could add mesenchymal elements to this preparation.

AGE-DEPENDENT CHANGES IN COLONOCYTE
GROWTH AND FUNCTION

A simple morphological explanation for the susceptibility of the neonate to diarrhea may be that these animals have an increased crypt:surface cell, and therefore a secretory:absorptive cell distribution. However, this is not the case. On the contrary, we[19] and others[32] have noted that crypt depth increases with age, from the neonate to the weanling to the adult. Parallel with the morphological increase in crypt depth,

TABLE 1. Newborn, weanling and adult rabbit distal colonocytes: Isolation buffer, cell yield, attachment to plastic (PL) versus collagen IV (Co IV) and transepithelial resistance (R^t)[19]

	Newborn (7 days)		Weanling (28 days)		Adult (6 months)	
Enzyme cocktail						
Pronase (%)	0.05		0.05		0.1	
Collagenase IV (%)	0.015		0.015		0.03	
DTT^a (%)	0.023		0.023		0.07	
Cell yield ($n = 5$) (cells/gm mucosa $\times 10^6$)	9.5 ± 0.5		22 ± 1.0		32 ± 1.5	
	PL	Co IV	PL	Co IV	PL	Co IV
% Attachment ($n = 5$)	21	88	8	78	4	15
R^t ($\Omega \cdot cm^2$; $n = 2$)	130		132		142	

^{a}Dithiothreitol.

and as shown in TABLE 1, the yield of distal colonocytes per gram mucosa increases with age.[19] These results confirm that in our preparation the pellet fraction is enriched for crypts, and this is the case for all three age groups. If there were a differential distribution with age, there should have been no correlation between crypt depth and numbers of cells harvested per gram mucosa.

As shown in TABLE 1, an enzymatic cocktail comprising pronase, collagenase IV, and dithiothreitol was optimal for the isolation of rabbit distal colonocytes, although the concentrations needed for the newborn and weanling[19] were lower than those required for the adult colon.[31] In all three age groups the isolated cells were epithelial in origin, as determined by cytokeratin staining,[19,31] and maintained >90% viability for 24 hours. When grown on plastic, the newborn and weanling colonocytes show a twofold increase in cell number, DNA, and protein content over 48 hours. By contrast, for all three parameters the adult colonocytes revealed only a 10% increase.

The colonocytes also show age-related differences in attachment to extracellular matrices, ranging from plastic to fibronectin, laminin, collagen I, collagen IV, and Matrigel.[19] Colonocytes of all age groups showed maximal attachment to Matrigel (data not shown) and collagen IV (TABLE 1). However, there was an age dependency to the degree of attachment, with newborn and weanling colonocytes showing >80% attachment and adult cells 15% attachment to collagen IV. Thus, the overall tendency of colonocytes to attach to matrices declines as the age advances, and colonocytes of older animals require more complex biomatrices for attachment. Matrigel was also seen to cause a small (15%) but significant increase in cell proliferation in the newborn and weanling colonocytes.[19]

In a preliminary study we determined whether these colonocytes grown on a collagen IV matrix were capable of forming resistive monolayers. By six days postplating, the transepithelial resistance (R^t) in all three age groups was approximately 120 ohm$\cdot$cm^2, as shown in TABLE 1. These values are similar to those reported for the neonatal colon mounted in Ussing chambers.[11] The functional responses of these monolayers remains to be studied.

To determine if there were any age-related differences in Cl⁻ transport manifested in the cultured colonocytes, Cl⁻ transport was studied using the fluorescent probe MQAE. Cells were studied in 24-hour cultures, either in suspension or when attached to a collagen IV matrix. In both preparations, at all age groups, both DPC-sensitive and furosemide-sensitive Cl⁻ transport could be measured, suggesting that the colonocytes possess both Cl⁻ channels and NKCC at an early age.[17] More recently, we have confirmed the presence of CFTR and NKCC-1 transcripts in newborn, weanling, and adult distal colonocytes.[18] Although there were no qualitative differences, the magnitude of basal Cl⁻ transport was the least in the newborn and progressively increased in the weanling and adult-attached colonocytes.[17] This age dependence was not as apparent in cells in suspension,[21] thereby underscoring the importance of the extracellular matrix in influencing cell function.

Thus, there are distinct age-related differences in the structural architecture of the colon. Clearly the intrinsic properties of the colonocytes and the underlying matrix must be contributing to these differences. It is of interest that some of these inherent, age-dependent features are retained when the colonocytes are maintained in short-term cultures. Over a period of 24–72 hours, the neonatal > weanling >>> adult colonocytes show an ability to proliferate and attach to extracellular matrices.[19] The adult colonocytes require more complex biomatrices for attachment. Although the matrix clearly influences the magnitude of growth, attachment, and function of the colonocytes, it does not appear to have qualitative influences; that is, even colonocytes in suspension have an ability to transport Cl⁻ and proliferate in culture.[17,19,20,21,30,33] In our characterization of the developing colon, neither the innate properties of the colonocytes nor the structural architecture of the neonatal colon would indicate that there is a preponderance of secretory machinery. On the contrary, the neonatal colon seems to be geared to lower levels of secretion. Is there perchance a greater decrease in the absorptive processes as well, thereby shifting the balance towards net secretion? In the rat distal colon Na^+ absorption shows a dramatic shift from being conductive in the neonate to an electroneutral process in the adult.[34] In the rabbit, Na^+ transport needs to be studied at the cellular level, inasmuch as studies in colonic epithelial sheets are conflicting, suggesting either an enhancement[9] or no change[11] in Na^+ absorption in the neonate as compared to the adult. There is also conflicting evidence as to whether neonatal Na^+ absorption is amiloride sensitive[9] or amiloride insensitive.[11] It is tempting to postulate that overall the basic characteristics of the neonatal and weanling colon are geared towards protecting the tissue from excess secretion.

REGULATION OF COLONIC CHLORIDE TRANSPORT
AT DIFFERENT AGES

Colonic Cl⁻ secretion is stimulated by neurohumoral, bacterial, or pharmacological secretagogues.[12,26] Secretory stimuli may activate one or more of the following: the apical Cl⁻ channel(s) directly; basolateral K^+ efflux, thereby hyperpolarizing the cell; and/or NKCC.[12,35] The accepted dogma[26] is that secretagogues act via second messengers, and their specific protein kinases, to modulate epithelial ion transport.[36,37] The picture is complicated by "cross talk" between the cascades, the net outcome of which controls the duration and amplitude of the final response. In the rabbit distal colon, agents acting via either cAMP [e.g., cholera toxin, prostaglandin E1 (PGE_1), PGE_2, and forskolin] or the Ca^{2+}/phospholipid/PKC pathway (e.g., histamine, muscarinic agonists, serotonin, and neurotensin) elicit Cl^- secretion.[12,35] Bile acids and hydroxy fatty acids are potent secretagogues in the adult rabbit colon, and both cAMP and Ca^{2+} are implicated in their action.[38] In the rat[39] and human[40] distal colon, but not in the rabbit, agents that increase cGMP, such as the heat-stable enterotoxin STa,[41] also stimulate Cl^- transport. To counterbalance the effects of the secretagogues, modulators such as PYY, norepinephrine, and somatostatin decrease net Cl^- secretion.[26] Evidence for multiple cascades, their cross talk interactions, and for tissue and cell-type differences is growing exponentially, and their implications as sites of developmental regulation remain to be explored. We therefore first examined whether the different signaling pathways were patent in the distal colon of the rabbit from an early age.

Effects of cAMP-Dependent Secretagogues

Cyclic AMP is a potent stimulus of Cl^- secretion in most, if not all, epithelia. The cAMP phosphodiesterase inhibitor theophylline stimulated Cl^- transport in intact colonic tissue from neonatal and adult rabbits.[11] These results suggested that a cAMP-responsive system was present at an early age but did not provide any evidence whether the proximal steps in cAMP signaling—namely, the receptor and cyclase—were also patent in the newborn. Therefore, using our 24-hour cell culture model, we examined if there were age-related changes in the cAMP signal transduction cascade in the neonate, weanling, and adult colonocytes.[17] We tested the effects of the following agents at concentrations known to elicit maximal responses in adult rabbit distal colonocytes: PGE_1, as a receptor-mediated activator of cAMP; forskolin, as a direct activator of adenylate cyclase; and 8-Br-cAMP, as a direct activator of cAMP-dependent protein kinase. An example of this is shown in FIGURE 2. As in the adult colonocytes, 1 μM forskolin elicited maximal transport responses in the neonatal and weanling colonocytes.[17] In data not shown here, PGE_1 and 8-Br-cAMP also stimulated DPC- and furosemide-sensitive Cl^- transport in all three age groups.[17] Thus, all the major steps of the cAMP signal transduction cascade are operative from an early stage of development.

Effects of cGMP-Dependent Secretagogues

The most potent stimulators of the cGMP-dependent signal transduction pathway in the intestinal mucosa are STa[42] and its endogenous homologues guanylin and uroguanylin.[43] These peptides act by binding to their receptor, guanylate cyclase C (GCC), which is localized to the apical membrane of enterocytes and colonocytes. Small intestinal epithelia have little soluble guanylate cyclase, and therefore activators of those enzymes, such as nitric oxide, do not directly stimulate Cl^- transport in enterocytes.[41] Both STa and guanylin are potent stimulators of Cl^- secretion in the small and large intestine, and the distribution of GCC has been studied in a number of intestinal tissues and species.[43] Age-related differences have been observed in the cGMP signaling cascade in the rat, pig, and human colon.[15,16,44,45] In all three spe-

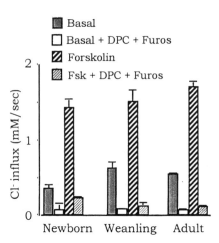

FIGURE 2. Basal and forskolin-stimulated chloride transport in newborn (7 days), weanling (28 days), and adult (6 months) distal colonocytes in suspension, in the presence or absence of inhibitors to Cl⁻ channel (DPC: 10 µM) and NKCC-1 (Furosemide (Furos): 50 µM) ($n = 4$).

cies, the neonate has been shown to have very high levels of STa receptors as determined by [¹²⁵I]STa binding. These values decline with age and in the human reach adult levels by two years of age.[45] Although this increase in STa receptors has not been directly linked to increased Cl⁻ transport, it has been postulated that the high incidence of STa receptors in the neonate may account for the susceptibility of the neonate to diarrhea. Despite the decline in STa receptors, the adult rat[39] and the human distal colon[40] secrete Cl⁻ in response to STa and/or cGMP.

In contrast to the cAMP signaling cascade, where an increase in intracellular cAMP results in an activation of protein kinase A (PKA), which in turn stimulates function, it appears that there may be at least three routes for cGMP activation of function in the colon (Fig. 3). The first is the direct "traditional" route of cGMP's activating a specific isoform of cGMP-specific protein kinase, PKGII, which then modulates transport. De Jonge and colleagues[39] have demonstrated this to occur in the rat small intestine and proximal colon. In the rat distal colon, cGMP employs a different route of activation. It appears to inhibit a cAMP-specific phosphodiesterase (PDE), which then allows cAMP concentrations to rise and thereby activate transport.[46] Studies on the human colon carcinoma cell line, T₈₄, reveal that cGMP can employ yet another route of activation. T₈₄ cells possess high levels of STa receptors, can generate high levels of cGMP, and show stimulated Cl⁻ transport in response to STa; but some batches fail to respond to 8-Br-cGMP.[47] Whereas 8-Br-cGMP is a fairly selective activator of PKGs, the endogenous cGMP, at sufficiently high concentrations, can activate PKAs. Indeed these T₈₄ cells were found to lack cGMP-specific protein kinases, and it was demonstrated that the cGMP levels generated by STa were indeed acting via PKA.[47] It must be noted that not all batches of T₈₄ respond in this fashion. Lin et al.,[48] and more recently our laboratory,[49] have observed that cGMP analogues, selective for cGMP and not cAMP-specific protein kinases, such

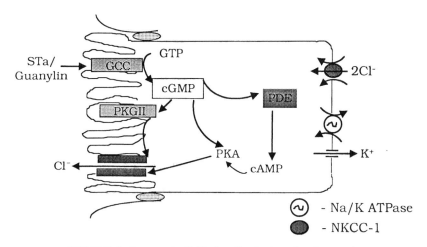

FIGURE 3. Routes of cGMP signaling in the colon (see text).

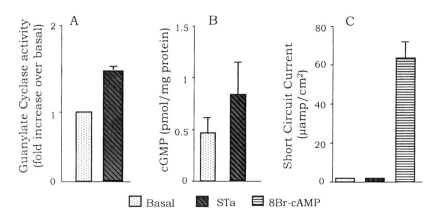

FIGURE 4. Effect of STa on (**A**) guanylate cyclase activity, (**B**) cGMP content, and (**C**) short circuit current in adult rabbit distal colonic epithelia. Data obtained from Ref. 41.

as 8-Br-cGMP and chloro-phenyl-thio-cGMP (cpt-cGMP), are capable of stimulating Cl⁻ transport in T_{84} cells. This underscores the variation encountered even in stable, transformed cell lines.

The adult rabbit distal colon has presented an interesting enigma with respect to the action of cGMP. As summarized in FIGURE 4,[41] the colonic mucosal membranes possess an STa-stimulable guanylate cyclase (FIG. 4, Panel A). This response is not the *in vitro* artifact of a cell-free preparation, inasmuch as epithelial tissue sheets exposed to STa also show an increase in tissue cGMP content (FIG. 4, Panel B). However, the tissue fails to secrete Cl⁻ in response to 8-Br-cGMP (not shown) or STa, as

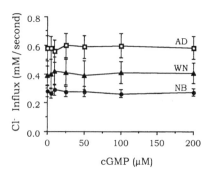

FIGURE 5. Effect of different concentrations of 8-Br-cGMP on chloride transport in newborn (NB), weanling (WN) and adult (AD) distal colonocytes ($n = 4$). The colonocytes were grown on a collagen IV matrix. (Reproduced with permission from Desai *et al.*[17])

determined by transepithelial short circuit current, while demonstrating a robust response to cAMP (FIG. 4, Panel C). Thus, although the adult distal colon has the machinery to generate cGMP, it appears to be lacking a critical step in cGMP signaling. We questioned whether this mechanism is absent in this tissue at all ages or whether, perhaps, as in the case of the rat and human, the neonatal rabbit has high levels of STa receptors that disappear rather than decline in the adult. Therefore, we compared the effects of different doses of STa and 8-Br-cGMP on Cl⁻ transport in the neonate, weanling, and adult colonocytes.[17] As shown in FIGURE 5, 8-Br-cGMP and STa (data not shown) did not stimulate Cl⁻ transport at any age in the distal colon. That the hydrolysis-resistant 8-Br-cGMP also did not elicit a response suggests that the lack of an effect of STa is not due to a rapid degradation of cGMP. It remains to be determined if the rabbit colon is lacking a cGMP-inhibitable PDE or a PKG.

Effects of Bile Acids

A fascinating example of coordinated regulation between the small and large intestine to combat the changing milieu of the developing gut is that of bile acids.[5,14,50] In the adult, the majority of bile acids are reabsorbed in the distal ileum by Na⁺-dependent bile acid transporters (N-BAT) and recycled to the liver via the enterohepatic circulation.[51] Excessive loss of bile acids to the colon results in salt secretion and choleretic diarrhea.[52] Depending on the species, the N-BAT in the ileum is expressed in the first few weeks after birth.[53] Yet the incidence of choleretic diarrhea is low in the newborn. Potter *et al.*[11] and O'Loughlin *et al.*[9] demonstrated that whereas bile acids stimulate electrogenic Cl⁻ secretion in the adult distal rabbit colon, they fail to do so in the neonate, accounting for the refractoriness.[5,14] These studies were conducted in intact distal colon preparations comprising the mucosa and underlying muscle. In such preparations, the bile acid taurodeoxycholate (TDC) was found to increase intracellular cAMP via a Ca²⁺-dependent mechanism in the adult, but not in the neonate.[54,55] However, the neonate exhibited cAMP-dependent Cl⁻ secretion and TDC-stimulated K⁺ secretion.[56] To determine if these age-related differences were due to the influence of subepithelial elements or if they were properties of the

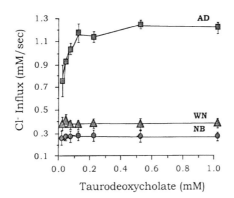

FIGURE 6. Effect of different concentrations of TDC on chloride transport in newborn (NB), weanling (WN), and adult (AD) distal colonocytes ($n = 3$). The colonocytes were grown on a collagen IV matrix. (Reproduced with permission from Desai *et al.*[17])

colonocytes per se, we examined the effects of TDC on Cl⁻ transport in our primary culture model. In addition to the adult and neonate we were particularly interested in examining the weanling animal, because the ileal bile acid transporters are present in the weanling and therefore excess bile acids in the colon should not be a potential threat.[17] As shown in FIGURE 6, TDC failed to stimulate Cl⁻ transport in weanling and neonatal colonocytes attached to a collagen IV matrix, over a concentration range of 25 µM–1 mM. By contrast, 50–100 µM TDC was sufficient to stimulate Cl⁻ transport in adult colonocytes.

The mechanisms of bile acid action have been studied in a variety of colonic models, including cell lines and intact tissue preparations. A number of pathways have been implicated, including cAMP, histamine, PGEs, intracellular Ca^{2+}, and PKC.[38,55,57–60] In the rabbit colon, Potter *et al.*[55] suggested a role for Ca^{2+} based on the inhibitory effects of TMB-8, an inhibitor of intracellular Ca^{2+} mobilization. However, intracellular Ca^{2+} was not directly measured. In a preliminary report we have demonstrated that TDC was capable of increasing intracellular Ca^{2+} in the adult, but not in the neonatal and weanling colonocytes.[20]

Effects of Activation of the Protein Kinase C Pathway

The PKC signal transduction pathway is known to be involved in secretagogue action. In the colon carcinoma cell line HT-29, PKC has been implicated in the action of bile acids.[59] Therefore we determined whether the inability of bile acids to stimulate Cl⁻ transport in the neonate and weanling may be due to an immaturity in the PKC cascade. Tumor promotors such as phorbol dibutyrate (PDB) are potent activators of PKC; its effects on adult, weanling, and newborn distal colonocytes were examined. As shown in FIGURE 7, PDB stimulated both DPC and furosemide-sensitive Cl⁻ transport in all three age groups, indicating that the PKC arm of signaling was present from an early age.

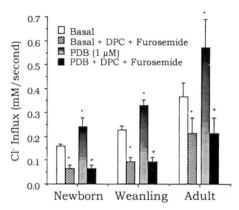

FIGURE 7. Basal and PDB-stimulated chloride transport in newborn, weanling, and adult distal colonocytes. Values represented as mean ± SEM ($n = 6$). *$p < 0.05$, different from basal; +$p < 0.05$, different from PDB treated.[17]

CONCLUSIONS AND PERSPECTIVES

Impaired fluid and salt absorption, enhanced water and salt secretion, and increases in colonic mucosal permeability are associated with diseases ranging from toxigenic secretory diarrheas to complex multifactorial diseases, such as inflammatory bowel diseases and necrotizing enterocolitis. Although significant inroads have been made in understanding how enterotoxins, neuromodulators, and immunomodulators act in the adult,[3,61,62] similar information in the developing mammal is lacking. Our observation that Cl⁻ transport regulation is age dependent underscores the importance of exploring these pathways.

From our studies at least two major regulatory pathways of Cl⁻ transport are worthy of pursuit, both for segmental and age-related differences: first the cGMP signaling cascade, and second the bile acid–signaling cascade. Elucidation of these pathways will provide a key framework for dissecting the complexities underlying normal development and disease states.

In data not shown here,[18,20,21] we have extended our primary culture methodology to studying proximal colonocytes and have found some intriguing ontogenic changes with the cGMP signaling cascade in that tissue. The physiological relevance of the GCC-cGMP signaling cascade is not fully understood, but it may play a larger role in colonic health and disease than that suggested by travelers' diarrhea. Bile acid–stimulated diarrhea is strictly a colonic phenomenon, and the whole question of age-related differences in stimulation of Cl⁻ transport by Ca^{2+}-dependent secretagogues is unexplored. Interestingly regional differences in Ca^{2+}-dependent Cl⁻ secretion have been reported in a nutrient-deprivation model, an oft-used paradigm for development. In the starved mouse, Ca^{2+}-dependent, but not cAMP-dependent, Cl⁻ secretion is greatly enhanced in the proximal and distal colon but not in the midcolon.[63] In the neonatal feline, gastric smooth muscle contraction in response to cholinergic agonists is attenuated. This appears to be due to lower IP_3 receptor numbers and therefore an impairment in the release of $[Ca^{2+}]_i$.[64] Our preliminary findings suggest that pharmacologically increasing $[Ca^{2+}]_i$ with the ionophore A23187 stimulates Cl^- transport in the distal colon at all ages,[20] suggesting that the distal steps in the Ca^{2+} signaling cascade are patent.

With the caveat that animal models are, at best, paradigms rather than replicas of the human and with the known limitations of performing developmental studies in the human, animal models such as the rabbit are excellent for defining ontogeny of colonic Cl^- transport from the level of the colonocyte to the intact mucosa. Our primary culture model is a powerful tool that allows us to delineate the cellular signaling mechanisms underlying these differences. Future studies need to explore whether the longer-term cultures, which form resistive monolayers, retain the age-related differences observed in the freshly isolated tissue and in the 24-hour cultures. If so, complex questions, such as manipulation of hormonal milieu to achieve gain/loss of age-related functions, can be addressed. An intriguing observation in the intact neonatal colon is that there is increased unidirectional Cl^- flux when compared to the adult.[11] This suggests an increase in paracellular permeability and, if this is so, needs to be examined in detail. Another major avenue to be pursued is to examine the ontogenic changes occurring in Na^+ transport in the surface cells. Clearly, although these reductionist models will help delineate the cellular and molecular events, ultimately they need to be examined with the backdrop of mesenchymal interactions. This could be achieved by coculture studies and finally related back to intact epithelial preparations. Such a systematic elucidation will provide a key framework for dissecting the complexities underlying normal development and disease states.

ACKNOWLEDGMENTS

This work was supported by National Institutes of Health grants (R01DK-46910; F32DK-08849) and the College Research Board, University of Illinois at Chicago.

REFERENCES

1. HENNING, S.J., D.C. RUBIN & R.J. SHULMAN. 1994. Ontogeny of the intestinal mucosa. *In* Physiology of the G. I. Tract. L.R. Johnson, Ed.: 571–610. Raven Press. New York.
2. SIMON, T.C. & J.I. GORDON. 1995. Intestinal epithelial cell differentiation: new insights from mice, flies and nematodes. Curr. Opin. Genet. Dev. **5:** 577–586.
3. CHANG, E.B. & M.C. RAO. 1994. Intestinal water and electrolyte transport: mechanisms of physiological and adaptive responses. *In* Physiology of the G.I. Tract. L.R. Johnson, Ed.: 2027–2081. Raven Press. New York.
4. GHISHAN, F.K. 1989. Electrolyte fluxes in the small intestine during development. *In* Human Gastrointestinal Development. E. Lebenthal, Ed.: 503–519. Raven Press. New York.
5. POTTER, G.D. 1989. Development of colonic function. *In* Human Gastrointestinal Development. E. Lebenthal, Ed.: 545–558. Raven Press. New York.
6. HENNING, S.J., D.C. RUBIN & R.J. SHULMAN. 1994. Ontogeny of the intestinal mucosa. *In* Physiology of the G. I. Tract. L.R. Johnson, Ed.: 571–610. Raven Press. New York.
7. HEATH, A. & P.J. MILLA. 1983. Development of colonic transport in early childhood: implications for diarrhoeal disease. Gut **24:** A977.

8. JENKINS, H.R. & P.J. MILLA. 1988. The development of colonic transport mechanisms in early life: evidence for reduced anion exchange. Early Hum. Dev. **16:** 213–218.
9. O'LOUGHLIN, E.V., D.M. HUNT & D. KREUTZMANN. 1990. Postnatal development of colonic electrolyte transport in rabbits. Am. J. Physiol. **258:** G477–G453.
10. JENKINS H.P., T.R. FENTON, M. SAVAGE *et al.* 1987. Development of colonic chloride transport processes in infancy: the influence of aldosterone. Gastrenterology **92:** A1453.
11. POTTER, G.B. & S.M. BURLINGAME. 1986. Ion transport by neonatal rabbit distal colon. Am. J. Physiol. **250:** G754–G759.
12. BINDER, H.J. & G.I. SANDLER. 1994. Electrolyte transport in the mammalian colon. *In* Physiology of the G.I. Tract. L.R. Johnson, Ed.: 2133–2171. Raven Press. New York.
13. NOCERINO, A., M. IAFUSCO & S. GUANDALINI. 1995. Cholera toxin–induced small intestine secretion has a secretory effect on the colon of the rat. Gastroenterology **108:** 287–288.
14. POTTER, G.D. 1990. Intestinal development and regeneration. Hosp. Pract. **25:** 131–144.
15. COHEN, M.B., A. GUARINO & R.A. GIANELLA. 1988. Age-related differences in receptors of *Escherichia coli* heat-stable enterotoxin in the small and large intestine of children. Gastroenterology **94:** 367–373.
16. MEZOFF, A.G., N.J. JENSEN & M.B. COHEN. 1991. Mechanism of increased susceptibility of immature and weaned pigs to *Escherichia coli* heat-stable enterotoxin. Pediatric Res. **29:** 424–428.
17. Desai, G.N., J. Sahi, P.M. Reddy *et al.* 1996. Chloride transport in primary cultures of mammalian colonocytes at different developmental stages. Gastroenterology **111:** 1541–1550.
18. NATARAJA, S., J. VENKATASUBRAMANIAN, D. VIDYASAGAR & M.C. RAO. 1998. Ontogeny of cGMP-stimulated chloride transport in rabbit colon. Gastroenterology **114:** G1636.
19. REDDY, P.M., J. SAHI, G.N. DESAI *et al.* 1996. Altered growth and attachment of rabbit colonocytes isolated from different developmental stages. J. Pediatric Res. **39:** 287–294.
20. VENKATASUBRAMANIAN, J., M. CARLOS, S.G. NATARAJA *et al.* 1998. Regulation of colonic chloride transport by bile acids (BA). Gastroenterology **114:** G1744.
21. VENKATASUBRAMANIAN, J., N. NEELAKANTAM, S. SKALUBA *et al.* 1998. Segmental differences in the regulation of colonic chloride transport by calcium mediators. Gastroenterology **114:** G1745.
22. FIELD, M. 1980. Regulation of small intestinal ion transport by cyclic nucleotides and calcium. *In* Secretory Diarrhea. M. Field J. Fordtran & S.G. Schultz, Eds.: 21–30. American Physiological Society. Bethesda, MD.
23. WELSH, M.J., P.L. SMITH, M. FROMM & R.A. FRIZZELL. 1982. Crypts are the site of intestinal fluid and electrolyte secretion. Science **281:** 1219–1221.
24. GREGER, R., M. BLEICH, J. LEIPZIGER *et al.* 1997. Regulation of ion transport in colonic crypts. News Physiol. Sci. **12:** 62–66.
25. ECKE, D., M. BLEICH & R. GREGER. 1996. The amiloride inhibitable Na^+ conductance of rat colonic cells is suppressed by forskolin. Pflugers Arch. **431:** 984–986.
26. CHANG, E.B. & M.C. RAO. 1991. Intestinal mediators of intestinal electrolyte transport. *In* Diarrheal Diseases. M. Field, Ed.: 49–72. Elsevier. New York.
27. FRIZZELL, R.A., M.A. KOCH & S.G. SCHULTZ. 1976. Ion transport by rabbit colon. 1. Active and passive components. J. Membr. Biol. **27:** 297–316.
28. MANDEL, K.G., K. DHARMSATHAPHORN & J.A. McROBERTS. 1986. Characterization of a cAMP-activated Cl^- transport pathway in the apical membrane of a human colonic epithelial cell line. J. Biol. Chem. **261:** 704–712.
29. FRIZZELL, R.A. 1995. Function of the cystic fibrosis transmembrane conductance regulator protein. Am. J. Respir. Crit. Care Med. **151:** S54–S58.
30. SAHI, J., J. GOLDSTEIN, T.J. LAYDEN & M.C. RAO. 1994. Cyclic AMP and phorbol ester regulated chloride permeabilities in primary cultures of human and rabbit colonocytes. Am. J. Physiol. **266:** G846–G855.

31. BENYA, R.V., L.N. SCHMIDT, J. SAHI *et al.* 1991. Isolation, characterisation and attachment of rabbit distal colon epithelial cells. Gastroenterology **101**: 692–702.
32. QUARRINO, A. & R.J. MAY. 1980. Establishment and characterization of intestinal epithelial cell cultures. Methods Cell Biol. **21B**: 403–427.
33. SAHI, J., G. BISSONNETTE, J.L. GOLDSTEIN *et al.* 1996. Effect of Ca^{2+}-dependent regulators in Cl^- transport and protein phosphorylation in human colon. FASEB J. **10**: A544.
34. FINKEL, Y. & A. APERIA. 1986. Role of aldosterone for control of colonic Na^+/K^+/ATPase activity in weaning rats. Pediatr. Res. **20**: 242–245.
35. RAO, M.C. & M. FIELD. 1983. Role of calcium and cyclic nucleotides in the regulation of intestinal ion transport. *In* Intestinal Transport: Fundamental and Comparative Aspects. M. Gilles-Baillien & R. Gilles, Eds.: 227–239. Springer Verlag. New York.
36. DE JONGE, H.R. & M.C. RAO. 1990. Cyclic nucleotide–dependent kinases. *In* Textbook of Secretory Diarrhea. E. Lebenthal & M.E. Duffey, Eds.: 191–207. Raven Press. New York.
37. RAO, M.C. & H.R. DE JONGE. 1990. Ca- and phospholipid-dependent protein kinases. *In* Textbook of Secretory Diarrhea. E. Lebenthal & M.E. Duffey, Eds.: 191–207. Raven Press. New York.
38. DEVOR, D.C., M.C. SEKAR, R.A. FRIZZELL & M.E. DUFFEY. 1993. Taurodeoxycholate activates potassium and chloride conductances via an IP3-mediated release of calcium from intracellular stores in a colonic cell line (T-84). J. Clin. Invest. **92**: 2173–2181.
39. VAANDRAGER, A., A. BOT & H. DE JONGE. 1997. Guanosine 3', 5'- cyclic monophosphate–dependent protein kinase II mediates heat-stable enterotoxin-provoked chloride secretion in rat intestine. Gastroenterology **112**: 437–443.
40. GOLDSTEIN, J.L., J. SAHI, M. BHUVA *et al.* 1994. *Escherichia coli* heat-stable enterotoxin-mediated colonic C1 secretion is absent in cystic fibrosis. Gastroenterology **107**: 950–956.
41. RAO, M.C., S. GUANDALINI, P.L. SMITH & M. FIELD. 1980. Mode of action of heat-stable *Escherichia coli* enterotoxin. Biochim. Biophys. Acta **632**: 35–60.
42. RAO, M.C. 1984. Toxins which activate guanylate cyclase: heat-stable enterotoxins. *In* Microbial Toxins and Diarrhoeal Disease. Ciba Found. Symp. **112**: 74–93.
43. FORTE, L.R. & M.G. CURRIE. 1995. Guanylin: a peptide regulator of epithelial transport. FASEB J. **9**: 643–650.
44. COHEN, M.B., M.S. MOYER, M. LUTTRELL & R.A. GIANELLA. 1986. The immature rat small intestine exhibits an increased sensitivity and response to *Escherichia coli* heat-stable enterotoxin. Pediatr. Res. **20**: 555–560.
45. GUARINO, A., M.A. COHEN & R.A. GIANNELLA. 1987. Small and large intestinal gunaylate cyclase activity in children: effect of age and stimulation by *E. coli* heat-stable enterotoxin. Pediatr. Res. **21**: 551–555.
46. NOBLES, M., M. DIENER & W. RUMMEL. 1991. Segment-specific effects of the heat-stable enterotoxin of *E. coli* on electrolyte transport in the rat colon. Eur. J. Pharmacol. **202**: 201–211.
47. FORTE, L.R., P.K. THORNE, S.L. EBER *et al.* 1992. Stimulation of intestinal Cl^- transport by heat-stable enterotoxin: activation of cAMP-dependent protein kinase by cGMP. Am. J. Physiol. **32**: C607–C615.
48. LIN, M, A.C. NAIRN & S.E. GUGGINO. 1992. cGMP-dependent protein kinase regulation of chloride channel in T-84 cells. Am. J. Physiol. **262**: C1304–1312.
49. NATARAJA, S., J.L. GOLDSTEIN & M.C. RAO. 1998. Expression of cGMP-dependent protein kinase II in T84 and primary human colonocytes (PHC). FASEB J. **10**: A712.
50. SETCHELL, K.D.R. & D.W. RUSSELL. 1994. Ontogenesis of bile acid synthesis and metabolism. *In* Liver Diseases in Children. F. Suchy, Ed.: 81–104. Mosby. St. Louis, MO.
51. SUCHY, F.J. 1994. Bile formation: mechanisms and development. *In* Liver Diseases in Children. F. Suchy, Ed.: 57–80. Mosby. St. Louis, MO.
52. KAPLAN, M. 1989. Medical treatment of primary biliary cirrhosis. Semin. Liver Dis. **9**: 138–143.

53. HEUBI, J.E. & J.L. FELLOWS. 1985. Postnatal development of intestinal bile salt transport. J. Lipid Res. **26:** 797–805.
54. POTTER, G.D, R. LESTER, S.M. BURLINGAME *et al.* 1987. Taurodeoxycholate and the developing rabbit distal colon: absence of secretory effect. Am. J. Physiol. **253:** G483–G488.
55. POTTER, G.D., J.H. SELLIN & S.M. BURLINGAME. 1991. Bile acid stimulation of cyclic AMP and ion transport in developing rabbit colon. Pediatr. Gastroenterol. Nutr. **13:** 335–341.
56. POTTER, G.D., J.H. SELLIN, S.M. BURLINGAME & R.C. DESOIGNIE. 1989. Potassium secretion in response to taurodeoxycholic acid in the newborn rabbit colon. J. Pediatr. Gastroenterol. Nutr. **9:** 365–370.
57. KARLSTROM, L., J. CASSUTO, M. JODAL & O. LUNDGREN. 1983. The importance of the enteric nervous system for the bile salt–induced secretion in the small intestine of the rat. Scand. J. Gastroenterol. **18:** 117–123.
58. FREEL, R.W. 1987. Dihydroxy bile salt–induced secretion of rubidium ion across the rabbit distal colon. Am. J. Physiol. **252:** G554–561.
59. HUANG, X., X. FAN, J. DESJEUX & M. CASTAGNA. 1992. Bile acids, non-phorbol ester–type tumor promoters, stimulate the phosphorylation of protein kinase C substrates in human platelets and colon cell line HT29. Int. J. Cancer **52:** 444–450.
60. GELBMANN, C.M., C.D. SCHTEINGART, S.M. THOMPSON *et al.* 1995. Mast cells and histamine contribute to bile acid–stimulated secretion in the mouse colon. J. Clin. Invest. **95:** 2831–2839.
61. CHANG, E.B. M.W. MUSCH. 1992. Immune regulation of intestinal arachidonic acid metabolism: effects on intestinal water and electrolyte transport. *In* Immunophysiology of the Gut. B.M. Squibb & M. Johnson, Eds. Academic Press. Orlando, FL.
62. FIELD, M. & C.E. SEMRAD. 1993. Toxigenic diarrheas, congenital diarrheas, and cystic fibrosis: disorders of intestinal ion transport. Annu. Rev. Physiol. **55:** 631–655.
63. SAGMANLIGIL, V. & R.J. LEVIN. 1993. Electrogenic ion secretion in proximal, mid and distal colon from fed and starved mice. Comp. Biochem. Physiol. **106:** 449–456.
64. DEUTSCH, D.E., K.N. BITAR & A.C. HILLEMEIER. 1998. Access to intracellular calcium during development in the feline gastric antrum. Pediatr. Res. **43:** 369–373.

Index of Contributors